COMPLEX PLANETARY SYSTEMS II: LATEST METHODS FOR AN INTERDISCIPLINARY APPROACH

IAU SYMPOSIUM 382

COVER ILLUSTRATION:

Artist's concept of an exoplanet (Credits : ESA)

IAU SYMPOSIUM PROCEEDINGS SERIES

INTERNATIONAL ASTRONOMICAL UNION

UNION ASTRONOMIQUE INTERNATIONALE

International Astronomical Union

IAU

COMPLEX PLANETARY SYSTEMS II: LATEST METHODS FOR AN INTERDISCIPLINARY APPROACH

PROCEEDINGS OF THE 382nd SYMPOSIUM OF THE INTERNATIONAL ASTRONOMICAL UNION NAMUR, BELGIUM 3–7 JULY 2023

Edited by

ANNE LEMAITRE
University of Namur, Belgium

and

ANNE-SOPHIE LIBERT
University of Namur, Belgium

CAMBRIDGE
UNIVERSITY PRESS

CAMBRIDGE UNIVERSITY PRESS
University Printing House, Cambridge CB2 8BS, United Kingdom
1 Liberty Plaza, Floor 20, New York, NY 10006, USA
10 Stamford Road, Oakleigh, Melbourne 3166, Australia

First published 2024

Printed in Great Britain by Henry Ling Limited, The Dorset Press, Dorchester, DT1 1HD

Typeset in System LaTeX 2ε

A catalogue record for this book is available from the British Library of Congress Cataloguing in Publication data

This journal issue has been printed on FSC^{TM}-certified paper and cover board. FSC is an independent, non-governmental, not-for-profit organization established to promote the responsible management of the world's forests. Please see www.fsc.org for information.

ISBN 9781009399050 hardback
ISSN 1743-9213

Table of Contents

Preface

The Symposium IAUS-382 entitled *Complex Planetary Systems II* (CPSII) took place at the University of Namur, in Belgium, from July 3 to 7, 2023. It was the second edition of CPS, the first one being organized in Namur in July 2014 (IAUS-310). The subtitle of this second edition was *Latest methods for an interdisciplinary approach.* The Kavli label was attributed to CPSII, it contributed to reinforce the interdisciplinarity of the Symposium and gave the opportunity to enlarge the scope of the Symposium, with additional contributions of scientists from close fields, as geologists, biologists, geographers, applied mathematicians, observers, and engineers.

By definition, complex systems are systems composed of interacting parts / agents whose local behavior, resulting from the interactions between them, cannot provide a complete understanding of the global behavior, i.e., when the system is considered as a whole, on a macroscopic scale. Several levels of modeling of the system should be considered at the same time. This forces complex systems to be studied by transdisciplinary teams, able to understand the whole construction and critically analyze the connections among the description levels. This vision is really efficient and challenging for space and astronomy sciences.

The huge number of available observations (from ground and space) and their accurate precision, as well as the computational power and speed of our present-day computers, have spectacularly changed the nature of the dynamical models, especially for planetary evolution studies. Planetology, celestial mechanics, and space geodesy have considerably evolved in that direction during the last decade, and have reinforced the need for crossing experiences and methods. Let us mention several examples: the concept of habitability of an extrasolar planet; the dynamical history of the Solar System and other planetary systems; the rotation of planets and satellites linked to their internal structure; the motion of natural satellites needing astrometry, tides and dissipations; the thermal effects on the evolution of the rotating small asteroids; the long-term evolution of space debris and satellites, including drags, shadowing effects, and collisional chain reactions. All these problems as well as many others were tackled at CPSII, showing how the formal historical border between analytical and numerical approaches has now disappeared.

The Symposium was a real opportunity to show the power of interdisciplinary collaboration and was a unique occasion to gather astronomers of many disciplines together. While the global thematic remained the planetary systems as in the first edition, CPSII highlighted that first, the tools and the methods have considerably evolved, and second, the interdisciplinarity has touched more communities. The Symposium opened new doors and created collaborations, exchanges of ideas, combinations of techniques, sometimes unexpected, to solve complex planetary systems.

CPSII was attended by 130 scientists from 30 different countries and proposed a rich program with 19 invited speakers, 57 talks, and 33 posters. Moreover, the organization of the meeting was innovative, with a priority to young promising speakers on interdisciplinary topics, two round tables on space awareness and habitability, poster flash talks ('one slide, one minute') with different prizes awarded to the best poster presentations. Cultural events, namely the visit of the citadel underground and a dance show about the astrophysicist Vera Rubin, completed the program.

The scientific communities of the Solar system and the exoplanets, were particularly happy to meet again after the pandemic. It was a great success in terms of contacts, discussions, and future projects.

CPSII was supported by the International Astronomical Union, Kavli Foundation, F.R.S.-FNRS, UNITER doctoral school, naXys Research Institute, UNamur, and the *Fonds Jacques Cox pour les générations futures.*

Anne Lemaitre and Anne-Sophie Libert (co-chairs)

Editors

Lemaitre Anne, University of Namur, Belgium
Libert Anne-Sophie, University of Namur, Belgium

SOC

LOC

List of Participants

Alnajjarine	Aya
Alves Silva	Raphael
Anghel	Simon
Baland	Rose-Marie
Balsalobre Ruza	Olga
Batygin	Konstantin
Baù	Giulio
Baycroft	Thomas
Benet	Luis
Bodart	Justine
Bolmont	Emeline
Căliman	Alexandru
Carletti	Timoteo
Celletti	Alessandra
Charalambous	Carolina
Christiaens	Valentin
Correia	Alexandre
Courtot	Ariane
Coyette	Alexis
Crespi	Samuele
Crida	Aurélien
Daquin	Jerome
Dardenne	Charline
Davoudi	Fatemeh
Delisle	Jean-Baptiste
Dermott	Stanley
Desoubrie	Baptiste
Di Ruzza	Sara
Dogkas	Anargyros
Dorn	Caroline
Doukhanin	Nikita
Duarte	Gladston
Efroimsky	Michael
Esseldeurs	Mats
Fabrycky	Daniel
Farhat	Mohammad
Fayolle	Marie
Füzfa	André
Gales	Catalin
Georgakarakos	Nikolaos
Ghosh	Tuhin
Gillon	Michaël
Gomes	Sérgio
Granvik	Mikael
Grassi	Clara
Griveaud	Philippine
Gronchi	Giovanni Federico

Habibi	Farhang
Hadden	Sam
Haghighipour	Nader
Hakim	Kaustubh
Hayashi	Toshinori
Hernandez	David
Hestroffer	Daniel
Huang	Shuo
Ida	Shigeru
Javaux	Emmanuelle
Johansen	Anders
Jones	Daniel
Knežević	Zoran
Kokubo	Eiichiro
Konings	Thomas
Korth	Judith
Kotoulas	Thomas
Kudryavtsev	Sergey
Kumar	Bhanu
Kuwahara	Ayumu
Kwok	Leon Ka-Wang
Lai	Dong
Lainey	Valery
Lari	Giacomo
Laskar	Jacques
Lau	Tommy Chi Ho
Lee Man	Hoi
Legnaro	Edoardo
Leleu	Adrien
Lemaitre	Anne
Lhotka	Christoph
Libert	Anne-Sophie
Liu	Xiang
Mah	Jingyi
Mastroianni	Rita
Minglibayev	Mukhtar
Mogavero	Federico
Morbidelli	Alessandro
Moursi	Ahmed
Nadabaica	Gabriela-Ana
Ochiai	Yoko
Okamoto	Tamami
Parviainen	Hannu
Perhirin	Goulven
Petit	Antoine
Petit	Alexis
Petrovich	Cristobal
Pichierri	Gabriele
Pilat-Lohinger	Elke
Prieur	Alexandre

Prokopenya	Alexander
Pucacco	Giuseppe
Quirrenbach	Andreas
Ragozzine	Darin
Rajkumar	Anitha Raj
Rambaux	Nicolas
Rekier	Jérémy
Renner	Stéfan
Revol	Alexandre
Robutel	Philippe
Rodriguez	Adrian
Romano	Matteo
Rosengren	Aaron Jay
Rossi	Mattia
Sadeh	Matan
Saillenfest	Melaine
Salnikova	Tatiana
Scheeres	Daniel
Sefilian	Antranik
Sicardy	Bruno
Sidorenko	Vladislav
Suto	Yasushi
Tealib	Shafeeq
Todorović	Nataša
Tommei	Giacomo
Torii	Naoya
Triaud	Amaury
Valente	Ema
Vavilov	Dmitrii
Vienne	Alain
Woo	Jason Man Yin
Xi	Xiao jin
Yseboodt	Marie
Zimmermann	Max
Zoppetti	Federico

Complex Planetary Systems II: Latest Methods for an Interdisciplinary Approach
Proceedings IAU Symposium No. 382, 2024
A. Lemaitre & A.-S. Libert, eds.
doi:10.1017/S1743921323004398

The obliquity of Mercury: Models and interpretation

Rose-Marie Baland

Royal Observatory of Belgium
email: rose-marie.baland@oma.be

Abstract. Mercury is locked in an unusual 3:2 spin-orbit resonance and as such is expected to be in a state of equilibrium called Cassini state. In that state, the angle between the spin axis and orbit normal, called obliquity, remains almost constant while the spin axis remains almost in the plane, also called Cassini plane, defined by the normal to the Laplace plane and the normal to the orbital plane. The spin axis and the orbit normal precess together with a period of about 300 kyr. The orientation of the spin axis of Mercury has been estimated using different approaches: (i) Earth-based radar observations, (ii) Messenger images and altimeter data, and (iii) Messenger radio tracking data. The different estimates all tend to confirm that Mercury occupies the Cassini state. The observed obliquity is small and close to 2 arcmin. It indicates a normalized polar moment of inertia of about 0.34. This information, combined with the existence of a liquid iron core, as evidenced by the librations, allows to constrain the interior structure of Mercury. However, the different estimates of the orientation of the spin axis locate the spin axis somewhat behind or ahead of the Cassini plane, and it is difficult to reconcile and interpret them coherently in terms of detailed interior properties. We review recent models for the obliquity and spin orientation of Mercury, which include the effects of complex orbital dynamics, tidal deformations and associated dissipation, and internal couplings related to the presence of fluid and solid cores. We discuss some implications regarding the interpretations of the orientation estimates in term of interior properties.

Keywords. planets and satellites: individual (Mercury), celestial mechanics, methods: analytical, obliquity, librations, geodesy, dynamics, rotation, planetary interiors

1. Introduction

1.1. *An unusual rotation*

Three important letters concerning the rotation of Mercury were published in 1965. First, using radar observations obtained at the Arecibo Ionospheric Observatory in Puerto Rico, Pettengill and Dyce (1965) determined that the rotation of Mercury is direct with a sidereal period of 59 ± 5 days which differs from the orbital period (88 days). They could not determine the direction of the pole, which is approximately normal to the planetary orbit. The fact that Mercury's rotation is not synchronous with its revolution was an unexpected discovery. Secondly, Peale and Gold (1965) noted that while for a planet in a circular orbit, tidal friction should synchronise its rotation with its revolution, the large eccentricity of Mercury (0.206) allows faster rotation because the tidal torque at pericenter, where the orbital angular velocity is maximum, exceeds that at other times. Finally, Colombo (1965) indicated that a uniform rotation with a period of 2/3 of the orbital period can be a stable state of rotation, because the axis of minimum moment of inertia (the largest axis in terms of dimension) is aligned with the Sun-Mercury direction at each pericenter passage, where the interaction between the two bodies is the strongest.

Table 1. Values of the orientation parameters of Mercury, obtained from different observational techiques. Based on Baland *et al.* (2017a) and Bertone *et al.* (2021).

Reference	**Technique**	$\alpha_0(^\circ)$	$\delta_0(^\circ)$
Margot *et al.* (2007)	Earth-based radar observations	281.0097	61.4143
Margot *et al.* (2012)	Earth-based radar observations	281.0103 ± 0.0015	61.4155 ± 0.0013
Mazarico *et al.* (2014)	MESSENGER radio tracking data	281.00480 ± 0.0054	61.41436 ± 0.0021
Stark *et al.* (2015)	Images and laser altimeter data from MESSENGER	281.00980 ± 0.00088	61.4156 ± 0.0016
Verma and Margot (2016)	MESSENGER radio tracking data	281.00975 ± 0.0048	61.41828 ± 0.0028
Genova *et al.* (2019)	MESSENGER radio tracking data	281.0082 ± 0.0009	61.4164 ± 0.0003
Konopliv *et al.* (2020)	MESSENGER radio tracking data	281.0138 ± 0.0025	61.4161 ± 0.0017
Bertone *et al.* (2021)	Images and laser altimeter data from MESSENGER	281.0093 ± 0.00063	61.4153 ± 0.00048

This part of the history of the study of Mercury's rotation is nicely documented in Sections 2 and 4 of Goldreich and Peale (1968).

1.2. *Cassini state*

Because the rotation rate of Mercury is commensurate with its orbital mean motion, the planet is assumed to be locked in a *Cassini state* and to follow Cassini's second and third laws (Peale 1969). In this state, the spin axis and the orbit normal precess together about the normal to the Laplace plane at the same rate (period of ~ 300 kyr), which means that all three axes are coplanar and the obliquity (the angle from the orbit normal to the spin axis) remains constant over time. The Laplace plane is the inertial plane that minimizes the variation in orbital inclination. In what follows, this particular state is referred to as the *Classical Cassini state*, as opposed to the more recent and advanced models which indicate that the spin axis can deviate by of a few arcsec from the Cassini plane (the plane defined by the Laplace and orbit normal axes), see Section 3.

2. Orientation measurements

2.1. *Published studies*

Using three different observational techniques, the orientation of Mercury's spin axis has been estimated no less than eight times since 2007, see Table 1. Each measurement is provided in terms of right ascension α_0 and declination δ_0 with respect to the International Celestial Reference Frame (ICRF) at the J2000 epoch, the format used by the IAU Working Group on Cartographic Coordinates and Rotational Elements of the Planets and Satellites. Almost all estimates listed in Table 1 are compatible with coplanarity/occupancy of the Classical Cassini state within the limits of measurements precision (see Fig. 1). The epoch spin location of Margot *et al.* (2012) is currently adopted by the IAU (Archinal *et al.* 2018). To those readers who desire more details on the evolution of the IAU adopted values for the rotation angles of Mercury, we recommend the articles by Margot (2009) and Stark *et al.* (2018).

2.2. *Obliquity/deviation representation*

Each pair (α, δ) corresponds to a pair of obliquity and deviation $(\varepsilon, \tilde{\delta})$, see Section 4.3 of Baland *et al.* (2017a) for the mathematical relationships. For example, the orientation of Stark *et al.* (2015) corresponds to an obliquity of about 2 arcmin and a deviation of 1.7 arcsec.

The obliquity/deviation representation is used in theoretical models to relate the estimated orientation to the interior properties of the planet, see Section 3. In particular,

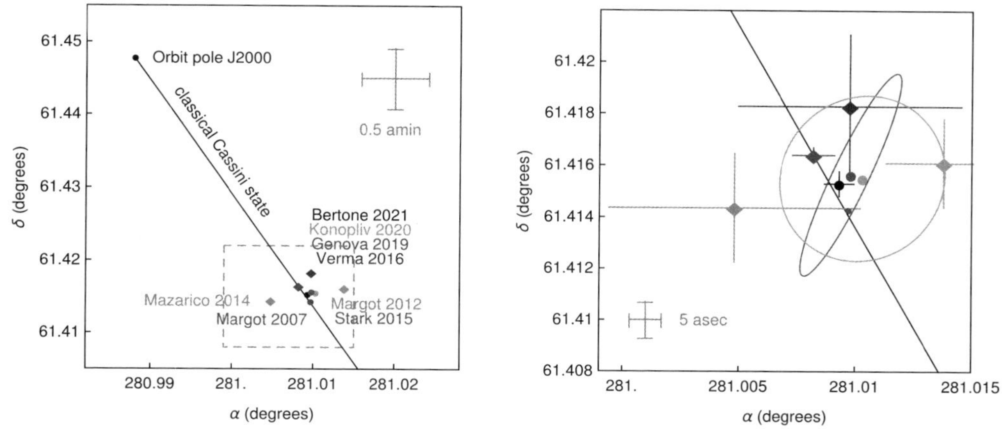

Figure 1. Orientation (right ascension α and declination δ w.r.t the ICRF) of the orbit pole and of the spin axis of Mercury at the J2000 epoch for the studies listed in Table 1. The solid line represents the location of the Cassini plane (Clasical Cassini state, see Section 1.2). The dashed rectangle on the left panel represents the plot limits used in the right panel (figure adapted from Baland *et al.* (2017a)).

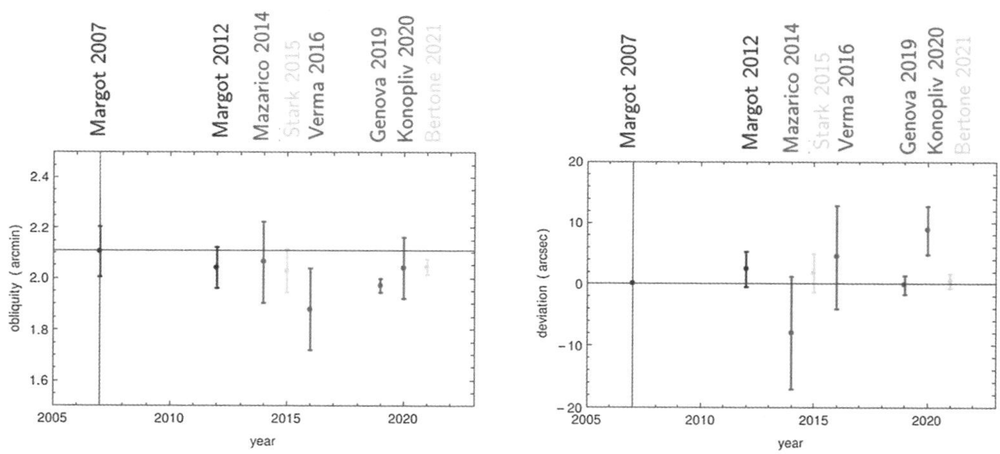

Figure 2. The different values for obliquity and deviation, according to the observational technique (Earth-based radar observations in blue - Messenger images and altimeter data in green - Messenger radio tracking data in red) and publication date.

the deviation $\tilde{\delta}$ represents the offest of the spin axis with respect to the Cassini plane and is zero in the Classical Cassini state.

From Margot *et al.* (2007) to Bertone *et al.* (2021), the estimated value for the obliquity has followed a downward trend (Fig. 2, left panel). As we will see in Section 3.2, this result has implication in terms of interior interpretation, and in particular in terms of the presence of a solid inner core. The estimated deviation started close to 0 with Margot *et al.* (2007). With the exception of Mazarico *et al.* (2014) and their negative deviation, $\tilde{\delta}$ is either close to 0, or follows an upward trend in subsequent studies (Fig. 2, right panel).

In some respects, the situation is reminiscent of the process of the determination of the elementary charge. Millikan's 1909 oil drop experiment, the initial direct determination of the magnitude of the elementary charge, yielded a value that we currently know slightly inaccurate due to his use of an incorrect value for the viscosity of the air. Although Millikan's first estimate was not correct, subsequent experimental attempts tended to be close to Millikan's, until they finally stabilized at a higher, correct value. To quote Feynman (1985): *"It's apparent that people did things like this: When they got a number that was too high above Millikan's, they thought something must be wrong and they would look for and find a reason why something might be wrong. When they got a number close to Millikan's value they didn't look so hard."* Were successive spin-axis orientation estimates influenced by the first obliquity estimate of Margot *et al.* (2007) and/or the authors' preconceived ideas about deviation? Since the discovery of the 3:2 spin orbit resonance of Mercury, the planet is reasonably considered to be in the Cassini state. Naturally, some authors did not imagine that the deviation could differ from zero, as appears clearly in Margot *et al.* (2007) † and Genova *et al.* (2019)‡. The possibility of a non zero deviation, and the implications in terms of energy dissipation due to solid-body tides and core-mantle interactions, was first recognized in Margot *et al.* (2012).

The authors sometimes explain the dispersion of their results by the different observation techniques used. On the one side, both the Earth-based radar (Margot *et al.* 2007, 2012) and laser altimeter (Stark *et al.* 2015; Bertone *et al.* 2021) determinations are specifically tied to the rotation of the solid surface layer, since Mercury has an at least partially liquid core. And indeed, all four orientation determinations (especially the last three) are in very good agreement with each other. They are characterized by similar obliquities despite differences in observational techniques. Despite a possible bias towards 0 at the beginning, they ultimately tend to indicate a small deviation. On the other side, the radio tracking determinations of Mazarico *et al.* (2014); Verma and Margot (2016); Genova *et al.* (2019); Konopliv *et al.* (2020) are allegedly tied to the orientation of the gravity field of the planet, and so we might expect to see them grouped together in Fig. 1. However, they are more scattered than the determinations tied to the surface, and marginally consistent with each other. Verma and Margot (2016) invoke the use of different ephemerides and range data to explain the difference between their estimate and that of Mazarico *et al.* (2014). The subsequent determinations of Genova *et al.* (2019) and Konopliv *et al.* (2020) did not resolved the differences. Topping it all off, the deviation is either negative, or positive but compatible with 0, or almost 0, or positive and not compatible with zero within the $1-\sigma$ limit.

3. Orientation modeling and interior interpretation

3.1. *Classical model*

The Classical Cassini state model for an entirely rigid Mercury comes with a relationship between the small obliquity ε and the polar moment of inertia C (see Eq. (4) of Peale 1981 and Eq. (12) of Yseboodt and Margot 2006):

$$\varepsilon = \frac{-C\dot{\Omega}\sin i}{C\dot{\Omega}\cos i + 2nM_{☿}R^2G_{201}(e)C_{22} - nM_{☿}R^2G_{210}(e)C_{20}} \quad (1)$$

† In Margot *et al.* (2007): *"The confidence regions for the spin orientation fall precisely on the locus of pole positions that satisfy the Cassini condition."* The Cassini condition referred to here is in fact the Classical Cassini state, with no deviation.

‡ Genova *et al.* (2019) argue that the spin pole orientations of Margot *et al.* (2012) and Stark *et al.* (2015) showed a *"substantial"* offset from the Cassini state, but that in contrast, their own estimate is in *"full agreement with the Cassini state"*, and by that, they mean the Classical Cassini state with a zero deviation.

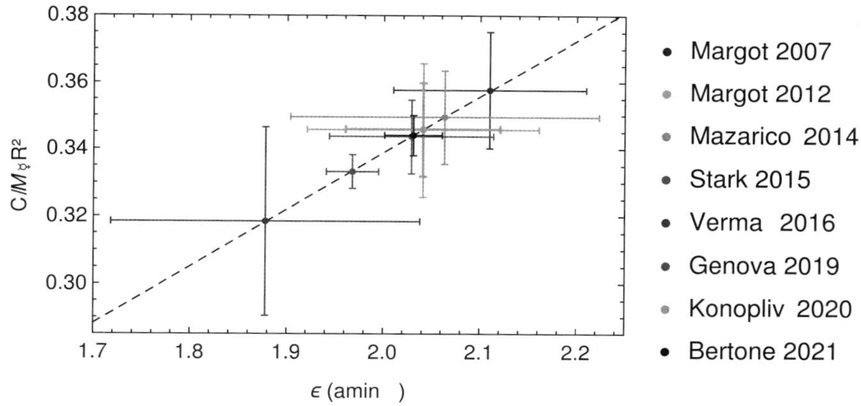

Figure 3. Normalized polar moment of inertia $C/M_{☿}R^2$ as a function of the obliquity ε. The solid line shows the quasi linear relationship resulting from Eq. (1) for the Classical Cassini state model.

where

$$G_{210}(e) = \left(1 - e^2\right)^{-3/2} \tag{2}$$

$$G_{201}(e) = \frac{7}{2}e - \frac{123}{16}e^3 + \frac{489}{128}e^5 + \mathcal{O}(e^7) \tag{3}$$

are eccentricity (e) functions, n is the mean motion, Ω and i are the longitude of the ascending node and orbital inclination with respect to the Laplace plane, $M_{☿}$ and R are the mass and mean radius of Mercury, and C_{20} and C_{22} are second-degree gravity field coefficients. Note that Eq. (1) has been derived independently in Baland *et al.* (2017a) to clarify the situation in the litterature, where uncorrect versions of the equation were in used, due to confusion regarding sign conventions (here the obliquity angle $\varepsilon > 0$ and the precession rate $\dot{\Omega} < 0$). Eq. (1) is obtained by assuming that Mercury behaves rigidly and that coplanarity is satisfied, so that no deviation is associated with the classical model.

Each estimate of the obliquity ($1.88 - 2.11$ arcmin) corresponds to an estimate of the normalized polar moment of inertia ($0.318 - 0.358$), see Fig. 3, which, together with the estimated mean density, indicates that Mercury is a differentiated body with a large iron core.

3.2. *Peale experiment*

To characterize the size and state of the core of Mercury, Peale (1976, 1981) proposed an observational procedure, since known as the *Peale experiment.* In the following equation

$$\left[\frac{(B-A)}{M_{☿}R^2}\right]\left[\frac{M_{☿}R^2}{C}\right]\left[\frac{C_m}{(B-A)}\right] = \frac{C_m}{C} \leq 1, \tag{4}$$

(i) the first factor is obtained from the second-degree and -order gravity coefficient C_{22}, (ii) the second factor is derived from the obliquity, using Eq. (1), and thus assuming that the core must follow the mantle on the time scale of forced precession, and (iii) the third factor is obtained from the forced libration at 88 days, the annual variation in the rotation due to the periodically reversing solar torque on the asymmetric figure, assuming that a liquid core would not follow the mantle's longitudinal oscillations. After multiplying these three factors, we obtain the ratio of the polar moment of inertia of the outer solid layer C_m over the total polar moment of inertia C. The ratio is 1 if the core is entirely solid, but is smaller if at least a part of the core is fluid. This procedure has been applied for example in Margot *et al.* (2007, 2012); Stark *et al.* (2015); Genova *et al.* (2019) and shows that $C_m/C \simeq 0.4$, indicating that the core is at least partially molten. The presence of a liquid core has important implications, as it provides a possible explanation based on a dynamo mechanism for the magnetic field observed by Mariner 10, helps to sort between thermal evolution models, and possibily explains the capture into resonance thanks to dissipation at the core-mantle boundary.

Considering a simple interior model consisting of two layers of uniform density, the ratio C_m/C implies that the size of the core ($\simeq 2000$ km) is $\simeq 80\%$ of the total radius and that the core and mantle densities are $\simeq 7250$ kg/m^3 and $\simeq 3200$ kg/m^3, respectively (Margot *et al.* 2012). Using more elaborate and realistic interior models and accurate estimates of the gravity coefficients by MESSENGER, Smith *et al.* (2012) or Hauck *et al.* (2013) estimated that the core radius is 2040 ± 37 or 2020 ± 30 km. Because their estimate of obliquity is lower than previous ones by Margot *et al.* (2007, 2012), Genova *et al.* (2019) derived a lower value for the polar moment of inertia and therefore concluded that Mercury has a solid inner core with a radius of at least 600 km (see also Steinbrügge *et al.* (2021)).

3.3. *Improving the experiment ?*

Annual libration

It has been shown that the relationship between the measured libration and the third factor of Eq. (4) is not valid (i) in the presence of a solid inner core, because of the gravitational-pressure coupling between the internal layers, and (ii) when the solid layers deform periodically. Tides decrease the forced annual libration amplitude, by about $1-2$ m, below the present and future expected observational precision, whereas an inner core larger than 1000 km could have a noticeable effect (Van Hoolst *et al.* 2012; Dumberry *et al.* 2013).

To avoid possible biases in the interpretation of rotation measurements in terms of interior structure, due to simplistic modeling of librations, Rivoldini and Van Hoolst (2013) has somewhat redesigned the *Peale experiment*. As the annual libration amplitude cannot be expressed as a function of the equatorial moment of inertia difference $(B-A)$ of the whole body (compare Eqs. 8 and 11 of Rivoldini and Van Hoolst (2013)), the improved experiment, unlike Eq. (4), can no longer be written in an elegant form to first obtain the moment of inertia of the silicate shell, which will then be interpreted in terms of interior properties. Instead, Rivoldini and Van Hoolst (2013) used as data the mean moment of inertia and the annual libration amplitude of Margot *et al.* (2012) in combination with the libration model of Van Hoolst *et al.* (2012) to directly constrain Mercury's interior structure.

Considering elaborate interior models that include a crust and an inner core, non uniform layers, and plausible chemical compositions and temperature profiles, see also Rivoldini *et al.* (2009) for additional details, they placed a lower constraint on the core radius (2004 ± 39 km) than Smith *et al.* (2012); Hauck *et al.* (2013), due to the mantle-core coupling taken into account in the libration model. They also determined the average density of the core (7233 ± 267 kg/m^3) and the fraction of sulfur in the core if sulfur is the only light element in the core ($4.5 \pm 1.8 wt\%$). They could not however confirm or infirm the existence of an inner core or impose constraints on the mantle density. According to Dumberry and Rivoldini (2015), which also use the rotation measurements of Margot *et al.* (2012), the largest inner core compatible with the rotation observations has a radius of 1325 ± 250 km, and geodetic observations suggest the formation of Fe-snow within the fluid core that could contribute to the magnetic field. The conclusions drawn by Rivoldini and Van Hoolst (2013); Dumberry and Rivoldini (2015) depend strongly on the chosen value of the polar moment of inertia, derived from the measured obliquity.

Obliquity

The relationship between obliquity and interior structure as provided by Eq. (1), and used to obtain the second factor of Eq. (4), needs to be questioned. Baland *et al.* (2017a) have listed various approximations in the modeling process that affect the determination of the polar moment of inertia from the measured orientation of the rotation axis. Among these, the most important are neglecting (i) the nutations due to the precession of the pericenter, (ii) the periodic tidal deformations, and (iii) the presence of a solid inner core (see details below). The use of an improved Cassini state model instead of Eq. (1) may, as with improved libration models, modify the interpretation of rotation measurements in terms of interior structure.

Using other observations

Finally, Peale's experiment can be extended to observations other than obliquity and annual libration, although in this case we can no longer really speak of Peale's experiment. For example, the long period libration (11.864 years) due to the orbital perturbations of Mercury's orbit by Jupiter could be resonantly amplified and affect the instantaneous spin rate, so that it deviates from the resonant rate (Stark *et al.* 2015, Yseboodt *et al.* 2013 and references within). In addition to rotation, the observed tidal Love numbers (see Mazarico *et al.* (2014); Verma and Margot (2016); Genova *et al.* (2019); Konopliv *et al.* (2020) for k_2 and Bertone *et al.* (2021) for h_2) and deviation of the spin axis from coplanarity (see below) can also contribute to a detailed study of Mercury's interior structure and rheology (e.g. Rivoldini *et al.* 2009; Padovan *et al.* 2014; Margot *et al.* 2018; Steinbrügge *et al.* 2021; Goossens *et al.* 2022). Note that the measured k_2, larger than expected for an entirely solid Mercury, confirms the presence of a liquid core, whereas the measured h_2 suggests the presence of a solid inner core.

3.4. *Cassini state: Effect of pericenter precession*

According to Baland *et al.* (2017a), assuming that Mercury rotates as a solid rigid body on very long timescales, the angular momentum equation governing the motion of its spin axis in space can be written in an inertial reference frame attached to the Laplace plane and centered at the center of mass of the planet as

$$\frac{d\vec{L}}{dt} = \vec{\Gamma}^{prec} + \vec{\Gamma}^{nut} \tag{5}$$

with $\vec{L}$ the angular momentum and $\vec{\Gamma}^{prec}$ and $\vec{\Gamma}^{nut}$ the two parts of the torque exerted by the Sun on Mercury. The solution for $\vec{L}$ is associated to time-varying obliquity and deviation:

$$\varepsilon(t) \simeq \varepsilon_\Omega + \varepsilon_\omega \cos 2\omega(t), \tag{6}$$

$$\tilde{\delta}(t) \simeq \varepsilon_\omega \sin 2\omega(t). \tag{7}$$

The main torque $\vec{\Gamma}^{prec}$ is caused by the precession of the orbital node (Ω, with a period of $\sim$325 000 years) and, alone, leads to a spin precession characterized by a constant obliquity ε_Ω given by Eq. (1) and a zero deviation. The classically neglected secondary part $\vec{\Gamma}^{nut}$ is caused by the precession of the orbital pericenter (ω, period of $\sim$134 000 years) and the large eccentricity e, and drives a small and very slow nutation about the

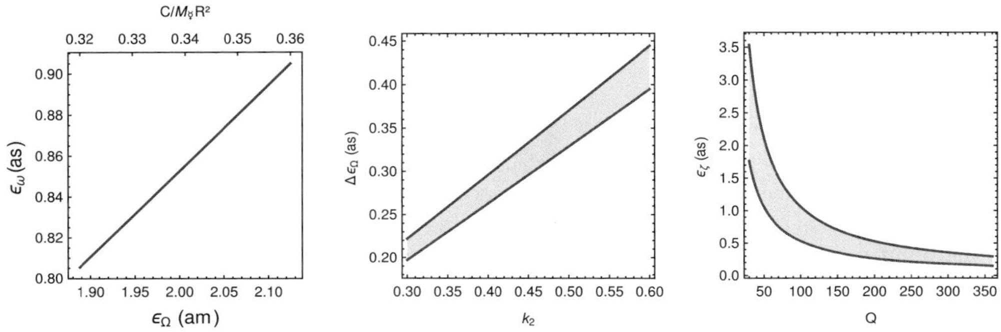

Figure 4. From left to right: nutation amplitude ε_ω as a function of the precession amplitude ε_Ω, tidal shift in obliquity $\Delta\varepsilon_\Omega$ as a function of k_2 ($C/M_m R^2 = 0.32$ and 0.36 for the bottom and top curves, respectively), and tidal shift in deviation as a function of Q ($k_2 = 0.3$ and 0.6 for the bottom and top curves, respectively), taken from Baland *et al.* (2017a).

main precession with the amplitude ε_ω, leading to periodic variations in obliquity and deviation with respect to the coplanarity. Due to the nutation, coplanarity is therefore very unlikely.

The amplitude ε_ω of the nutation is proportional to the amplitude ε_Ω of the main precession. For $C/M_☿ R^2$ ranging from 0.32 to 0.36, ε_ω ranges between 0.8 and 1 arcsec, see Fig. 4 (left panel). Since $\omega(0) = 50.379554°$, $\varepsilon(0)$ is reduced by 0.15 to 0.19 arcsec ($\sim 0.14\%$) with respect to ε_Ω and the deviation $\tilde{\delta}(0)$ is positive and ranges between 0.8 and 1 arcsec.

The nutation induced by the pericenter precession is essential to explain and interpret any observed positive deviation. The deviation related to the nutation is about half the observed deviation of Stark *et al.* (2015) for instance. The nutation results from the orbital dynamics and certainly does exist, even though it was classically neglected. In Peale (1974), the Hamiltonian includes one term that would lead to nutation induced by the pericenter precession, if it had not been neglected on the grounds that it is of third order in eccentricity. Peale *et al.* (2016) numerically identifies the nutation in their model.

3.5. *Cassini state: Effet of solid tides and associated dissipation*

Even assuming that Mercury rotates as a solid body on very long timescales, it deforms tidally at the orbital period, which consequently modifies the solar torque through (i) the tidal Love number k_2, which describes changes in the external gravitational potential, and (ii) the quality factor Q associated with tidal dissipation and corresponding to a phase shift $\zeta = 1/Q$ in the mean anomaly in the torque expression. Due to solid tides, the angular momentum of Mercury is affected by a constant shift in orientation over time. The solution for the time-varying obliquity and deviation, including both the nutation due to the pericenter precession and the constant shift in orientation due to the solid tides reads, see Baland *et al.* (2017a):

$$\varepsilon(t) \simeq \varepsilon_\Omega + \Delta\varepsilon_\Omega + \varepsilon_\omega \cos 2\omega(t), \quad (8)$$

$$\tilde{\delta}(t) \simeq \varepsilon_\zeta + \varepsilon_\omega \sin 2\omega(t). \quad (9)$$

The estimated k_2 of Mercury varies between 0.45 and 0.57 (see Mazarico *et al.* (2014); Verma and Margot (2016); Genova *et al.* (2019); Konopliv *et al.* (2020), whereas Q, as yet unestimated, probably lies in the interval [30, 360] that covers current estimates for the Earth, the Moon and Mars (Lainey 2016). The shift in obliquity $\Delta\varepsilon_\Omega$ is proportional to k_2, positive, and varies from ~ 0.30 to ~ 0.45 arcsec ($\sim 0.30\%$) for k_2 in the interval

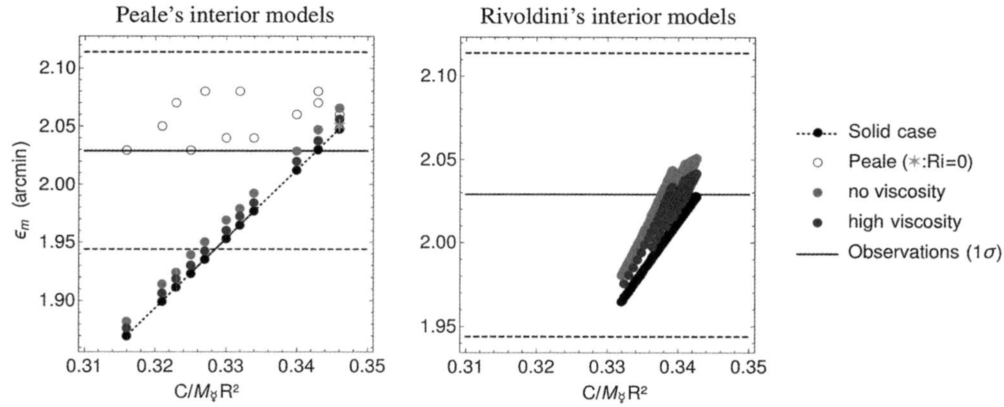

Figure 5. Mantle obliquity of Mercury as a function of the normalized polar moment of inertia for the set of 13 interiors structure models defined in Peale *et al.* (2016) (left panel) and for a large set of realistic interior models consistent with the annual libration amplitude and based on Rivoldini and Van Hoolst (2013). The black dots indicate the solid case obliquities according to Eq. (1), the clear dots represent the results of Peale *et al.* (2016) (their Table 1), the purple and red dots represent the results of Baland *et al.* (2017b), and the solid and dashed line represent the measured obliquity of Margot *et al.* (2012). Note that the radius of the inner core R_i increases with decreasing MOI and that one of Peales's interior model, with $R_i = 0$, has no solid inner core and the largest MOI. Figure adapted from from Baland *et al.* (2017b).

$[0.45, 0.60]$, see Fig. 4 (middle panel). The shift in deviation ε_ζ is inversely proportional to Q, positive, and of the order of 1 arcsec (Fig. 4, right panel).

Note that the tidal Love number k_2 cannot be inferred from an obliquity measurement, because it appears in $\Delta\varepsilon_\Omega$ and would be correlated with the polar moment if inertia which appears in ε_Ω. This is why it is useful to analyze tidal and rotational measurements together. The shift in deviation due to solid body tides, also identified analytically in MacPherson and Dumberry (2022), along with the deviation associated with pericenter precession-induced nutation, could explain any observed positive deviation of a few arcsec and help to constrain Mercury's interior structure and rheology.

3.6. *Cassini state: Effect of the fluid outer core and the inner core*

As shown by the measured annual librations and the tidal Love number k_2, the core of Mercury is at least partially molten. The presence of a solid inner core is suspected, although the conclusion seems to depend on the value considered for the polar moment of inertia: Genova *et al.* (2019), thanks to their small obliquity, claim to have confirmed its presence, while Rivoldini and Van Hoolst (2013) using the larger obliquity of Margot *et al.* (2012) were unable to settle the question. In the presence of a solid inner core, internal gravitational-pressure and magnetic-viscous couplings between the internal layers could significantly influence the obliquity of the planet, thus affecting its interpretation in terms of interior. However, contradictory results coexist on this subject in the literature.

Peale *et al.* (2016) computed the mantle obliquity ε_m for a limited set of 13 interior structure models, see Fig. 5 (left panel). For each interior model, ε_m is larger (up to $5 - 10\%$) than the solid case obliquity ε of Eq. (1). It should be noted that no clear relationship between ε_m and $C/M_☿R^2$ emerges from these results, which is intriguing. Baland *et al.* (2017b) tried to reproduce the results for the same set of interior models and found other values for ε_m, which depend linearly on $C/M_☿R^2$. They also found that the modeled obliquity ε_m is only 0.5 to 1% larger than the solid case obliquity.

Dumberry (2021), for a different set of interior models which satisfy exactly the value of $C/M_☿ R^2 = 0.3455$, concluded that the presence of a solid inner core affects the mantle obliquity by as much as 0.5%.

The problem with the results discussed above is that they are based on a set of interior models that is unrepresentative of the current knowledge of Mercury's internal structure. Baland *et al.* (2017b), using a set of plausible interior structure models based on Rivoldini and Van Hoolst (2013), found that the modeled obliquity may be up to 2% larger than the solid case obliquity, see Fig. 5 (right panel). An additional increase in mantle obliquity arises in the presence of a solid inner core, due to the elastic deformations caused by the misalignment of the solid layers, which perturb the obliquity as much as the tidal deformations due to gravitational potential of the Sun (MacPherson and Dumberry 2022).

All in all, interpreting a measured obliquity with a model which includes the effect of pericenter precession, solid body tides, and the internal couplings between the internal layers, instead of the Classical Cassini state model, may lead to a polar moment of inertia decreased by up to $\sim 2\%$ and therefore to a more differentiated interior more likely to include a solid inner core. For now, the actual precision ($3-4\%$) on the orientation precludes further constraints on Mercury's interior.

4. Conclusion

Mercury is in a 3:2 spin-orbit resonance and is most likely locked in a Cassini state, as different estimates of the orientation of its spin axis tend to confirm. Precise modeling of the rotation is required to interpret these measurements in terms of the planet's inner properties. The most advanced theoretical models predict a slightly larger obliquity than the classical Peale's equation, making the presence of a solid inner core slightly more likely. The advanced models also predict a deviation of a few arcsec from coplanarity, in agreement with most observations. On the basis of theoretical modelling, it is practically impossible for the actual deviation to be zero. Long-period librations, tidal Love numbers, and deviation of the spin axis can be used in combination with measured obliquity, annual libration, and gravity field and with advanced rotation models to constrain Mercury's interior structure and rheology more precisely than with Peale's experiment.

5. Acknowledgment

I am grateful to my colleagues on this topic: Marie Yseboodt, Attilio Rivoldini and Tim Van Hoolst. This work was financially supported by the Belgian PRODEX program managed by the European Space Agency in collaboration with the Belgian Federal Science Policy Office.

References

Archinal, B. A. and A'Hearn, M. F. and Bowell, E., *et al.* 2011, *Celestial Mechanics and Dynamical Astronomy*, 109, 101–135

B.A. Archinal, C.H. Acton, M. F. A'Hearn, *et al.* 2018, *Celestial Mechanics and Dynamical Astronomy*, 130, 22

Baland, R.-M., Yseboodt, M., Rivoldini, A., *et al.* 2017a, *Icarus*, 291, 136–159

Baland, R.-M., Yseboodt, M., Rivoldini, A., *et al.* 2017b, *European Planetary Science Congress*, EPSC2017–521

Bertone, S., Mazarico, E., Barker, M. K., *et al.* 2021, *Journal of Geophysical Research (Planets)*, 126, e06683

Colombo, G. 1965, *Nature*, 208, 575

Dumberry, M., Rivoldini, A., Van Hoolst, T., Yseboodt, M. 2013, *Icarus*, 225, 62–74

Dumberry, M., Rivoldini, A., 2015, *Icarus*, 248, 254–268

Dumberry, M. 2021, *Journal of Geophysical Research (Planets)*, 126, e06621
Feynman, Richard P. 1985, Surely you're joking Mr. Feynman!, New York: W.W.Norton
Genova, A., Goossens, S., Mazarico, E., *et al.* 2019, *Geophysical Research Letters*, 46, 3625–3633
Goldreich, Peter and Peale, Stanton J. 1968, *Annual Review of Astronomy and Astrophysics*, 6, 287
Goossens, S., Renaud, J. P., Henning, W. G., *et al.* 2022, *The Planetary Science Journal*, 3, 37
Hauck, S. A., Margot, J.-L., Solomon, S. C., *et al.* 2013, *Journal of Geophysical Research (Planets)*, 118, 1204–1220
Konopliv, A. S., Park, R. S., Ermakov, A. I. 2020, *Icarus*, 335, 113386
Lainey, V. 2016, *Celestial Mechanics and Dynamical Astronomy*, 126, 145–156
MacPherson, I., Dumberry, M. 2022, *Journal of Geophysical Research (Planets)*, 127, e07184
Margot, J. L., Peale, S. J., Jurgens, R. F., *et al.* 2007, *Science*, 316, 710–714
Margot, J. L. 2009, *Celestial Mechanics and Dynamical Astronomy*, 105, 329–336
Margot, J. L., Peale, S. J., Solomon, S. C., *et al.* 2012, *Journal of Geophysical Research (Planets)*, 117, E00L09
Margot, J.-L., Hauck, S. A., Mazarico, E., *et al.* 2018, Mercury's Internal Structure, Mercury The View after MESSENGER: Cambridge University Press
Mazarico, E., Genova, A., Goossens, S., *et al.* 2014, *Journal of Geophysical Research (Planets)*, 119, 2417–2436
Padovan, S., Margot, J.-L., Hauck, S. A., *et al.* 2014, *Journal of Geophysical Research (Planets)*, 119, 850–866
Peale, S. J. and Gold, T. 1965, *Nature*, 206, 1240–1241
Peale, S. J. 1969, *Astronomical Journal*, 74, 483–489
Peale, S. J. 1974, *Astronomical Journal*, 79, 722
Peale, S. J. 1976, *Nature*, 262, 765–766
Peale, S. J. 1981, *Icarus*, 48, 143–145
Peale, S. J, Margot, J.-L., Hauck, S. A., Solomon, S. C. 2016, *Icarus*, 264, 443–455
Pettengill, G. H. and Dyce, R. B. 1965, *Nature*, 206, 1240
Rivoldini, A., Van Hoolst, T., Verhoeven, O. 2009, *Icarus*, 201, 12–30
Rivoldini, A., Van Hoolst, T. 2013, *Earth and Planetary Science Letters*, 377, 62–72
Smith, D. E., Zuber, M. T., Phillips, R. J., *et al.* 2012, *Science*, 336, 214
Stark, A., Oberst, J., Preusker, F., *et al.* 2015, *Science*, 117, 64–72
Stark, A., Oberst, J., Preusker, F., *et al.* 2018, *Journal of Geodesy*, 92, 949–961
Steinbrügge, G., Dumberry, M., Rivoldini, A., *et al.* 2021, *Geophysical Research Letters*, 48, e89895
Van Hoolst, T., Rivoldini, A., Baland, R.-M., Yseboodt, M. 2012, *Earth and Planetary Science Letters*, 333, 83–90
Verma, A. K., Margot, J.-L. 2016, *Journal of Geophysical Research (Planets)*, 121, 1627–1640
Yseboodt, M., Margot, J.-L. 2006, *Icarus*, 181, 327–337
Yseboodt, M., Rivoldini, A., Van Hoolst, T., Dumberry, M. 2013, *Icarus*, 226, 41–51

Complex Planetary Systems II: Latest Methods for an Interdisciplinary Approach
Proceedings IAU Symposium No. 382, 2024
A. Lemaitre & A.-S. Libert, eds.
doi:10.1017/S1743921323004787

Dynamics of Circumstellar Planets in Binary Star Systems

Man Hoi Lee[1,2], Ka Ho Wong[1], Ho Wan Cheng[1], Trifon Trifonov[3], Sabine Reffert[4] and Andreas Quirrenbach[4]

[1]Department of Earth Sciences, The University of Hong Kong, Pokfulam Road, Hong Kong
email: mhlee@hku.hk

[2]Department of Physics, The University of Hong Kong, Pokfulam Road, Hong Kong

[3]Max-Planck-Institut für Astronomie, Königstuhl 17, 69117 Heidelberg, Germany

[4]Landessternwarte, Zentrum für Astronomie der Universität Heidelberg, Königstuhl 12, 69117 Heidelberg, Germany

Abstract. Circumstellar planets in binary star systems provide unique constraints on the formation and dynamical evolution of planets. We present an empirical formula for the stability boundary of coplanar retrograde orbits, similar to the classic one for coplanar prograde orbits. We discuss two of the tightest binaries with circumstellar planets: HD 59686 and ν Octantis. For HD 59686, dynamical fitting of the radial velocity data and stability analysis show that the planet must be either on a nearly coplanar retrograde orbit or in one of the narrow regions of prograde orbits stabilized by secular apsidal alignment. For ν Octantis, a nearly coplanar retrograde planetary orbit is the only option for dynamical stability. We also discuss the mysterious case of ϵ Cygni. It shows short-period radial velocity variations that closely resemble the signal of a Jupiter-mass planet, but the period and amplitude change over time and dynamical stability analysis rules out a planet.

Keywords. planets - binary stars - stability - HD59686 - ν Octantis

1. Planets in Binary Star Systems

Planets in binary star systems provide unique constraints on the formation and dynamical evolution of planets. They can be orbiting around one of the stars in a circumstellar (S-type) orbit or around the binary in a circumbinary (P-type) orbit (Dvorak 1986). To date, the number of detected planets in binary star systems is only about 440, even though nearly 1/3 of nearby solar-type stars are in binary systems (Duquennoy and Mayor 1991; Raghavan et al. 2010) and the number of confirmed exoplanets is over 5500 (see https://exoplanetarchive.ipac.caltech.edu/). The small detection number of these planets is in part due to observational bias. But observations have also shown that the planet occurrence rate in binary systems with orbital semimajor axis $a_B \lesssim 47$ AU is only $\sim 34\%$ of that for wider binaries or single stars (Wang et al. 2014a,b; Kraus et al. 2016).

Figure 1 shows the orbital semimajor axis of the outermost planet versus the projected separation or semimajor axis (if orbit is known) of the binary for all binaries of separation < 500 AU with circumstellar planets. The filled square and triangle are the ν Octantis and HD 59686 systems, respectively, with the horizontal bars indicating the periastron and apastron binary separations. They are two of the tightest binaries ($a_B \lesssim 20$ AU) with circumstellar planets on relatively wide orbits ($a_p \gtrsim 1$ AU), and in both cases, the planet is likely on retrograde orbit relative to the binary's. We examine these systems in this

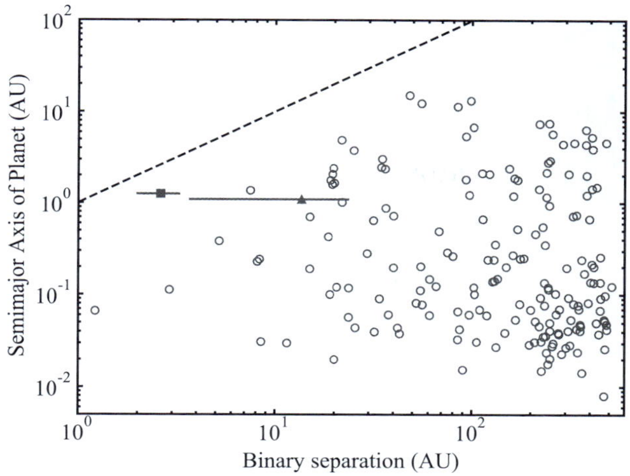

Figure 1. Orbital semimajor axis of the outermost planet versus the projected separation or semimajor axis (if the orbit is known) of the binary for all binaries of separation < 500 AU with circumstellar planets. Data obtained from `http://exoplanet.eu/planets_binary/`.

paper, after we discuss the stability boundary of circumstellar orbits in binary systems and the mysterious case of ϵ Cygni.

2. Stability of Circumstellar Orbits in Binary Systems

In the classic work by Holman and Wiegert (1999), the stability of both circumstellar and circumbinary planetary orbits was studied by performing numerical simulations for a wide range of binary mass ratio $\mu_B = M_B/(M_A + M_B)$ and eccentricity e_B (see also Rabl and Dvorak 1988; Quarles and Lissauer 2016; Quarles et al. 2020). The system consists of a primary star of mass M_A and a secondary star of mass M_B in an orbit of semimajor axis a_B and a test particle in a coplanar, prograde, and initially circular orbit (i.e., orbital angular momentum of the test particle is parallel to that of the binary) at semimajor axis a_p. For each combination of μ_B and e_B, the stability boundary for circumstellar planets is defined as a_p where all orbits with $a_p < a_{c,\text{pro}}$ are stable for 10^4 binary periods. By fitting an empirical formula to the numerical results, Holman and Wiegert (1999) obtained the following formula for the stability boundary of coplanar prograde orbits:

$$a_{c,\text{pro}}/a_B = 0.464 - 0.380\mu_B - 0.631e_B + 0.586\mu_B e_B + 0.150e_B^2 - 0.198\mu_B e_B^2. \tag{1}$$

It should be noted that the boundary between stable and unstable regions is not sharp and that there is a transition region between $a_{c,\text{pro}}$ and a second boundary at $a_{c,2,\text{pro}} > a_{c,\text{pro}}$ where some orbits are stable.

It is well known since the work by Harrington (1977) that retrograde orbits are more stable than prograde ones (see also Morais and Giuppone 2012; Quarles and Lissauer 2016), but as far as we know, an empirical formula similar to Equation (1) had not been constructed. We have performed numerical simulations using the Wisdom-Holman symplectic integrator in the SWIFT package with setup similar to those of Holman and Wiegert (1999), but for coplanar, retrograde, and initially circular test particle orbits (see Wong & Lee, in preparation, for details). The results are shown in the left panel of Figure 2. As expected, systems with a more massive perturber (i.e., larger μ_B) and/or larger e_B have smaller $a_{c,\text{retro}}$. We apply a least-squares fit to the results and

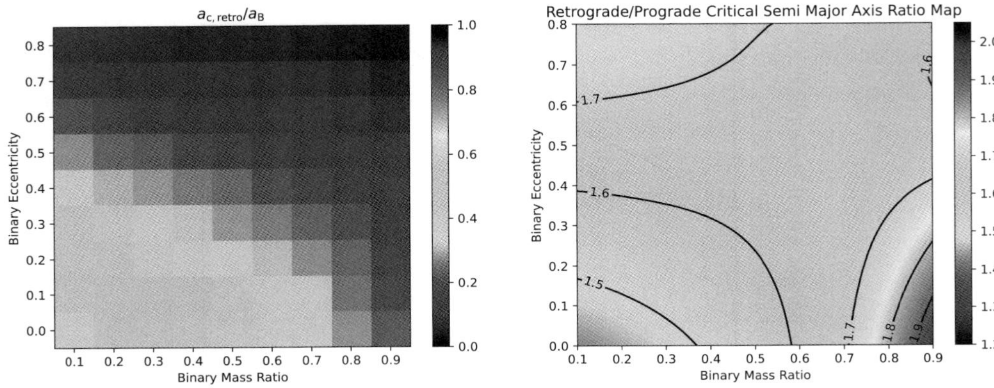

Figure 2. Stability boundary $a_{c,\rm retro}/a_B$ (left panel) and the ratio $a_{c,\rm retro}/a_{c,\rm pro}$ (right panel) in the parameter space of the binary mass ratio μ_B and eccentricity e_B.

obtain the following formula for the stability boundary of coplanar retrograde orbits:

$$a_{c,\rm retro}/a_B = 0.654 - 0.456\mu_B - 0.684e_B + 0.418\mu_B e_B - 0.0182e_B^2 + 0.0855\mu_B e_B^2. \quad (2)$$

The ratio $a_{c,\rm retro}/a_{c,\rm pro}$, using Equations (1) and (2), is shown in the right panel of Figure 2. The ratio $a_{c,\rm retro}/a_{c,\rm pro}$ is larger than one over the whole region, which means that retrograde orbits are more stable than prograde orbits for general binary systems. The ratio $a_{c,\rm retro}/a_{c,\rm pro}$ is conveniently nearly constant (≈ 1.5–1.7) over a wide range of μ_B and e_B relevant for circumstellar planets in binary star systems. In the Hill's regime where $\mu = 1 - \mu_B = M_A/(M_A + M_B) \ll 1$ and $e_B = 0$, we find that $a_{c,\rm retro}/a_{c,\rm pro} \approx 2$. Further analysis shows that both $a_{c,\rm pro}$ and $a_{c,\rm retro}$ scale with the Hill radius $r_H = (\mu/3)^{1/3} a_B$, with $a_{c,\rm pro} \approx 0.45 r_H$ and $a_{c,\rm retro} \approx 0.92 r_H$ (see also Innanen 1980; Hamilton and Burns 1991, 1992).

3. A Cautionary Tale: ϵ Cygni

Heeren et al. (2021) have analyzed in detail the radial velocity (RV) observations of ϵ Cygni, which is a K0 III giant star of $M_A = 1.10 M_\odot$. ϵ Cygni was observed by the Lick G and K giants RV survey (Frink et al. 2001; Reffert et al. 2015). The Lick RV measurements and earlier RV measurements by McMillan et al. (1992) show a long term trend due to a distant binary companion and short-period ($\sim 290\,$d) variations that could be due to a circumstellar planet orbiting the primary star. A large set of RVs was subsequently obtained with the SONG telescope from 2015 to 2018. The stellar companion happened to go through periastron passage during this time, which allows the properties of the highly eccentric binary to be accurately determined: $a_B = 15.8\,$AU, $e_B = 0.93$, and the minimum mass of the secondary $M_B \sin i = 0.265 M_\odot$. If the short-period RV variations were due to a planet, the best double-Keplerian fit indicates that the planet would have an orbital period of about 291 d, $a_p = 0.89\,$AU, and a minimum mass of $1.0 M_J$, where M_J is Jupiter mass. However, a more careful analysis shows that both the period and amplitude of the short-period variations change over time. In addition, a planet at $a_p \sim 0.89\,$AU should be unstable according to Equations (1) and (2), and it would come within the Hill radius of the stellar companion at periastron distance $q_B \sim 1.1\,$AU. A more detailed stability analysis of this system confirms that no stable orbits could be found in a large region of parameter space around the best fit. So it is highly unlikely that the short-period RV variations are due to a planet. The most likely explanation for the short-period variations of ϵ Cygni is the heartbeat phenomenon, which are stellar oscillations excited by a stellar companion passing close to the primary. But it would be a rather extreme example of a heartbeat system, because of its large periastron separation (> 20 stellar radii) and large stellar radius ($\approx 11 R_\odot$).

4. HD 59686

HD 59686 is another star observed by the Lick G and K giants RV survey. It is a K2 III giant star of $M_A = 1.9 M_\odot$. The 88 RV measurements over 12 years clearly show a circumstellar planet on a nearly circular orbit around the primary star of an eccentric binary, with the double-Keplerian fit giving $M_B \sin i = 0.530 M_\odot$, $P_B = 11700$ d, $a_B = 13.6$ AU, $e_B = 0.729$, $M_p \sin i = 6.9 M_J$, $P_p = 299$ d, $a_p = 1.09$ AU, and $e_p = 0.05$ (Ortiz et al. 2016). HD 59686 is the most eccentric close binary to harbor a circumstellar planet. The high binary eccentricity means that the stars are only separated by ~ 3.7 AU at periastron, leading to strong gravitational perturbations of the planet at $a_p = 1.09$ AU by the secondary. Indeed, a_p is just outside the stability boundary for coplanar prograde orbits, which is $a_{c,\mathrm{pro}} \approx 0.97$ AU according to Equation (1). On the other hand, a_p is well inside the stability boundary for coplanar retrograde orbits, which is $a_{c,\mathrm{retro}} \approx 1.67$ AU according to in Equation (2).

Trifonov et al. (2018) have performed a detailed orbital and stability analysis of this system. The best coplanar prograde dynamical fit is unstable on timescales of about 42 kyr. The best coplanar retrograde dynamical fit is slightly better than the prograde dynamical fit in χ^2 value, and it is stable for at least 10 Myr. A bootstrap analysis finds that a few % of the edge-on prograde dynamical fits are stable, while almost all of the edge-on retrograde dynamical fits are stable for at least 10 Myr. The stable prograde fits fall in narrow regions of the parameter space, and they are stabilized by secular apsidal alignment, with the difference in the longitude of periapse $\Delta\varpi = \varpi_p - \varpi_B$ librating about 0°

4.1. *Stability of Planetary Orbits in HD 59686*

Chan & Lee (in preparation) have explored systematically the parameter space of planetary orbits in the HD 59686 system to investigate its stability behaviors, particularly those related to secular apsidal alignment and mean-motion resonances (MMR). The planet is treated as a test particle in a circumstellar orbit around the primary star. We use the Wisdom-Holman symplectic integrator in the SWIFT package for the numerical simulations, and the maximum integration time is 10^5 binary periods.

Figure 3 shows the numerical results on the survival time of the particles in the (a_p, e_p) parameter space for coplanar prograde orbits with initial $\Delta\varpi = 0°$ and mean anomalies $\mathcal{M}_p = 0°$ and $\mathcal{M}_B = 180°$. There is a bulk stable region in the lower left corner that shrinks toward smaller semimajor axis at higher eccentricity, which is expected because more eccentric orbits are generally less stable. The stability boundary, which is the boundary of the bulk stable region at $e_p = 0$, is $a_{c,\mathrm{pro}} = 0.975$ AU, which is in good agreement with 0.97 AU from Equation (1). Other than the bulk stable region, we see two new stability features. The first feature is the set of stable islands at $e_p \sim 0.1$ that extend from $a_p \approx 0.95$ AU to $a_p \approx 1.1$ AU. They exist at distances beyond the circular stability boundary $a_{c,\mathrm{pro}}$. They fall on a nearly horizontal line but break apart at the positions of MMR. These stable islands are due to secular apsidal alignment. The second feature is the set of stable islands at the odd-order MMR such as 1:40, 1:42, 1:44, and 1:46 at 1.072, 1.038, 1.006, and 0.977 AU, respectively. These stable islands also exist at distances beyond the circular stability boundary $a_{c,\mathrm{pro}}$, and they are at higher eccentricities than the secular apsidal alignment islands. Figure 4 is similar to Figure 3, but for simulations with the planet initially at its apocenter ($\mathcal{M}_p = 180°$). The islands stabilized by secular apsidal alignment are similar, but the MMR stable islands are at even orders (1:39, 1:41, etc.).

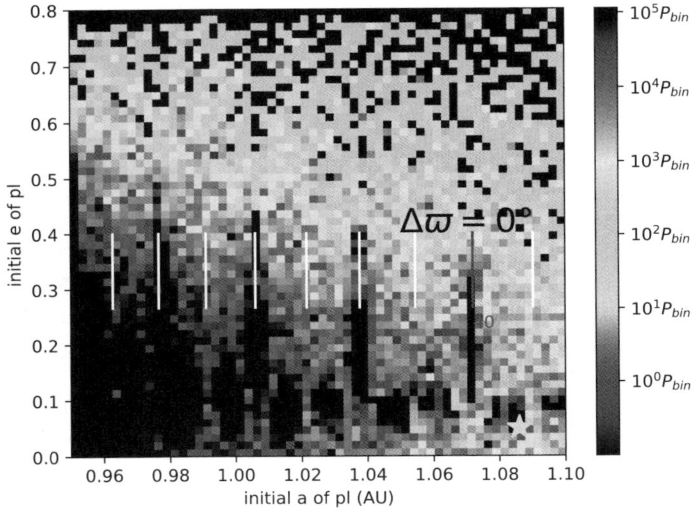

Figure 3. Stability map of the HD 59686 system showing the time at which a test particle becomes unstable as a function of initial a_p and e_p for coplanar prograde orbits with initial $\Delta\varpi = 0°$, $\mathcal{M}_p = 0°$, $\mathcal{M}_B = 180°$. The best-fit (a_p, e_p) of the observed planet is indicated by the star. Positions of MMR are indicated by the vertical lines, with the red one for 1:40.

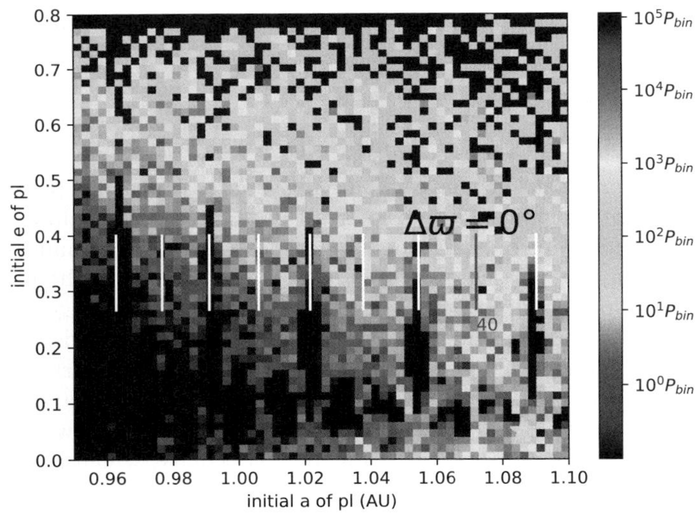

Figure 4. Same as Figure 3, but with initial $\Delta\varpi = 0°$, $\mathcal{M}_p = 180°$, $\mathcal{M}_B = 180°$.

The different patterns are due to the different initial $\mathcal{M}_p$, which leads to different initial MMR angles.

Figure 5 is similar to Figure 3 but for coplanar retrograde planetary orbits. The planetary orbits are initially aligned with the binary orbit ($\Delta\varpi = 0°$), with $\mathcal{M}_p = 0°$ and $\mathcal{M}_B = 180°$. The bulk stable region in the lower left corner is significantly larger than that for prograde orbits in Figure 3, with the stability boundary $a_{c,\mathrm{retro}} = 1.5$ AU, which is slightly smaller than 1.67 AU from Equation (2). Besides the bulk stable region, there is a series of islands stabilized by secular apsidal alignment that extend from the bulk region to the right along a nearly horizontal line at $e_p \sim 0.25$, and there are also stable

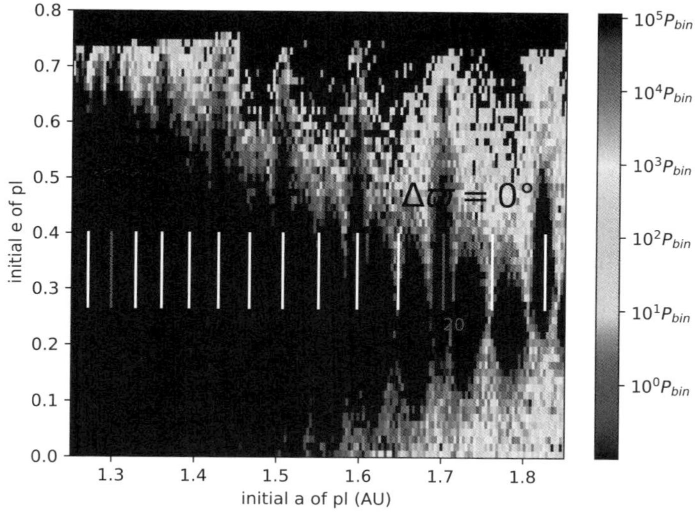

Figure 5. Same as Figure 3, but for coplanar retrograde orbits with initial $\Delta\varpi = 0°$, $\mathcal{M}_p = 0°$, $\mathcal{M}_B = 180°$. The red vertical lines indicate the 1:20 and 1:30 MMR.

islands at higher e_p at the odd-order MMR such as 1:18, 1:20, 1:22, etc. As in the case of prograde orbits, simulations with the same initial $\Delta\varpi = 0°$ and $\mathcal{M}_B = 180°$ as Figure 5 but $\mathcal{M}_p = 180°$ show similar stable islands due to secular apsidal alignment but MMR stable islands at even orders. The range of a_p simulated and shown in Figure 5 is chosen such that it covers well the stability boundary ($a_{c,\mathrm{retro}}$) and one can see the details of the transition from the stable region to the unstable region. The observed planet at $a_p = 1.09$ AU and $e_p = 0.05$ sits well inside the bulk stable region to the left of the region shown in Figure 5.

Based on these results, we can conclude that the real planet is more likely to be in a retrograde configuration, because coplanar retrograde orbits have a much larger stable region around the real planet's position than coplanar prograde orbits. If the real planet is in a prograde configuration, it must be in one of the narrow islands stabilized by secular apsidal alignment (in particular, the ones just inside and outside the 1:39 MMR, which are within the 1σ errors of the observed parameters), and it must not be in MMR with the binary companion, because the stable islands due to MMR have much higher e_p than the observed value.

5. ν Octantis

The planet in the ν Octantis system is undoubtedly one of the most dynamically challenging circumstellar planets in binary star systems. It was originally discovered by Ramm et al. (2009) based on ~ 220 cross-correlation (CCF) RV measurements taken with the HERCULES spectrograph on the 1-m telescope at Mt John Observatory. The data show evidence for a circumstellar planet and a binary companion around the K1 III giant star of $M_A = 1.6 M_\odot$, with $M_B \sin i \approx 0.50 M_\odot$, $P_B \approx 1050$ d, $a_B \approx 2.5$ AU, $e_B \approx 0.23$, $M_p \sin i \approx 2.4 M_J$, $P_p \approx 417$ d, $a_p \approx 1.2$ AU, and $e_p \approx 0.12$. The planetary orbit is located at almost twice the distance of the prograde stability boundary of $a_{c,\mathrm{pro}} \approx 0.65$ AU from Equation (1). Thus, the planet could not possibly be in a nearly coplanar and prograde orbit: it is in a highly unstable region of parameter space, which is unlikely to have any orbits stabilized by secular apsidal alignment or MMR (as in the case of HD 59686). On the other hand, a_p is comparable to the retrograde stability boundary of

$a_{c,\rm retro} \approx 1.0\,\rm AU$ from Equation (2), and Eberle and Cuntz (2010), Quarles et al. (2012), and Goździewski et al. (2013) have shown that the planet could be in a nearly coplanar and retrograde orbit.

Ramm et al. (2016) have subsequently obtained an additional ~1200 RV measurements with higher precision in 2009–2013 using an iodine cell with the HERCULES spectrograph. The RV data continues to show the short-period signal. A careful analysis of the photospheric and chromospheric activity indicators has also found no significant RV-correlated variability (Ramm et al. 2021). We have recently obtained new HARPS RV data, which together with the previous data, show that the short-period signal has been coherent for over 16 periods, which strongly suggests that the planet is real.

Finally, the mass of the secondary star is $M_B = 0.57 M_\odot$ (with inclination $i_B = 71.8°$ from HIPPARCOS data), which means that it could be either a low-mass main-sequence star or a white dwarf. We have obtained VLT/SPHERE adaptive optics image that could help us answer this question. If the secondary star is a white dwarf, then it is initially the more massive star and the evolution of the binary would have significant implications for the origin and evolution of the planet. A detailed analysis of the RV and imaging data will be reported in a forthcoming article (Cheng et al., in preparation).

6. Summary

Although it is important to keep in mind that nearly periodic RV variations of a star may not be due to an orbiting planet, as in the case of ϵ Cygni, there is strong evidence that the RV signals for the circumstellar planets in the HD 59686 and ν Octantis binary systems are real. HD 59686 and ν Octantis are two of the tightest binaries with circumstellar planets, and the planets are most likely on retrograde orbits, which are stable out to larger orbital semimajor axis than prograde orbits. For HD 59686, although we cannot yet rule out that the planet is on a prograde orbit, one would have to explain how a planet could be formed in a narrow region of the parameter space stabilized by secular apsidal alignment. If the circumstellar planets in both of these systems are confirmed to be on retrograde orbit, they pose serious challenging questions to the theory of planet formation: Did these planets form in retrograde circumstellar disk? Were they captured from prograde circumbinary orbits into retrograde circumstellar orbits (Gong and Ji 2018)? Are they the result of binary evolution?

Acknowledgements. This work was supported in part by Hong Kong RGC grants 17305015 and 17306720.

References

Duquennoy, A. & Mayor, M. 1991, *A&A*, 248, 485–524.
Dvorak, R. 1986, *A&A*, 167, 379–386.
Eberle, J. & Cuntz, M. 2010, *ApJL*, 721, L168–L171.
Frink, S., Quirrenbach, A., Fischer, D., Röser, S., & Schilbach, E. 2001, *PASP*, 113, 173–187.
Gong, Y.-X. & Ji, J. 2018, *MNRAS*, 478, 4565–4574.
Goździewski, K., Słonina, M., Migaszewski, C., & Rozenkiewicz, A. 2013, *MNRAS*, 430, 533–545.
Hamilton, D. P. & Burns, J. A. 1991, *Icarus*, 92, 118–131.
Hamilton, D. P. & Burns, J. A. 1992, *Icarus*, 96, 43–64.
Harrington, R. S. 1977, *AJ*, 82, 753–756.
Heeren, P., Reffert, S., Trifonov, T., Wong, K. H., Lee, M. H., Lillo-Box, J., Quirrenbach, A., Arentoft, T., Albrecht, S., Grundahl, F., Andersen, M. F., Antoci, V., & Pallé, P. L. 2021, *A&A*, 647, A160.
Holman, M. J. & Wiegert, P. A. 1999, *AJ*, 117, 621–628.
Innanen, K. A. 1980, *AJ*, 85, 81–85.
Kraus, A. L., Ireland, M. J., Huber, D., Mann, A. W., & Dupuy, T. J. 2016, *AJ*, 152, 8.

McMillan, R. S., Smith, P. H., Moore, T. L., & Perry, M. L. 1992, *PASP*, 104, 1173.
Morais, M. H. M. & Giuppone, C. A. 2012, *MNRAS*, 424, 52–64.
Ortiz, M., Reffert, S., Trifonov, T., Quirrenbach, A., Mitchell, D. S., Nowak, G., Buenzli, E., Zimmerman, N., Bonnefoy, M., Skemer, A., Defrère, D., Lee, M. H., Fischer, D. A., & Hinz, P. M. 2016, *A&A*, 595, A55.
Quarles, B., Cuntz, M., & Musielak, Z. E. 2012, *MNRAS*, 421, 2930–2939.
Quarles, B., Li, G., Kostov, V., & Haghighipour, N. 2020, *AJ*, 159, 80.
Quarles, B. & Lissauer, J. J. 2016, *AJ*, 151, 111.
Rabl, G. & Dvorak, R. 1988, *A&A*, 191, 385–391.
Raghavan, D., McAlister, H. A., Henry, T. J., Latham, D. W., Marcy, G. W., Mason, B. D., Gies, D. R., White, R. J., & ten Brummelaar, T. A. 2010, *ApJS*, 190, 1–42.
Ramm, D. J., Nelson, B. E., Endl, M., Hearnshaw, J. B., Wittenmyer, R. A., Gunn, F., Bergmann, C., Kilmartin, P., & Brogt, E. 2016, *MNRAS*, 460, 3706–3719.
Ramm, D. J., Pourbaix, D., Hearnshaw, J. B., & Komonjinda, S. 2009, *MNRAS*, 394, 1695–1710.
Ramm, D. J., Robertson, P., Reffert, S., Gunn, F., Trifonov, T., Pollard, K., & Cantalloube, F. 2021, *MNRAS*, 502, 2793–2806.
Reffert, S., Bergmann, C., Quirrenbach, A., Trifonov, T., & Künstler, A. 2015, *A&A*, 574, A116.
Trifonov, T., Lee, M. H., Reffert, S., & Quirrenbach, A. 2018, *AJ*, 155, 174.
Wang, J., Fischer, D. A., Xie, J.-W., & Ciardi, D. R. 2014,a *ApJ*, 791a, 111.
Wang, J., Xie, J.-W., Barclay, T., & Fischer, D. A. 2014,b *ApJ*, 783b, 4.

Complex Planetary Systems II: Latest Methods for an Interdisciplinary Approach
Proceedings IAU Symposium No. 382, 2024
A. Lemaitre & A.-S. Libert, eds.
doi:10.1017/S1743921323004167

Long-term stability and dynamical spacing of compact planetary systems

Antoine C. Petit

Université Côte d'Azur, Observatoire de la Côte d'Azur, CNRS, Laboratoire Lagrange, France
email: antoine.petit@oca.eu

Abstract. Exoplanet detection surveys revealed the existence of numerous multi-planetary systems packed close to their stability limit. In this proceeding, we review the mechanism driving the instability of compact systems, originally published in (Petit et al. 2020). Compact systems dynamics are dominated by the interactions between resonances involving triplets of planets. The complex network of three-planet mean motion resonances drives a slow chaotic semi-major axes diffusion, leading to a fast and destructive scattering phase. This model reproduces quantitatively the instability timescale found numerically. We can observe signpost of this process on exoplanet systems architecture. The critical spacing ensuring stability scales as the planet-to star mass ratio to the power 1/4. It explains why the Hill radius is not an adapted measure of dynamical compactness of exoplanet systems, particularly for terrestrial planets. We also provide some insight on the theoretical tools developped in the original work and how they can be of interest in other problems.

Keywords. Planetary systems, Planetary system stability, Exoplanet systems architecture, Celestial Mechanics

1. Introduction

Exoplanets discoveries have shown that one of the most common type of planetary system is composed of multiple Super-Earths orbiting very close to their host star (*e.g.* Fabrycky et al. 2014). The orbits of these planets are almost circular and coplanar (Johansen et al. 2012; Xie et al. 2016; He et al. 2020) and the large majority of them are not in resonance (Izidoro et al. 2017). Moreover these systems have a strong intra-system uniformity: planet radii and spacings between the planets are correlated with their neighbours (Weiss et al. 2018; Murchikova & Tremaine 2020). The spacing between the planets is constrained by the dynamical stability as there is a lower limit to the distance between two orbits at around 10 Hill radii (Pu & Wu 2015).

Understanding the origin of this stability boundary has been the topic of many research works. Indeed, the mutual Hill radius $R_{\rm H} = (a_1 + a_2)/2((m_1 + m_2)/(3m_\star))^{1/3}$, where a_k, m_k are respectively the planet semi-major axis and mass, is the natural length unit in the planetary three-body problem, as it measures the distance where the planet-planet interactions start to dominate over the stellar gravity. In this problem, the Hill stability gives an absolute boundary (Marchal & Bozis 1982) preventing from planet close encounters and collisions (but not from ejections or star-planet collsions). This boundary, which can be expressed (Petit et al. 2018) as a condition on the total Angular Momentum Deficit (Laskar 1997; Laskar & Petit 2017) can be simplified to a minimum spacing between circular and coplanar orbits (Gladman 1993) such that $a_2 - a_1 > 2\sqrt{3}R_{\rm H}$ guarantees the longterm stability of the system.

This sharp boundary has no equivalent in systems with more than three planets. Chambers et al. (1996) found numerically that multiplanetary systems experience a long quiescent phase where the system dynamics are almost secular before a very rapid transition to collisional dynamics. The instability timescale also scales exponentially with the spacing between the planets. While Chambers et al. (1996) proposed to scale the spacing with the mutual Hill radius of the planets, they also showed that a scaling as $(m_{\rm pl}/m_\star)^{1/4}$ is more accurate over a large mass ratio regime. Many works further confirmed that pionneering result (*e.g.* Faber & Quillen 2007; Smith & Lissauer 2009; Obertas et al. 2017, see Petit et al. 2020 for a comprehensive review) but most authors sticked with the Hill radius as a way to rescale the problem.

In Petit et al. (2020), we proposed the first quantitative analytical model of the instability mechanism in compact multiplanetary systems. Building upon the work of Quillen (2011), we derived a criterion for the overlap of a mean motion resonance (MMR) network involving three planets. We also showed that the effective diffusion of the semi-major axis due to this resonant network has a timescale consistent with numerical results. Moreover, the critical spacing for stability between the planets scales as $(m_{\rm pl}/m_\star)^{1/4}$.

In this proceeding, we come back on the results from Petit et al. (2020) and discuss how the concepts developped in this study can be used beyond this case. In Sect. 2, we briefly recall the main findings that work. We show in Sect. 3.1 that the dynamical spacing measurement coming from our model provide a better descritpion than the Hill spacing to understand the instability limit of exoplanet systems. In particular, we show that exoplanets systems are clustered much closer to the instability than if the spacing is measured in Hill radii. In Sect. 3.2, we discuss in detail why the idea of resonant optical depth is so powerful and the conditions where such a tool can be useful, in particular to compute analytically diffusion timescales.

2. The path to instability in compact systems

This section summarizes the results from Petit et al. (2020), we refer to it for a more detailed description. We study a system of three planets of masses m_1, m_2, m_3 orbiting a star of mass $m_\star$. The orbits are assumed to be circular and coplanar (see Sect. 3.3 for a discussion on generalization to the non-idealized case). The system coordinates are the Delaunay coordinates (Λ_i, λ_i) where $\Lambda_i = m_i\sqrt{\mu a_i}$, a_i is the planet semi-major axis and $\mu = \mathcal{G}m_\star$. We consider systems out of resonant chains, whose instabilities are studied in (Pichierri & Morbidelli 2020). While most numerical studies on this problem focus on the equal spacing case, we do not make this assumption, simply assuming that the spacing is small with respect to the semi-major axes.

2.1. *Instability mechanism*

The key features the tightly packed system instability present can be summarized as

(a) The survival time $T_{\rm surv}$ has an exponential dependency on the orbital spacing, measured in units of $(m_{\rm pl}/m_\star)^{1/4}$. The fit is valid over 6 to 8 orders of magnitude for $T_{\rm surv}$ between 100 and 10^{10} orbits.

(b) Instabilities occur for spacings larger than the Hill stability criterion, it is an intrinsically multi-planet phenomenon. While there is a small quantitative change going from three planet to five planet systems, there is no dependency beyond that number (Chambers et al. 1996). Three planets are necessary but also sufficient to reproduce the effect.

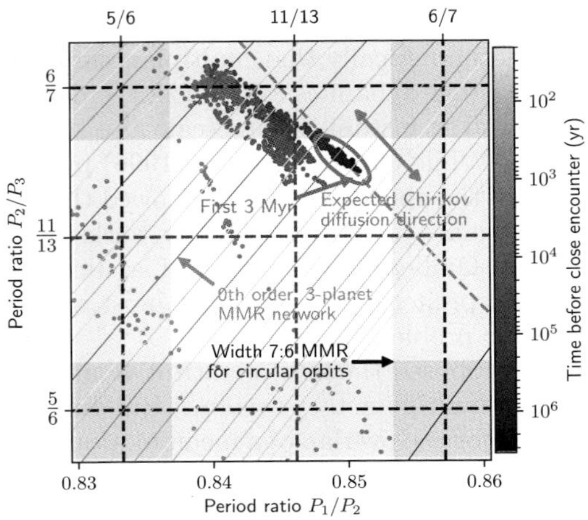

Figure 1. Evolution in the period ratios plane. The points are colour-coded according to their time before the close encounter. We note that the system spends almost all of its time very close to its starting location. Green oblique lines correspond the loci of the zeroth-order three-planet MMRs. The first-order two-planet MMRs, 7:6 and 6:5, are plotted in grey. The second-order resonance 13:11 is plotted in red. Adapted from Petit et al. (2020).

(c) The survival time distribution suggests that the evolution is driven by a diffusion process (Hussain & Tamayo 2020). The dips close to first-order two-planet MMRs indicate that these latter play a fundamental role in enabling the orbit crossing.
(d) Unstable systems spend most of their lifetime in a quiescent phase where they only experience weak diffusion of their semi-major axes but no AMD evolution. Once a system encounters a two-planet first-order MMR, a rapid chaotic phase leads to orbit crossing.

We illustrate the mechanism leading to instability on a typical numerical simulation. In this case, the planet masses are equal to $m_{\rm p} = 10^{-5}$ $\rm M_\odot$ and the period ratio is initially close to 1.175. We reproduce in Fig. 1, the Figure 3 of Petit et al. (2020). This Figure shows the evolution of the period ratios of both pair of planets P_1/P_2 and P_2/P_3 during the system lifetime. Planets start outside of the main neighbouring MMRs (the 6:5 and 7:6) and remain close to the initial position for 3 Myr. They diffuse towards the 7:6 MMR between the second and third planet perpendicularly to zeroth-order three planet MMRs (see below) as expected due to Chirikov (1979) diffusion. Once the 7:6 MMR is reached, a close encounter is almost immediate.

2.2. *Three-planet mean motion resonances overlap*

Based on Quillen (2011), we postulate that the slow diffusion is driven by the overlap of three-planet zeroth-order MMR. A zeroth-order three-planet MMR can be described by two integers p and q. The resonance equation is

$$pn_1 - (p+q)n_2 + qn_3 = 0 \quad \text{or equivalently} \quad \nu_{23} = 1 - \frac{p}{q}(\nu_{12}^{-1} - 1), \tag{2.1}$$

where $\nu_{ij} = P_i/P_j = n_j/n_i$, p and q are integers and, n_j is the planet mean motion. The resonance loci in the period ratio plane (ν_{12}, ν_{23}) correspond to hyperbola going through the point (1,1). and the full network is dense in that plane.

The resonance loci do not depend on the MMR index $p + q$, but only on the ratio of the integers p/q. An adapted set of coordinates to describe the period ratio plane can be defined to take advantage of this property. We define the resonance locator η and the generalised period ratio ν

$$\eta = \frac{1-\nu_{23}}{\nu_{12}^{-1}-\nu_{23}} \quad \text{and} \quad \nu = \frac{(1-\nu_{12})(1-\nu_{23})}{1-\nu_{12}\nu_{23}}. \tag{2.2}$$

The resonant locator η is a smooth function of the period ratio and $\eta = p/(p+q)$ on the resonance defined by (p,q). Constant ν lines are hyperbola along which the resonance strength is roughly comparable for a given $p+q$. Inspired by the generalized period ratio ν, we can also define a generalized spacing between the planets that the takes the form

$$\delta = \frac{\delta_{12}\delta_{23}}{\delta_{12}+\delta_{23}} = \left(\frac{1}{1-\alpha_{12}} + \frac{1}{1-\alpha_{23}}\right)^{-1} \simeq \frac{2}{3}\nu, \tag{2.3}$$

where $\delta_{ij} = 1-\alpha_{ij}$ and $\alpha_{ij} = a_i/a_j$.

When considering a single resonance defined by integers (p,q), we can show that there exist a canonical transformation from the Delaunay coordinates to a set of coordinates $(\Theta, \Gamma, \Upsilon, \theta_{\rm res}, \theta_\Gamma, \theta_\Upsilon)$ such that Γ, Υ are first integral of the motion after averaging over the fast angles to the second order in mass (Petit et al. 2020). In these coordinates, the motion is described by a classical pendulum Hamiltonian $\mathcal{H}_{\rm res} = -\mathcal{K}_2/2(\Theta-\Theta)^2 + \varepsilon^2 R_{pq}\cos\theta_{\rm res}$ where Θ_0 is the value of the action Θ at the resonance locus for given Γ and Υ, ε is a factor proportional to the planet-to-star mass ratio to remind the order of each term, $\mathcal{K}_2$ is the second-order expansion of the Keplerian part around the resonance center and $\varepsilon^2 R_{pq}$ is the factor in front of the resonant term. For these two last terms, we refer to Eqs. (42,47,49) of Petit et al. (2020). The width of the resonance and the small amplitude libration are straightforward to compute, we have respectively $\Delta\Theta = 2\varepsilon\sqrt{2R_{pq}/\mathcal{K}_2}$ and $\omega_{pq} = \varepsilon\sqrt{\mathcal{K}_2 R_{pq}}$.

The resonances have a clearer geometrical interpretation in the period ratio space, particularly when one needs to compare them. We compute the width of the resonances perpendicularly to the network, that is, the width in terms of the variable η

$$\Delta\eta_{pq} = \frac{\mathrm{d}\eta}{\mathrm{d}\Theta}\Delta\Theta = 6.55\varepsilon M \frac{(\eta(1-\eta))^{3/2}}{\delta^2} e^{-(p+q)\delta}, \tag{2.4}$$

where

$$\varepsilon M = \frac{\sqrt{m_1 m_3 + m_2 m_3 \eta^2 \alpha_{12}^{-2} + m_1 m_2 \alpha_{23}^2 (1-\eta)^2}}{m_\star} \tag{2.5}$$

describes the planet mass dependency. Importantly, the width $\Delta\eta_{pq}$ only depends on the sum $p+q$ such that the width is a smooth function for the subnetworks having the same resonant index $p+q$.

In order to find the region where the resonances overlap, we rely on the concept of resonance network optical depth or filling factor developped by Quillen (2011) and applied in the two-planet case by Hadden & Lithwick (2018). The idea is to consider the local proportion of the space that is occupied by the MMRs. If this proportion is larger than 1, it is a good approximation to consider that MMRs fully overlap and a diffusion of the actions is possible. We detail more in Sec. 3.2 the conditions where such concept is useful. Resonances in subnetworks with constant index $p+q$ have comparable size and a constant spacing in terms of η, such that the expression of the optical depth of the subnetwork is easy computed $\rho_{p+q} = (p+q)\Delta\eta$. The total optical depth of the network is bounded by the sum over $k = p+q$ of the optical depth of all subnetworks $\rho_{\rm tot} \leqslant \sum_k \rho_k$ as some resonances are counted with multiplicity. Yet, the exponential decay in k allow us

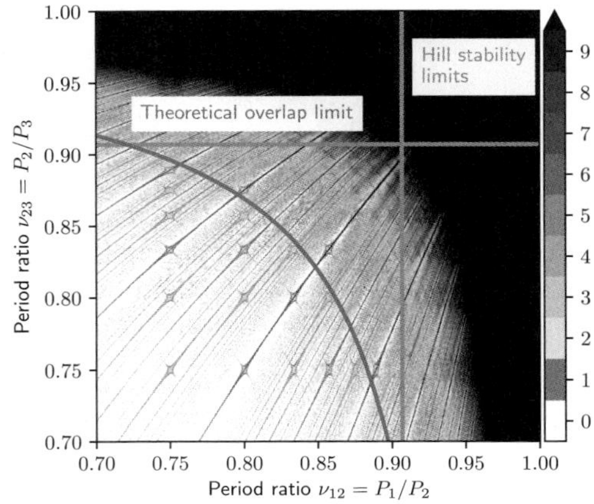

Figure 2. Number of resonances covering the period ratios space for equal-mass planets, with masses 10^{-5} $M_\odot$. At first order, the number of resonances can be compared to the resonance optical depth ρ_{tot}. We plot the two-planet circular Hill-stability limits (Petit et al. 2018) for both planet pairs in green and the predicted overlap limit for the three-planet MMRs (Eq. 2.7) in blue. Adapted from Petit et al. (2020).

to replace the inequality by an equality. Substituting the sum by an integral, the optical depth is

$$\rho_{tot} = 6.55\varepsilon M \frac{(\eta(1-\eta))^{3/2}}{\delta^2} \int_0^{+\infty} k e^{-k\delta} \mathrm{d}k = 6.55\varepsilon M \frac{(\eta(1-\eta))^{3/2}}{\delta^4}. \tag{2.6}$$

As Quillen (2011), we find that the filling factor depends linearly on the mass ratio and scales as δ^{-4}. We confirm that the natural spacing rescaling for the problem is not the Hill radius, which scales as $\varepsilon^{1/3}$, but rather a dependency on $\varepsilon^{1/4}$. We can define a critical spacing value δ_{ov} such that the zeroth-order three-planet MMR network fills the entire space. Taking $\rho_{tot} = 1$ and solving for δ, one obtains

$$\delta_{ov} = 1.59(\varepsilon M)^{1/4}(\eta(1-\eta))^{3/8}. \tag{2.7}$$

Here, δ_{ov} is a function of the masses and η. In the case of equal mass and spacing systems, Eq. (2.7) becomes $\delta_{ov,eq} = 1.0\,(m_p/m_0)^{1/4}$†.

We plot in Figure 2, the number of resonances that overlap at a given point in the plane (ν_{12}, ν_{23}) for three equal-mass planets as well as the two-planet Hill stability limits and the MMR overlap criterion derived in Eq. (2.6). The number of resonances is to first order a proxy for the filling factor ρ_{tot}. We see in Figure 2 that the region where the overlap of the three-planet MMR network takes place extends well beyond the Hill-stability limits, particularly for comparable spacings between the two neighbouring planet pairs.

2.3. *Diffusion model and instability timescale*

When resonances overlap, Chirikov (1979) predicts that the actions diffuse perpendically to the resonance hyperplane with a diffusion rate that depends on the resonance width and the period of the resonance such that the diffusion depends expoentially in the resonance index. As the resonance index increases, the associated diffusion rate vanishes, such that in the limit where the diffusion is dominated by smaller and smaller resonances, the timescale effectively tends to infinity. However, for $\delta < \delta_{ov}$, not all the resonances are

† There is a typo in (Petit et al. 2020) in Eq. (61) noticed by D. Tamayo (private communication).

necessary to cover the phase space. We therefore only need to consider the largest ones to compute the survival time.

The diffusion in the η direction is quite representative of the general Chirikov direction while giving a good parametrization of the resonant network. Since the diffusion rate depends on the index of the resonance, the diffusion is not homogeneous and the diffusion of η follows the equation

$$\frac{\mathrm{d}\eta}{\mathrm{d}t} = \sqrt{D(\eta)}b(t), \tag{2.8}$$

where $D(\eta)$ represent the diffusion coefficient due to the lowest index resonance at this location. Morbidelli & Vergassola (1997) give a solution to Eq. (2.8) providing that we can describe the function $D(\eta)$. However, in our case, this function is highly variable and we instead use a probabilistic approach. Instead of following the random walk by looking at which exact resonances are crossed until the first close encounter, we obtain a mean survival time by averaging over all the possible combinations of resonances that can contribute to the diffusion process (*i.e* the MMR with index lower than k_{ov}). In particular, we can define an effective diffusion coefficient, taking into account the contribution of all the resonances necessary to locally cover the phase space

$$D_{\mathrm{eff}} = \left(\int_0^{k_{\mathrm{ov}}} \frac{\rho_k}{\sqrt{D_k}}\mathrm{d}k\right)^{-2} = \left(\int_0^{k_{\mathrm{ov}}} \frac{k}{\sqrt{\omega_k}}\mathrm{d}k\right)^{-2}. \tag{2.9}$$

While the chaotic diffusion will be much more heterogenous (with fast and slow phases), on average the system will behave the same way if it where subject to a constant diffusion process of parameter D_{eff}. Taking into account the distance that the system needs to travel until it reaches a two-planet MMR that destabilizes it for good, we obtain a mean survival time for compact systems where three-planet zeroth-order MMR overlap

$$\left\langle \log \frac{T_{\mathrm{surv}}}{P_1} \right\rangle \simeq -\log\left(26.2\varepsilon M\sqrt{\eta(1-\eta)}\frac{1-(\delta/\delta_{\mathrm{ov}})^4}{(\delta/\delta_{\mathrm{ov}})^6}\right) + \sqrt{\left|\ln 1 - \left(\frac{\delta}{\delta_{\mathrm{ov}}}\right)^4\right|}. \tag{2.10}$$

This expression can be decomposed into a prefactor that mainly depends on the planet-to-star-mass ratio and a function that only depends on $\delta/\delta_{\mathrm{ov}}$. At first glance, this expression is not linear in δ, however the linear approximation is correct in the regime of interest, that is, for Hill-stable planet pairs not too close to the overlap limit. As Quillen (2011), we predict that beyond the overlap limit, the survival time is likely much larger since the Chirikov diffusion is not an efficient process on its own.

To compare with previous numerical studies, we compute the estimated survival time for equal-mass and equally spaced planets $\log_{10}(T_{\mathrm{surv}}/P_1) \simeq -\log_{10}\varepsilon_p - 6.51 + 3.56\Delta/\varepsilon_p^{1/4}$, where $\varepsilon_p = m_p/m_\star$. The slope coefficient 3.56 is very close to values obtained in previous works and as already discussed the spacing is rescaled by $\varepsilon_p^{1/4}$. The prefactor proportional to $1/\varepsilon_p$ is consistent with numerical simulations and the numerical constant is very close to the one obtained by Faber & Quillen (2007). We show on Fig. 3 a direct comparison to numerical simulations. We see that the analytical expression captures most of the trends observed, including the change of behaviour where the network stops overlapping. A more in-depth comparison to numerical results is done in Petit et al. (2020), from planet mass of 10^{-7} $\mathrm{M}_\odot$to 10^{-3} $\mathrm{M}_\odot$, unequal planet masses or spacing showing the very good agreement of the analytical expression.

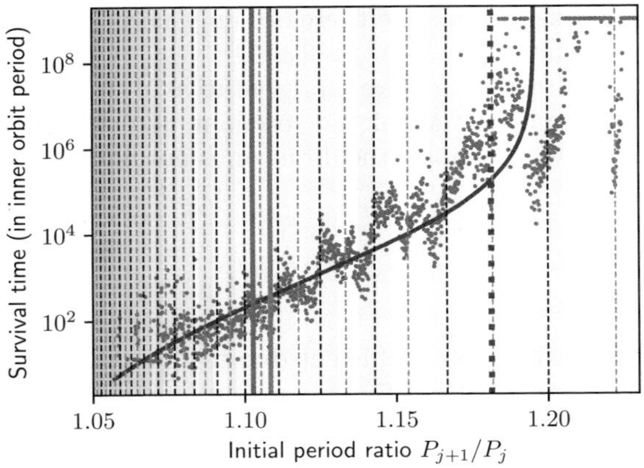

Figure 3. Survival time in numerical simulations for a equal spacing and 10^{-5} $M_\odot$ mass three-planet system as a function of the initial period ratio. The red curve corresponds to the survival time estimate (eq. 2.10), the blue vertical line to the Hill-stability limit (Petit et al. 2018), the orange vertical line to the two-planet MMR overlap criterion (Wisdom 1980). The dashed black (resp. green) lines are the two-planet first(resp. second)-order MMR. The light orange rectangles show an estimate of the width of the two-planet MMR (Petit et al. 2017). We refer to Petit et al. (2020) (from which this firgure is adapted) for the description of the numerical setup.

3. Discussion: key concepts and generalization

3.1. *Dynamical spacing: implication for observation and theoretical models*

Until now, most of the studies focused on the spacing of exoplantary systems used the Hill radius as a way to measure compactness. However, exoplanet to star mass ratio ε spans from a few 10^{-7} to more than 10^{-3}, such that the difference between a scaling in $\varepsilon^{1/3}$ or $\varepsilon^{1/4}$ is significant. Smaller exoplanets tend to have wider spacing when expressed in Hill radii but there should be no significant mass dependency when expressed in the unit of the MMR overlap δ_{ov} (Eq. 2.7).

We select from the Exoplanet Archive† all the systems containing more than three planets to study how close to the stability limits these systems are. For systems with only radius know, we use the mass-radius relationship from Marcy et al. (2014). Due to the weak exponents at play, there is no need for a more refined relationship. We plot on Fig. 4 the dynamical spacing δ (Eq. 2.3) for triplets of adjacent exoplanets rescaled by the critical value δ_{ov} (left plot), as a histogram and as a function of the triplet mass factor εM. We also plot the spacing between adjacent pairs of planets measured in Hill radii as a histogram and as a function of the planet mass ratio. Solar System triplets and pairs are plotted as stars. While the two mass ratio units are not the same, they allow for a meaningful comparison of the two spacing metrics. The black continuous lines give the three-planet stability criterion for circular orbits. On the right plot, the Hill stability limit is plotted in blue.

We see on the Hill spacing plot that a clear mass ratio dependency is visible. As the mass ratio decreases, spacings close to the two planet limit become very rare and almost all systems are consitstent with the three planet stability criterion. Importantly, terrestrial planets all show Hill spacing larger than 17. On the left plot, there is no obvious mass dependency of the lower limit in spacing, terrestrial planets are consistent with the spacing value found for the bulk of the other exoplanets.

Expressed in terms of dynamical spacing δ, exoplanetary systems are much closer to the stability limit than expected from their spacing in Hill radius. Indeed, the Hill spacing

† https://exoplanetarchive.ipac.caltech.edu/, consulted on the 30/06/2023.

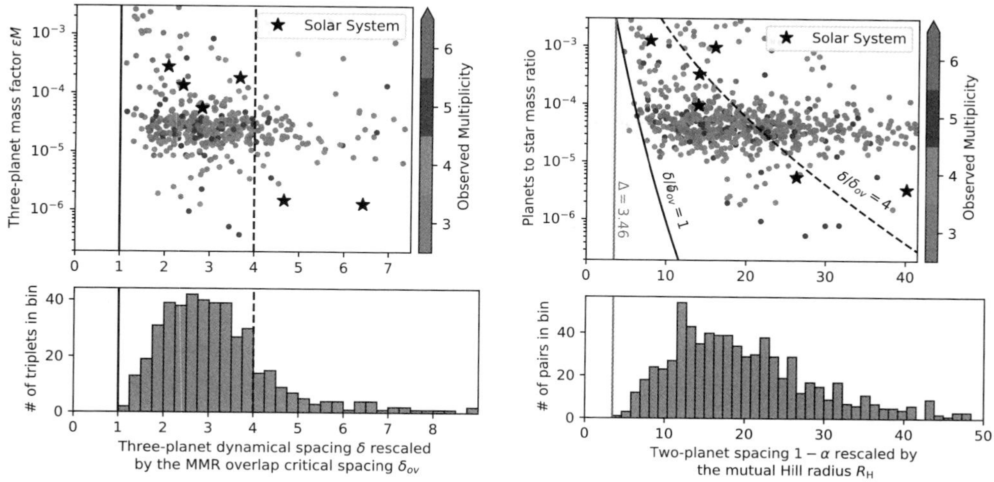

Figure 4. Left: Three-planet dynamical spacing δ scaled by the MMR overlap critical spacing as a function of the triplet mass factor εM (top panel) and binned (bottom panel). The continuous black line is the stability condition, 80% of the triplets have $\delta \leqslant 4\delta_{\rm ov}$ (dashed line). Right: Separation in mutual Hill radii for pairs of planets in 3+ planetary systems as a function of the planets-to-star mass ratio (top panel) and binned (bottom panel). The blue line at 3.46 is the Hill stability criterion, the two black lines represent the equivalent criteria from the left figure (assuming equal spacing).

distribution starts from the two-planet stability, have a maximum between 10 and 20 and 80% of the pairs have a spacing lower than 30. So the spread in spacing spans roughly an order of magnitude. On the other hand the dynamical spacing spread is only a factor 4.

Our sample selection is subject to selection biases, for instance detected systems tend to be closer together as wide spacing leads to larger period for the outer planets (Zhu 2020). Yet, the general pattern is strong enough that the conclusions should be robust in general. Future studies of dynamical packing should use the measure of spacing defined in this proceeding. For pair of planets, an alternative to the Hill spacing was proposed by Tamayo et al. (2021).

3.2. *Using the resonance optical depth*

The optical resonance depth has been shown to be an extremely powerfool tool to approach stability and diffusion problem for overlapping resonances (Quillen 2011; Hadden & Lithwick 2018; Petit et al. 2020). This section reviews the conditions which allow the use of this comcept.

As it relies on Chirikov (1979) theory, the phase space should be shaped by a resonance network having at least some partial overlapping. The network can be composed of very diverse resonances in terms of width, index, order, and, resonant direction. However, an important feature is that each individual resonance can be described by a simple model when taken in isolation. Works on generic resonances such as Hadden (2019); Petit (2021); Batygin et al. (2021) remain thus important to future studies on instabilities.

The resonant optical depth shines when one can exploit the symmetries of subnetworks where resonances have similar widths, spacing so that partial overlap criteria can be derived for the subnetworks. It is then straightforward to sum over the subnetworks to obtain a general criterion. Summing over the different subnetworks is in general possible because while resonance locii are not uniformely spread, the rational numbers are dense.

In a sense, we use to our advantage the main hurdle in the search of invariant tori that real numbers with a bad Diophantine approximation are rare.

Diffusion rates obtained under the resonant optical depth approach also have requirements for a direct application. For a general network, each resonance defines an hypersurface in phase space and the associated chaotic diffusion takes place perpendicular to it. As a result, the diffusion is in general anisotropic and multidimensional. Diffusion rate can be estimated simply if the motion can be reduced to a simple direction or if one can project the random walk onto a single coordinate such that the other variables affect only weakly the diffusion rate. In the case of compact system, this direction is η as the diffusion due to the three-planet MMR zeroth-order network forces the orbits to stay circular and the conservation of the total angular momentum limits the diffusion in the parallel diffusion. In the case of a two-planet system, the conservation of angular momentum and energy binds the change in period ratio and AMD such that the motion in the (α, AMD) is unidimensional (Petit et al. 2018).

3.3. *More general systems: non-circular orbits, more than three planets*

Stability criteria for more general systems based on three-planet MMR overlap are still an active research direction. Tamayo et al. (2021) proposed a first step in that direction by summing the optical depths for two-planet MMR (Hadden & Lithwick 2018) for adjacent pairs of planets. They showed that while taking into account the secular evolution was necessary, the problem was solvable. In Bodart et al. (*in prep.*) we will propose an extension of this result accounting for both the two-planet and three-planet results.

We proposed in (Petit et al. 2020) a tentative extension of the three planet result to four and more than five planet systems. We showed that increasing the number of MMR in the network provide a heuristic result consistent with simulations. A more in-depth study remains necessary.

4. Conclusion

To summarise, we estimate the instability time of compact planetary systems by modelling the diffusion rate along the zeroth-order three-planet MMR network. We show that the complex random walk along the resonance network can be represented by a diffusion process with an effective locally constant diffusion coefficient thanks to the powerful concept of resonance optical depth. Our estimate survival time is exponential in planet spacing as fitted in numerical simulations for the range of times of interest. For longer times, we predict much longer survival times, which is hinted by the longest performed simulations (Obertas et al. 2017).

The mean survival timescales depends on the planet separation in units of the planet-to-star mass ratio $\varepsilon^{1/4}$ and not in units of Hill radii (scaling as $\varepsilon^{1/3}$). The critical spacing for stability thus scales as $\varepsilon^{1/4}$ too. In particular, considering systems of various masses allows both in numerical simulations and observed systems highlights the difference in scaling, proving the proposed mechanism. Future missions for exoplanet discoveries such as Plato will increase our number of known terrestrial planets which will allow to test definitively the mechanism driving planetary system instability.

References

Batygin, K., Mardling, R. A., & Nesvorný, D. 2021, The Astrophysical Journal, 920, 148
Chambers, J., Wetherill, G., & Boss, A. 1996, Icarus, 119, 261
Chirikov, B. V. 1979, Physics Reports, 52, 263
Faber, P. & Quillen, A. C. 2007, Monthly Notices of the Royal Astronomical Society, 382, 1823

Fabrycky, D. C., Lissauer, J. J., Ragozzine, D., et al. 2014, The Astrophysical Journal, 790, 146
Gladman, B. 1993, Icarus, 106, 247
Hadden, S. 2019, AJ, 158, 238
Hadden, S. & Lithwick, Y. 2018, The Astronomical Journal, 156, 95
He, M. Y., Ford, E. B., Ragozzine, D., & Carrera, D. 2020, The Astronomical Journal, 160, 276
Hussain, N. & Tamayo, D. 2020, Monthly Notices of the Royal Astronomical Society, 491, 5258
Izidoro, A., Ogihara, M., Raymond, S. N., et al. 2017, Monthly Notices of the Royal Astronomical Society, 470, 1750
Johansen, A., Davies, M. B., Church, R. P., & Holmelin, V. 2012, The Astrophysical Journal, 758, 39
Laskar, J. 1997, Astronomy and Astrophysics, 317, L75
Laskar, J. & Petit, A. C. 2017, Astronomy & Astrophysics, 605, A72
Marchal, C. & Bozis, G. 1982, Celestial Mechanics, 26, 311
Marcy, G. W., Isaacson, H., Howard, A. W., et al. 2014, The Astrophysical Journal Supplement Series, 210, 20
Morbidelli, A. & Vergassola, M. 1997, Journal of Statistical Physics, 89, 549
Murchikova, L. & Tremaine, S. 2020, The Astronomical Journal, 160, 160
Obertas, A., Van Laerhoven, C., & Tamayo, D. 2017, Icarus, 293, 52
Petit, A. C. 2021, Celestial Mechanics and Dynamical Astronomy, 133, 39
Petit, A. C., Laskar, J., & Boué, G. 2017, Astronomy and Astrophysics, 607, A35
Petit, A. C., Laskar, J., & Boué, G. 2018, Astronomy & Astrophysics, 617, A93
Petit, A. C., Pichierri, G., Davies, M. B., & Johansen, A. 2020, Astronomy and Astrophysics, 641, A176
Pichierri, G. & Morbidelli, A. 2020, Monthly Notices of the Royal Astronomical Society, 494, 4950
Pu, B. & Wu, Y. 2015, The Astrophysical Journal, 807, 44
Quillen, A. C. 2011, Monthly Notices of the Royal Astronomical Society, 418, 1043
Smith, A. W. & Lissauer, J. J. 2009, Icarus, 201, 381
Tamayo, D., Murray, N., Tremaine, S., & Winn, J. 2021, The Astronomical Journal, 162, 220
Weiss, L. M., Marcy, G. W., Petigura, E. A., et al. 2018, The Astronomical Journal, 155, 48
Wisdom, J. 1980, The Astronomical Journal, 85, 1122
Xie, J.-W., Dong, S., Zhu, Z., et al. 2016, Proceedings of the National Academy of Sciences, 113, 11431
Zhu, W. 2020, The Astronomical Journal, 159, 188

Complex Planetary Systems II: Latest Methods for an Interdisciplinary Approach
Proceedings IAU Symposium No. 382, 2024
A. Lemaitre & A.-S. Libert, eds.
doi:10.1017/S1743921323004775

Long-term evolution of exoplanet systems

Cristobal Petrovich[1,2]

[1]Instituto de Astrofisica, Pontificia Universidad Católica de Chile,
Av. Vicuña Mackenna 4860, 782-0436 Macul, Santiago, Chile.

[2]Millennium Institute for Astrophysics, Chile
email: petrovich.cristobal@gmail.com

Abstract. In this Review, I discuss recent developments on the long-term dynamical evolution of exoplanet systems, focusing on how distinctive dynamical processes may have shaped the orbital architectures of observed populations. I include three applications that highlight part of my own work. First, I examine the high-eccentricity tidal migration of hot Jupiters from a phase of dynamical instability and subsequent secular interactions in two-planet systems. Second, secular chaos as the origin of ultra-short-period planets with extreme period ratios. Third, secular resonance sweeping driven by a dispersing protoplanetary disk as the origin hot Neptunes residing in polar orbits. Finally, I discuss how upcoming observations will allow further constraining the prevalence of these dynamical processes.

Keywords. Planets and satellites: dynamical evolution and stability – Chaos – Planet-disk interactions

1. Introduction

The orbital architectures of exoplanet systems is diverse, often displaying vestiges of an active dynamical history. Examples include the population of cold Jupiters in eccentric orbits, hot Jupiters and hot Neptunes in tilted orbits relative to their host's star axis, close-in planets with extreme period ratios, among others (Winn and Fabrycky 2015; Zhu and Dong 2021). As a major goal in exoplanet science we aim to describe the history of these intricate systems and shed light on the relevant mechanisms that shaped their orbits and those that may have taken place in the Solar System.

In this review, I describe two classes of long-term dynamical processes and how they may account for the architectures of different exoplanet populations, highlighting my own work:

(1) *Dynamical instabilities* in multi-planet systems whose long-term evolution reduces the number of planets due to collisions and/or ejections. The remaining planets could reach relaxed, though dynamically excited, states that evolve secularly and lead to tidal decay after their eccentricities reach large values. Possible end results of this evolution path are the *hot Jupiters* (§2, Petrovich 2015) and the *ultra-short-period planets* (§3, Petrovich et al. 2019).

(2) *Sweeping secular resonances* due to gradual changes in the frequencies of the systems' secular modes leading to crossings of secular resonances and subsequent excitation of large eccentricities and/or inclinations. The driver of these changes could be due to a time-varying external perturbation (e.g., a dispersing protoplanetary disk or a decaying stellar quadrupolar moment) or an internal change in the system (e.g., a migrating planet or planetary engulfment). I discuss a possible end

result of this evolution history that may account for an emerging population of *polar Neptunes* (§4, Petrovich et al. 2020).

As the exoplanet sample with well-characterized orbital architectures continues to increase, we will better constrain the prevalence of these long-term dynamical processes and their conditions to manifest in nature.

2. Hot jupiters from dynamical instabilities

Soon after the discovery of the first gas giant planets by radial velocities, gravitational scattering was recognized as a promising mechanism to explain the observed large eccentricities and the existence of hot Jupiters by tidal captures (Weidenschilling and Marzari 1996; Rasio and Ford 1996). Subsequent population-level studies showed that the eccentricity distribution of the gas giants remaining in the system relaxes to a distribution that reproduces that of the observed cold Jupiters (Chatterjee et al. 2008; Jurić and Tremaine 2008). Similar studies including tidal dissipation found that a significant fraction of systems ($\sim 10\%$) are capable of shrinking the orbit of a gas giant and giving rise to a hot Jupiter (Nagasawa et al. 2008; Beaugé and Nesvorný 2012).

Next, I present my own numerical experiments as they were presented at the Kavli-IAU Symposium 382. Although these results have not been published elsewhere, they are broadly consistent with previous results from Beaugé and Nesvorný (2012).

2.1. *Numerical experiments*

I present N-body simulations of the evolution of the orbits of giant planets orbiting a solar-type star including extra forces to emulate the effects from tidal dissipation on the planets and relativistic precession (see Appendix A). The planets are assumed to have a Jupiter mass and radius. Planet-star and planet-planet collisions are assumed to result in momentum-conserving mergers with no fragmentation. Collisions are assumed to occur when the distance between two planets (or planet and star) becomes less than the sum of their physical radii.

We choose log of the semi-major axis to be uniformly distributed in the range $3-7$ au. Labeling the planets by subscripts i in order of increasing semi-major axis, we impose a minimum initial spacing of the orbits given by

$$\Delta a_{i,i+1} \equiv a_{i+1} - a_i > K R_{H,i,i+1}, \quad \text{where} \quad R_{H,i,i+1} = \left(\frac{M_i + M_{i+1}}{3M_\star}\right)^{1/3} \frac{a_i + a_{i+1}}{2}, \tag{1}$$

where $R_{H,i,i+1}$ is the mutual Hill radius of planets with masses M_i and M_{i+1}. The eccentricities and inclinations are drawn from a Rayleigh distribution with parameters σ_e and σ_i. The parameters can be found in Table 1.

In Figure 1 and Table 1, we show the outcomes of the simulation *3pl-K3*. Most of the systems ($\simeq 95\%$) lead to instability with a set of outcomes involving planetary collisions (C), ejections (E), stellar collisions (S), and hot Jupiter formation (HJ). These simulations have an initial Safronov number for parabolic orbits, defined as

$$\Theta \equiv \left(\frac{2GM_p}{R_p}\right)\left(\frac{a}{GM_\star}\right) \simeq \left(\frac{M_p}{M_J}\right)\left(\frac{M_\odot}{M_\star}\right)\left(\frac{R_J}{R_p}\right)\left(\frac{a}{0.25\text{ au}}\right), \tag{2}$$

in the range of $\simeq 10-30$ so that close encounters mostly lead to ejections relative to planetary collisions (Ford et al. 2001; Petrovich et al. 2014). These scattering encounters

Table 1. Summary of simulated systems and outcomes.

Name	K	$t_{\max}$ [yr]	$t_{\mathrm{V,pl}}$ [yr]	$\sigma_e = \sigma_i$	a [au]	N_{sys}	3 pl. (HJ, inactive)	2 pl. (E, C, S)	1pl (2E)
3pl-K3	3	10^7	0.1	0.01	3-7	1,000	221 (162, 49)	634 (444, 161, 29)	141

Notes: 2 pl. (3 pl.) means that two (three) planets remain in stable orbits for a time $t_{\max}$. C, E, and S stand for planet-planet collisions, planet ejections ($d > 10^5$ AU), and planet collisions with the star. 2E means that two planet ejections occur and HJ stands for hot Jupiter (one planet with $a < 0.1$ AU). All the planets in the simulations have radius R_J.

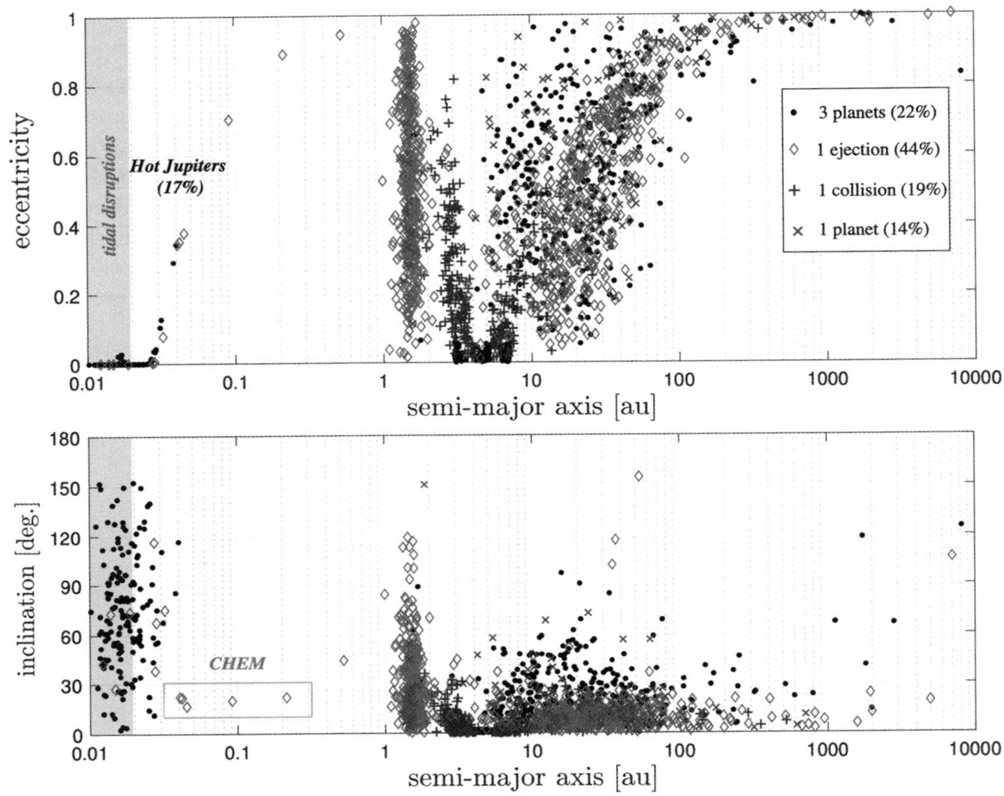

Figure 1. The final orbital elements after 10 Myrs for three-planet systems, where the majority ($\simeq 95\%$) undergoes a phase of strong gravitational scattering. The symbols indicate various outcomes (see Table 1 for more details). A significant population of hot Jupiters is formed, the majority in systems where all 3 planets remain in the system. Note, however, that these planets would have likely been disrupted as their pericenter distances reach inside $\simeq 2R_\odot$ (shaded region in red). In turn, hot Jupiters in two-planet systems tend to circularize to wider orbits, some of which are driven by Coplanar High-eccentricity Migration (CHEM; see green box in lower panel and an example in Figure 2).

are reflected as outcomes involving ejections from the system, collisions with the star, and hot Jupiter formation. Consistently, these outcomes largely dominate relative to collisions.

2.2. *Rate of hot Jupiter formation*

The integrations show that hot Jupiters are formed in $\simeq 17\%$ of the systems. This fraction falls between the $\sim 10\%$ quoted by Beaugé and Nesvorný (2012) and the $\sim 30\%$ from Nagasawa et al. (2008). My HJ rate may be larger than the former because my simulations included only equal-mass planets where the scattering is most efficient. In

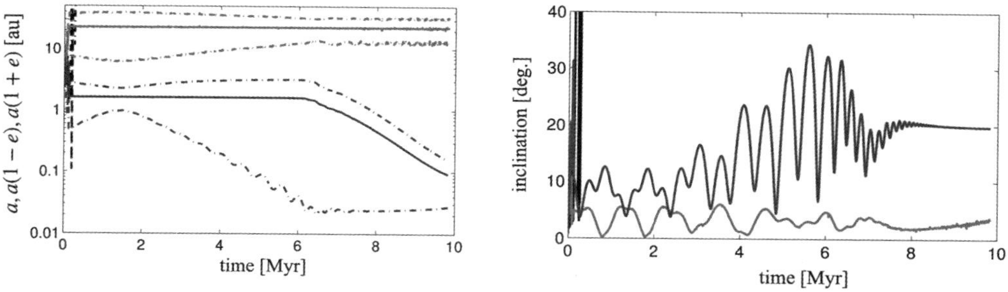

Figure 2. Orbital evolution the planet in the green box of Figure 1, undergoing *Coplanar High-eccentricity Migration* (CHEM; Petrovich 2015). A planet is ejected early on (black line), while the other two relax into a hierarchical configuration ($a_{\rm out}/a_{\rm in} \gtrsim 10$) with eccentric ($e_{\rm in}, e_{\rm out} \gtrsim 0.5$), and mildly inclined orbits ($I_{\rm in}, I_{\rm out} \lesssim 15°$). The subsequent evolution is driven by octupole-level secular perturbations reaching $e_{\rm in} \to 1$ ($a_{\rm in}[1-e_{\rm in}] \simeq 0.02$ au) after 6 Myrs, time at which the orbits start to lose orbital energy due to tides. On top of the octupole-level cycle, we observe shorter-term quadrupole-level oscillations, most noticeable on the inclinations when the eccentricities are large. The final state of the system would be a hot Jupiter at $a_{\rm in} \simeq 0.04$ au (3-day period) in a circular orbit and an inclination of $\sim 20°$ relative to the initial invariable plane.

turn, these may be lower than the latter, as these authors do not include planetary collisions that reduce the eccentricity excitation and also their efficient prescription for tides that largely prevent collisions with the star (taking place in $\sim 10\%$ of my simulations, mainly as E+S outcomes).

Regardless of the slight discrepancies, the fraction of hot Jupiters in these simulations may not be physical, since none of the above works included the possibility of tidal disruptions when the pericenter distance is $r_p = a(1-e) \lesssim 2R_\odot$†. Due to the conservation of orbital angular momentum during migration, the final semi-major axis is $a_{\rm final} = a(1-e^2) \simeq 2r_p$, implying that hot Jupiters with $a_{\rm final} \lesssim 4R_\odot \simeq 0.019$ au would have been disrupted (red-shaded region containing $\sim 70\%$ of the HJs).

2.3. *Coplanar High-eccentricity Migration (CHEM)*

The main outcome from the scattering experiments is the ejection of one planet (red diamonds) leading behind two planets in excited and well-separated orbits. We clearly distinguish an inner planet piled-up at $a_{\rm in} \gtrsim 1$ au as a result of energy conservation, and an outer planet following a track with $a_{\rm out}(1-e_{\rm out}) \sim 10$ au connecting its orbital path to the region scattering took place. As a result, most of these systems lead to a hierarchical configuration with $a_{\rm out}(1-e_{\rm out})/a_{\rm in} \sim 5-12$.

As recognized in previous work, if these hierarchical systems relax into orbits with large mutual inclinations, then the von Zeipel-Lidov-Kozai (ZLK) mechanism can be triggered and lead to subsequent migration (Nagasawa et al. 2008; Beaugé and Nesvorný 2012). Though this evolution track does happen in some systems, it is not the only possibility. Alternatively, two eccentric planets ($e_{\rm in}, e_{\rm out} \gtrsim 0.5$) with low to moderate mutual inclinations $I_{\rm mut.} \lesssim 20°$ may secularly drive the inner planet to $e_{\rm in} \simeq 1$ and its subsequent migration. A process termed Coplanar High-eccentricity Migration and proposed by Petrovich (2015).

† The tidal disruption radius for parabolic orbits is $r_t \simeq 2R_J(M_\star/M_i)^{1/3} \simeq 2R_\odot$ (e.g., Faber et al. 2005).

An example is shown in Figure 2, which corresponds to a system in the green box of Figure 1 (system at $a \sim 0.1$ au, $e \sim 0.7$ and $I \sim 20°$). Here, the dynamics is governed by the following dimensionless interaction potential:

$$\tilde{\phi}_{\rm oct} = \frac{e_{\rm in}^2 + 2/3}{2(1-e_{\rm out}^2)^{3/2}} - \frac{15}{16}\left(\frac{a_{\rm in}}{a_{\rm out}}\frac{e_{\rm out}}{1-e_{\rm out}^2}\right)\frac{e_{\rm in}(e_{\rm in}^2+3/4)}{2(1-e_{\rm out}^2)^{3/2}}\cos\Delta\varpi, \tag{3}$$

where the second term is the octupole-level coupling responsible for the eccentricity growth. Depending on its amplitude relative to the first quadrupole-level term and the relative apsidal orientations $\Delta\varpi = \varpi_{\rm in} - \varpi_{\rm out}$, the inner orbit can reach $e_{\rm in} \to 1$ (see also Li et al. 2014 for a condition in the test particle limit). The figure shows that this process takes ~ 4 Myr to excite the eccentricity to $e_{\rm in} \simeq 0.98$, corresponding to the octupole timescale. On top of this, we observe other oscillations that take place in quadrupole timescale due to the nonzero inclinations and become most prominent after ~ 5 Myr when the eccentricity is quite large.

2.4. *The paths to high-eccentricity migration*

Recently, Garzón (2022) performed similar scattering experiments to quantify the relative importance of various eccentricity excitation mechanisms, including secular chaos, CHEM, ZLK oscillations, and a combination of the last two (eccentric and inclined orbits). Their results show that secular excitation from both eccentric (octupole-driven like CHEM) and inclined (quadrupole-driven like ZKL) are most prevalent. In other words, after scattering, the orbits relax to inclined and eccentric orbits as shown in Figure 1. Thus, CHEM operates but generally shows extra inclinations wiggles driven by quadrupole perturbations. Similarly, ZKL typically displays strong octupolar modulations.

We can turn to observations to gauge the leading mechanism, if any. In particular the sky-projected spin-orbit angles λ that have been measured in around 200 systems with short-period gas giants, mainly hot Jupiters. In this sample, the majority has small values of λ, possibly indicative of CHEM or other channels to form hot Jupiters (in situ formation and disk-driven migration). However, such systems may have also started with large λ following other high-eccentricity channels, but have their large tilted damped out by tides (Winn and Fabrycky 2015).

Most recently, measurements of λ have been possible for a more pristine population of warm Jupiters ($a \gtrsim 0.1$ au), which are much less affected by tidal damping of λ compared to hot Jupiters. A notable example is TOI-3362b, a proto HJ with $a \simeq 0.15$ au and $e \simeq 0.7$ that would circularize down to $a(1-e^2) = 0.07$ au. Espinoza-Retamal et al. (2023a) measured $\lambda = 1.2^{+2.8}_{-2.7}$ deg, indicating with exquisite precision that this migrating planet is well aligned with the host star's equator. As such, the system is most consistent with CHEM with a history similar to that depicted in Figure 2.

3. Ultra-short-period planets from secular chaos

The *Kepler* mission revealed that sub-Neptunes inside ~ 100-day orbits abundantly revolve stars such as the Sun. Although in most systems a single planet is observed to transit, recent works show that this is merely due to geometric selection biases and that the underlying population consists of multiple planets (tipically ~ 3 planets inside ~ 100 days; Zhu and Dong 2021).

Using independent methods to constrain the presence of non-transiting planets, a novel trend arises: *systems with less planets tend to have larger mutual inclinations* (Zhu et al. 2018; He et al. 2020). This trend has been interpreted as indicative of a phase of giant impacts (Hansen and Murray 2013). Much like the case of giant planets discussed above,

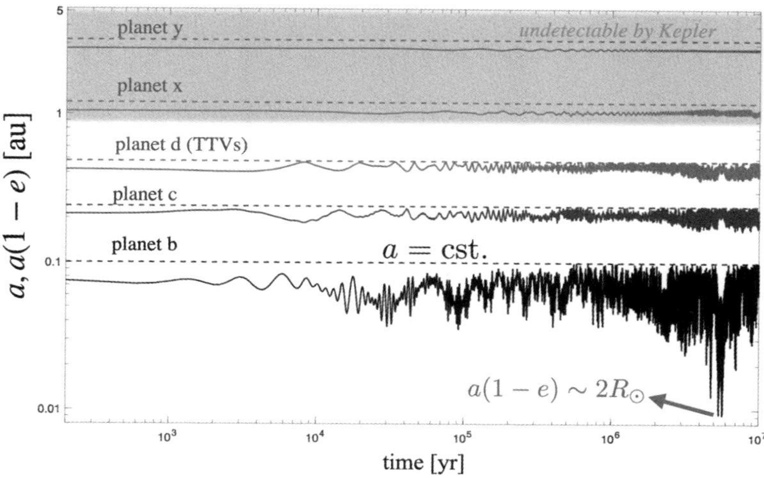

Figure 3. Proposed orbital evolution of the Kepler-10 system from secular chaos that might have led to the tidal migration of planet b from $a_b \sim 0.1$ au down to its current value of 0.016 au from Petrovich et al. (2019). The integration includes planets c and d, and two hypothetical planets (x and y) undetectable by Kepler. The simulation shows the secular and chaotic behavior of the eccentricity of planet b that after ~ 5 Myrs diffuses to large values such that $a_b(1-e_b) \lesssim 0.01$ au at which point the orbit can circularize to $a_b(1-e_b) \lesssim 0.018$ au. The initial eccentricities and inclinations of planets c to y are assumed to be $e = 0.12$ and $i = 7°$, while higher values ($e_b = 0.12$ and $i_b = 7°$) are assumed for Kepler-10b, to speed up its chaotic diffusion. Integrations are carried out with REBOUND-x (Rein and Liu 2012; Tamayo et al. 2020) and include relativistic precession and apsidal precesion from tidal bulges.

these planetary systems may have started with a higher planet multiplicity, and instabilities gradually reduced the number of planets. The main difference is that the Safronov-like number in Equation 2 is significantly reduced compared to that of the gas giants. For a super-Earth with $R_p = 2R_\oplus$ and $M_p = 5M_\oplus$ at 0.2 au, $\Theta \simeq 0.07$. Thus, we expect that instabilities result in planetary collisions, not ejections. As this process repeats, the planets grow and their orbital spacing and 'dynamical temperature' (eccentricities and/or inclinations) also increases (Tremaine 2015).

If the systems indeed settle in these excited states, their long-term orbital evolution may be dictated by secular chaos, much like the evolution of Mercury in our solar system (Laskar 2008). Next, I discuss a possible manifestation of this process in the exoplanet population.

3.1. *Ultra-short-period planets*

Planets with radii between $0.5 - 2R_\oplus$ and periods $\lesssim 1$ days, known as ultra-short-period planets (USPs), represent an extreme planet population. Their equilibrium temperature can exceed $\sim 2000K$ for a Sunlike host, which is hot enough to sublimate dust grains. It has been suggested that, as for hot Jupiters, these planets may have migrated to their locations (Zhu and Dong 2021). Furthermore, the orbits of USPs are also unusual compared to the *Kepler*-like population, as they exhibit significantly larger orbital spacing and higher mutual inclinations (Dai et al. 2018; Petrovich et al. 2019). This orbital architecture suggests a distinctive dynamical past.

A scenario proposed by Petrovich et al. (2019) is that these planets are the end-result of a secular chaos. A scaled-down version in planetary mass and orbital distance of a mechanism proposed for hot Jupiters by Wu and Lithwick (2011). Figure 3 shows an example motivated by Kepler-10 that harbors at least three planets and an extremely detached USP with $P_c/P_b \simeq 55$. Here, tidal migration due to secular chaos would naturally explain these extreme period ratios starting from a more representative value of $P_c/P_b \sim 4$.

But not all USPs are as extreme as Kepler-10, and their outer neighbors may reside in closer orbits, though still detached. As such, the migration may have started closer to their current orbits without the need of extreme eccentricity excitation. A low-eccentricity excitation model, put forward by Pu and Lai (2019), may fit these systems better.

In either case, the tidal migration provides an explanation for the orbital detachment and also to the increased orbital inclinations that result from these secular interactions and tidal migration.

4. Polar Neptunes from sweeping secular resonances

The long-term evolution of a planetary system can also be influenced by gradual changes in the perturbing potential. Examples include the decaying potential arising from the rotationally induced quadrupolar deformation of a host star (Ward et al. 1976) and that from dispersing protoplanetary disk (Heppenheimer 1980). The time-varying perturbation can gradually modify the secular frequencies of the planetary orbits and lead to passages through secular resonance—a process termed "scanning" or "sweeping" of secular resonances. These passages can drastically increase eccentricities and/or inclinations.

The role and relevance of sweeping secular resonances have been studied in the context of the asteroid belt (Heppenheimer 1980; Lemaitre and Dubru 1991), planet formation through planetesimal accretion (Nagasawa et al. 2005; Best et al. 2023), and even excitation of eccentricities in exoplanet systems (Nagasawa et al. 2003).

In these applications one can simplify the dynamics of the resonance sweeping and reduce it to the second fundamental model of the resonance (Henrard and Lemaitre 1983; Lemaitre and Dubru 1991):

$$\mathcal{H} = -3\Delta(t)R + R^2 - 2\sqrt{2R}\cos(r), \tag{4}$$

where $R \propto 1 - \sqrt{1-e^2} \simeq e^2/2$ (or $R \propto [1 - \cos I] \simeq I^2/2$). The sweeping is encapsulated in the evolution of the resonance distance $\Delta(t)$. Note that most previous works ignore the non-linear term R^2, thus only modeling "passages" through linear resonances and not resonant captures. The application described below shows one of the first applications where this distinction becomes crucial.

4.1. *Polar Neptunes*

A number of measurements of the Rossiter-Mclaughlin effect have recently been carried out for planets smaller than Jupiter. As pointed out by Petrovich et al. (2020), these observations reveal an emerging population of hot Neptunes that reside in nearly polar orbits relative to their host star's equator. I caution that the sample is still small with only 5 members, but so far they outnumber the systems residing in aligned orbits in stars smaller than the Sun (Stefànsson et al. 2022). Also, in 3 of these systems long-period Jovian companions are detected (e.g., HAT-P-11 described in Figure 4).

Using this information, Petrovich et al. (2020) proposed that these two-planet systems could engage in an inclination secular resonance capture driven by a dispersing disk. Thus, the angular momentum deficit

$$\mathcal{A} \equiv M_{\rm in} a_{\rm in}^{1/2}(1 - \cos I_{\rm in}) + M_{\rm out} a_{\rm out}^{1/2}(1 - \cos I_{\rm out}) \ , \tag{5}$$

conserved during this process, and largely stored in the outer orbit, is then significantly stored in the inner orbit as $\cos I_{\rm in} \to 0$.

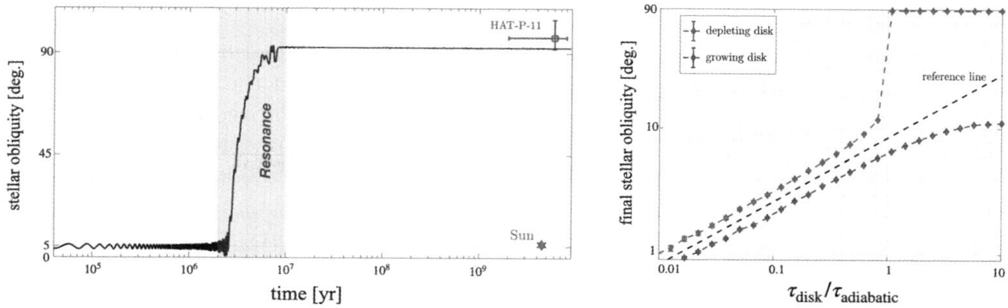

Figure 4. *Left panel*: Evolution of the stellar obliquity relative to the inner planet for a two-planet with parameters (masses, orbital separations) based on HAT-P-11. An initially low obliquity is driven to $\simeq 90°$ during resonance capture after ~ 2 Myrs (shaded region). This evolutionary path would reproduce the observed polar state of HAT-P-11b. *Right panel*: The final stellar obliquity as a function of the adiabaticity parameter $\tau_{\rm disk}/\tau_{\rm adia}$ (see Eq. 7). As expected, polar orbits are only reached when $\tau_{\rm disk}/\tau_{\rm adia} > 1$ (or slowly depleting disks). In order to illustrate the importance of the non-linear term in the resonance model in Equation (4), we include (unrealistic) integrations with a growing disk. Here, the sweep would lead to crossing with $\dot{\Delta} < 0$, not an adiabatic capture. Moreover, this crossing would fall below what is predicted from a linear resonance model due to a non-linear de-tuning (compare to reference dashed line) and saturating at obliquities of $\sim 10°$.

In this work, we reduce the dynamics to the Hamiltonian in Equation (4) with the resonance distance given by

$$\Delta(t) = \frac{2}{3}\left[\frac{1+\eta_\star}{I_{\rm out,0}}\right]^{2/3}\left[1 - \frac{a_{\rm out}^{9/2}}{a_{\rm in}^{3/2}R_{\rm in}^2 R_{\rm out}}\frac{M_{\rm disk}(t)}{(1+\eta_\star)M_{\rm out}}\right], \quad \text{where } \eta_\star = \frac{2J_2 M_\star}{M_{\rm out}}\frac{R_\star^2 a_{\rm out}^3}{a_{\rm in}^5} \tag{6}$$

and the disk of mass $M_{\rm disk}(t)$ has edges at $R_{\rm in}$ and $R_{\rm out}$ (both $> a_{\rm out}$). The star has a quadrupolar moment J_2 and $I_{\rm out,0}$ is the initial inclination of the outer planet. Similarly, we define the adiabatic time scale

$$\tau_{\rm adia} \simeq \frac{2P_{\rm in}}{3\pi}\frac{M_\star}{M_{\rm out}}\frac{a_{\rm out}^3}{a_{\rm in}^3} I_{\rm out,0}^{-4/3}(1+\eta_\star)^{1/3}, \tag{7}$$

such that the resonance capture is guaranteed if the disk dispersal timescale $\tau_{\rm disk} = |d\log M_{\rm disk}/dt|^{-1}$ longer than this. In other words, if the crossing is adiabatic.

In Figure 4 we show the evolution of a system that reproduced the orbital architecture of HAT-P-11, where inner hot Neptune has a nearly polar orbit. A key advantage of this proposal is that high inclinations are achieved starting from small values $\lesssim 5°$. This is simply a result of the much larger angular momentum of the outer orbit compared to that of the inner. The disk only facilitates, quite efficiently, the transfer of angular momentum deficit between the orbits.

Our model relies on a resonance capture, not a simply a crossing event as in most previous studies. To illustrate the importance of this distinction and the role of the nonlinear term R^2 in the dynamics, we included simulations with a range of adiabaticity parameters $\tau_{\rm disk}/\tau_{\rm adia}$ (right panel of Figure 4) and growing and depleting disks. As expected, polar orbits are only reached when $\tau_{\rm disk}/\tau_{\rm adia} > 1$ and for a depleting disk (crossing the resonance with $\dot{\Delta} > 0$). In turn, the unrealistic case of a growing disk would cross the resonance with $\dot{\Delta} < 0$, thus never leading to an adiabatic capture even if $\tau_{\rm disk}/\tau_{\rm adia} \gg 1$. This is a result of the nonlinear resonance detuning and stresses the importance of going beyond the often-used linear secular model which would predict a linear growth of the obliquity with $\tau_{\rm disk}/\tau_{\rm adia}$ (see dashed line as reference).

5. Future prospects

The numerical models presented above have a range of simplifications, ranging from the prescriptions, or absence, of tidal dissipation in the planets and the extent in which the integrations have been carried out. These aspects should be improved upon, and we have the required tools to do so, including new numerical integrators such as REBOUND-x (Tamayo et al. 2020) as well as ring or secular codes to efficiently model the very long-term dynamics (e.g., Sefilian et al. 2023).

Less clear, though, is how to improve our, somewhat arbitrary, choice of initial conditions. For the scattering of giant planets, how many giant planets participate in the scattering process and at what orbital distances? Was there residual gas when the instability ensued? The initial eccentricities, inclinations, and orbital spacing seem less relevant for this application, as the outcomes have little memory of their values after the orbits cross. Some constraints on the initial states of systems with giant planets are becoming possible with the advent of ALMA observations. Examples include direct detections in PDS-70 and indirect means in HD 163296. For the latter, recent work by Garrido-Deutelmoser et al. (2023) showed that combining resonant dynamics with images can place sensitive constraints on their orbital architectures. The subsequent evolution of systems like HD 163296 can dictate the timing of the instabilities.

Finally, a promising venue for moving forward is the better characterization of exoplanet systems. Radial velocity follow-ups with high-precision spectrographs such as ESPRESSO and NEID are pushing the observations of Rossiter-Maclaughlin to smaller planets. We should expect many more measurements of stellar obliquities for warm Neptunes. Moreover, the astrometric measurements from Gaia will likely provide with orbital orientations for cold Jupiters and also information on relative inclinations when combined with TESS data (Espinoza-Retamal et al. 2023b). Thus, opening a new window into the 3D architectures of exoplanet systems.

Acknowledgements. C.P. acknowledges support from the ANID Millennium Science Initiative-ICN12_009, CATA-Basal AFB-170002, ANID BASAL project FB210003, FONDECYT Regular grant 1210425, CASSACA grant CCJRF2105, and ANID+REC Convocatoria Nacional subvencion a la instalacion en la Academia convocatoria 2020 PAI77200076.

Appendix A. Numerical integrations of dynamically unstable gas giants

We use the publicly available integration packages of MERCURY6.2 (Chambers 1999). We use use MERCURY's Bulirsch-Stoer (BS) integration algorithm; we justify this choice because we are mostly interested in the evolution of dynamically active systems, where planets experience close encounters, and the BS algorithm handles close encounters better than the other integration algorithms in MERCURY. We include extra forces to model the effects of tides from the host star and general-relativistic apsidal precession due to the host star. These forces are explicitly included in the *mfo-user* routine in MERCURY. The first order post-Newtonian term is given by:

$$\mathbf{f}_{\mathrm{GR}} = -\frac{3G^2 M_\odot^2 a(1-e^2)}{c^2 r^4}\hat{\mathbf{r}}. \tag{8}$$

For the tides with the host star we consider a prescription for the equilibrium tidal bulge as in Wu and Lithwick (2011), where the acceleration associated with the delayed tidal bulge raised on the planet with mass m_p and radius R_p by the host star is written as:

$$\mathbf{f}_{\mathrm{TD}} = -\frac{GM_\odot^2}{\mu r^2}\left(\frac{R_p}{r}\right)^5 k_{L,p}\left[3 + 9\frac{\mathbf{v}\cdot\hat{\mathbf{r}}}{r}\tau_p\right]\hat{\mathbf{r}}, \tag{9}$$

where $k_{L,p}$ is the Love number of the planet chosen to be 0.52 (polytrope with $n=1$) and τ_p is the planet's time lag, which is related to the planets' viscous time $t_{V,p} = 3(1+k_{L,p})R^3/(Gm\tau_p)$ (Eggleton et al. 1998). This simple prescription ignores the component of the tidal force that is tangential to the planet's orbital motion, which is equivalent to saying that the planet is rapidly synchronized with its orbital mean motion. Such simplification is justified by the negligible spin angular momentum of the planet relative to its orbital angular momentum. This prescription also ignores the dissipation inside the star because for the values of τ_p we use, dissipation is dominated by that inside the planet. We ignore the evolution of the spin of the star and, therefore, the effect from the rotational bulges.

References

Beaugé, C. & Nesvorný, D. 2012, Multiple-planet Scattering and the Origin of Hot Jupiters. *ApJ*, 751(2), 119.

Best, S., Sefilian, A. A., & Petrovich, C. 2023, The influence of cold Jupiters in the formation of close-in planets. I. planetesimal transport. *arXiv e-prints*,, arXiv:2304.02045.

Chambers, J. E. 1999, A hybrid symplectic integrator that permits close encounters between massive bodies. *MNRAS*, 304(4), 793–799.

Chatterjee, S., Ford, E. B., Matsumura, S., & Rasio, F. A. 2008, Dynamical Outcomes of Planet-Planet Scattering. *ApJ*, 686(1), 580–602.

Dai, F., Masuda, K., & Winn, J. N. 2018, Larger Mutual Inclinations for the Shortest-period Planets. *ApJL*, 864(2), L38.

Eggleton, P. P., Kiseleva, L. G., & Hut, P. 1998, The Equilibrium Tide Model for Tidal Friction. *ApJ*, 499(2), 853–870.

Espinoza-Retamal, J. I., Brahm, R., Petrovich, C., Jordán, A., & Stefánsson, Guðmundur, e. a. 2023,a The Aligned Orbit of the Eccentric Proto Hot Jupiter TOI-3362b. *arXiv e-prints*,, arXiv:2309.03306.

Espinoza-Retamal, J. I., Zhu, W., & Petrovich, C. 2023,b Prospects from TESS and Gaia to constrain the flatness of planetary systems. *arXiv e-prints*,, arXiv:2309.08665.

Faber, J. A., Rasio, F. A., & Willems, B. 2005, Tidal interactions and disruptions of giant planets on highly eccentric orbits. *Icarus*, 175(1), 248–262.

Ford, E. B., Havlickova, M., & Rasio, F. A. 2001, Dynamical Instabilities in Extrasolar Planetary Systems Containing Two Giant Planets. *Icarus*, 150(2), 303–313.

Garrido-Deutelmoser, J., Petrovich, C., Charalambous, C., Guzmán, V. V., & Zhang, K. 2023, A Gap-sharing Planet Pair Shaping the Crescent in HD 163296: A Disk Sculpted by a Resonant Chain. *ApJL*, 945(2), L37.

Garzón, H. e. a. 2022, Production of hot Jupiter candidates from high-eccentricity mechanisms for different initial planetary mass configurations. *MNRAS*, 517(4), 4986–5002.

Hansen, B. M. S. & Murray, N. 2013, Testing in Situ Assembly with the Kepler Planet Candidate Sample. *ApJ*, 775(1), 53.

He, M. Y., Ford, E. B., Ragozzine, D., & Carrera, D. 2020, Architectures of Exoplanetary Systems. III. Eccentricity and Mutual Inclination Distributions of AMD-stable Planetary Systems. *AJ*, 160(6), 276.

Henrard, J. & Lemaitre, A. 1983, A Second Fundamental Model for Resonance. *Celestial Mechanics*, 30(2), 197–218.

Heppenheimer, T. A. 1980, Secular resonances and the origin of eccentricities of Mars and the asteroids. *Icarus*, 41(1), 76–88.

Jurić, M. & Tremaine, S. 2008, Dynamical Origin of Extrasolar Planet Eccentricity Distribution. *ApJ*, 686(1), 603–620.

Laskar, J. 2008, Chaotic diffusion in the Solar System. *Icarus*, 196(1), 1–15.

Lemaitre, A. & Dubru, P. 1991, Secular Resonances in the Primitive Solar Nebula. *Celestial Mechanics and Dynamical Astronomy*, 52(1), 57–78.

Li, G., Naoz, S., Kocsis, B., & Loeb, A. 2014, Eccentricity Growth and Orbit Flip in Near-coplanar Hierarchical Three-body Systems. *ApJ*, 785(2), 116.

Nagasawa, M., Ida, S., & Bessho, T. 2008, Formation of Hot Planets by a Combination of Planet Scattering, Tidal Circularization, and the Kozai Mechanism. *ApJ*, 678(1), 498–508.

Nagasawa, M., Lin, D. N. C., & Ida, S. 2003, Eccentricity Evolution of Extrasolar Multiple Planetary Systems Due to the Depletion of Nascent Protostellar Disks. *ApJ*, 586(2), 1374–1393.

Nagasawa, M., Lin, D. N. C., & Thommes, E. 2005, Dynamical Shake-up of Planetary Systems. I. Embryo Trapping and Induced Collisions by the Sweeping Secular Resonance and Embryo-Disk Tidal Interaction. *ApJ*, 635(1), 578–598.

Petrovich, C. 2015, Hot Jupiters from Coplanar High-eccentricity Migration. *ApJ*, 805(1), 75.

Petrovich, C., Deibert, E., & Wu, Y. 2019, Ultra-short-period Planets from Secular Chaos. *AJ*, 157(5), 180.

Petrovich, C., Muñoz, D. J., Kratter, K. M., & Malhotra, R. 2020, A Disk-driven Resonance as the Origin of High Inclinations of Close-in Planets. *ApJL*, 902(1), L5.

Petrovich, C., Tremaine, S., & Rafikov, R. 2014, Scattering Outcomes of Close-in Planets: Constraints on Planet Migration. *ApJ*, 786(2), 101.

Pu, B. & Lai, D. 2019, Low-eccentricity migration of ultra-short-period planets in multiplanet systems. *MNRAS*, 488(3), 3568–3587.

Rasio, F. A. & Ford, E. B. 1996, Dynamical instabilities and the formation of extrasolar planetary systems. *Science*, 274, 954–956.

Rein, H. & Liu, S. F. 2012, REBOUND: an open-source multi-purpose N-body code for collisional dynamics. *A&A*, 537, A128.

Sefilian, A. A., Rafikov, R. R., & Wyatt, M. C. 2023, Formation of Gaps in Self-gravitating Debris Disks by Secular Resonance in a Single-planet System. II. Toward a Self-consistent Model. *ApJ*, 954(1), 100.

Stefànsson, G., Mahadevan, S., Petrovich, C., Winn, J. N., Kanodia, S., & Millholland, S. C. e. a. 2022, The Warm Neptune GJ 3470b Has a Polar Orbit. *ApJL*, 931(2), L15.

Tamayo, D., Rein, H., Shi, P., & Hernandez, D. M. 2020, REBOUNDx: a library for adding conservative and dissipative forces to otherwise symplectic N-body integrations. *MNRAS*, 491(2), 2885–2901.

Tremaine, S. 2015, The Statistical Mechanics of Planet Orbits. *ApJ*, 807(2), 157.

Ward, W. R., Colombo, G., & Franklin, F. A. 1976, Secular Resonance, Solar Spin Down, and the Orbit of Mercury. *Icarus*, 28(4), 441–452.

Weidenschilling, S. J. & Marzari, F. 1996, Gravitational scattering as a possible origin for giant planets at small stellar distances. *Nature*, 384(6610), 619–621.

Winn, J. N. & Fabrycky, D. C. 2015, The Occurrence and Architecture of Exoplanetary Systems. *ARA&A*, 53, 409–447.

Wu, Y. & Lithwick, Y. 2011, Secular Chaos and the Production of Hot Jupiters. *ApJ*, 735(2), 109.

Zhu, W. & Dong, S. 2021, Exoplanet Statistics and Theoretical Implications. *ARA&A*, 59, 291–336.

Zhu, W., Petrovich, C., Wu, Y., Dong, S., & Xie, J. 2018, About 30% of Sun-like Stars Have Kepler-like Planetary Systems: A Study of Their Intrinsic Architecture. *ApJ*, 860(2), 101.

Complex Planetary Systems II: Latest Methods for an Interdisciplinary Approach
Proceedings IAU Symposium No. 382, 2024
A. Lemaitre & A.-S. Libert, eds.
doi:10.1017/S1743921323004088

Derivation and Properties of the Angular Momentum Relative Amended Potential

Daniel J. Scheeres

Smead Department of Aerospace Engineering Sciences University of Colorado Boulder
email: scheeres@colorado.edu

Abstract. The paper concerns the determination of the Angular Momentum Relative Amended Potential (AMR potential) in the framework of the Full n-Body problem and some of its basic properties are discussed. The AMR potential is derived using two different approaches : first using a Routh reduction of the system relative to rotation about the total angular momentum, second as a variation of the Sundman Inequality, using the Cauchy inequality.

Keywords. Amended Potential, Full Body Problem, Routh Reduction

1. Introduction

In this contribution the "Angular Momentum Relative Amended Potential" (AMR potential) is derived and some of its basic properties are discussed. This amended potential was specifically developed for the Full n-Body problem [Scheeres (2012)], which concerns the dynamics of n rigid bodies as they attract each other gravitationally, and as they impact and rest on each other. In the current contribution this problem is stated using Jacobi coordinates as derived in [Scheeres & Brown (2023)], with a new discussion on the Lagrangian dynamics of this statement. The AMR potential is derived using two different approaches. First it is derived using a Routh reduction of the system relative to rotation about the total angular momentum, which shows that it is directly related to the forces acting within the system. Then it is derived as a variation of the Sundman Inequality, using the Cauchy inequality. It is compared with a few other amended potentials in common usage, derived as related versions of the Sundman Inequality or through geometric mechanics as the Smale Amended Potential [Simo et al. (1991)]. The results provide a rigorous and multi-faceted derivation of the angular momentum relative amended potential, establishing its use for evaluation of equilibria and their energetic stability.

2. Model

Consider the finite-density n-body problem where all of the component bodies are rigid and can thus rotate and rest on each other as well as orbit about each other, as derived and discussed in [Scheeres (2012, 2016)]. To pose the problem define $N+1$ finite density bodies P_i, each of which has a mass distribution $\mathcal{B}_i$, where $i = 0, 1, 2, ..., N$. Each body has a mass $m_i = \int_{\mathcal{B}_i} dm$, a rigid body inertia dyadic $\bar{\mathbf{I}}_i^R = -\int_{\mathcal{B}_i} \tilde{\rho} \cdot \tilde{\rho}\, dm$, a center of mass position $\mathbf{r}_i$, an orientation dyadic $\bar{\mathbf{T}}_i$, a velocity $\mathbf{v}_i$, and an angular velocity $\boldsymbol{\Omega}_i$, all generally specified with respect to an inertial frame, except that the orientation dyadic will orient the body frame in inertial space and the rigid body inertia dyadic is specified in a body-fixed frame. The attitude definition of each body can use Euler Angles, which allow a traditional Lagrangian approach to specifying the equations of motion. With this assumption, the orientation dyads and the angular velocities for each body are functions of its Euler Angles and their rates, $\boldsymbol{\Theta}_i = \left[\theta_i^1, \theta_i^2, \theta_i^3\right]$ and $\dot{\boldsymbol{\Theta}}_i = \left[\dot{\theta}_i^1, \dot{\theta}_i^2, \dot{\theta}_i^3\right]$, leading to the functional form $\bar{\mathbf{T}}(\boldsymbol{\Theta}_i)$ and $\boldsymbol{\Omega}(\boldsymbol{\Theta}_i, \dot{\boldsymbol{\Theta}}_i)$.

One important implication of the finite density rigid body assumption is that any two bodies have a constraint on how close their centers of mass can come to each other, denoted as $\|\mathbf{r}_{ij}\| \geqslant d(\mathbf{r}_{ij}, \mathbf{T}_{ij})$, where $\mathbf{r}_{ij} = \mathbf{r}_j - \mathbf{r}_i$ and $\bar{\mathbf{T}}_{ij} = \bar{\mathbf{T}}_j^T \cdot \bar{\mathbf{T}}_i$ and equality occurs when the two bodies touch. This form of the constraint implicitly implies that the bodies are mutually convex, although this assumption can be relaxed. As discussed in [Scheeres (2019)] this formulation can accommodate the holonomic and non-holonomic constraints that arise when the components are in contact. As a final note, the mutual potential between any two bodies i and j is $\mathcal{U}_{ij}(\mathbf{r}_{ij}, \bar{\mathbf{T}}_{ij}) = -\mathcal{G} \int_{\mathcal{B}_i} \int_{\mathcal{B}_j} dm_i \; dm_j / \left|\mathbf{r}_{ij} + \rho_j - \rho_i\right|$ where $\mathcal{G}$ is the gravitational constant. This leads to the total gravitational potential $\mathcal{U} = \sum_{0 \leqslant i < j \leqslant N} \mathcal{U}_{ij}$.

2.1. *Jacobi Coordinates*

Given the standard form of the problem, transform the positions and velocities of the bodies into Jacobi Coordinates, which makes them all relative to each other in a recursive way. Note that the rigid body components are not affected by this transformation, which only operates on the centers of mass of each body.

The Jacobi formulation defines a sequence of transformations where the position and velocity of each body P_i is measured from the collective center of mass of the bodies P_j, $j = 0, 1, \ldots, i-1$ [Scheeres & Brown (2023)]. To start the sequence define: $M_0 = m_0, \mathbf{R}_0 = \mathbf{V}_0 \;=\; \mathbf{0}, \mathbf{R}_0^C = \mathbf{r}_0$, and $\mathbf{V}_0^C = \mathbf{v}_0$. Then the remaining Jacobi coordinates are defined as

$$\mathbf{R}_i = \mathbf{r}_i - \mathbf{R}_{i-1}^C \tag{2.1}$$

$$\mathbf{V}_i = \mathbf{v}_i - \mathbf{V}_{i-1}^C \tag{2.2}$$

$$M_i = M_{i-1} + m_i \tag{2.3}$$

$$\mathbf{R}_i^C = \frac{1}{M_i}\left[M_{i-1}\mathbf{R}_{i-1}^C + m_i \mathbf{r}_i\right] \tag{2.4}$$

$$\mathbf{V}_i^C = \frac{1}{M_i}\left[M_{i-1}\mathbf{V}_{i-1}^C + m_i \mathbf{v}_i\right] \tag{2.5}$$

all for $i = 1, 2, \ldots, N$. The center of mass position and velocity vectors $\mathbf{R}_i^C$ and $\mathbf{V}_i^C$ are computed accounting for all bodies with that index and lower. The relative position and velocity vectors $\mathbf{R}_i$ and $\mathbf{V}_i$ are of body i relative to the center of mass of all bodies with index $i-1$ and lower. The vector $\mathbf{R}_N^C$ is the total center of mass of the entire system.

Finally, note that the difference of two body positions $\mathbf{r}_j$ and $\mathbf{r}_i$ where $j > i$ can be stated as

$$\mathbf{r}_j - \mathbf{r}_i = \mathbf{R}_j - \mathbf{R}_i + \mathbf{R}_{j-1}^C - \mathbf{R}_{i-1}^C \;=\; \mathbf{R}_j - \mathbf{R}_i + \sum_{k=i}^{j-1} \frac{m_k}{M_k}\mathbf{R}_k \tag{2.6}$$

Thus, when considering the potential energy, the mutual potential between bodies j and i, $\mathbf{r}_{ij}$, is only a function of $\mathbf{R}_k$ from $k = i, i+1, \ldots, j$.

2.2. *Angular Momentum, Energy and Moments of Inertia*

A key advantage of the Jacobi coordinates is that they explicitly decouple the quantities of angular momentum, moment of inertia and kinetic energy at a given order from each other and are only a function of the lower order indices. Define the angular momentum, kinetic energy and inertia tensor at each order, respectively,

$$\mathbf{H}_i = \frac{M_{i-1}m_i}{M_i}\mathbf{R}_i \times \mathbf{V}_i + \bar{\mathbf{T}}_i \cdot \bar{\mathbf{I}}_i^R \cdot \boldsymbol{\Omega}_i \tag{2.7}$$

$$T_i = \frac{1}{2}\frac{M_{i-1}m_i}{M_i}\left(\mathbf{V}_i \cdot \mathbf{V}_i\right) + \frac{1}{2}\boldsymbol{\Omega}_i \cdot \bar{\mathbf{I}}_i^R \cdot \boldsymbol{\Omega}_i \tag{2.8}$$

$$\bar{\mathbf{I}}_i = \frac{M_{i-1}m_i}{M_i}\left[R_i^2\bar{\mathbf{U}} - \mathbf{R}_i\mathbf{R}_i\right] + \bar{\mathbf{T}}_i \cdot \bar{\mathbf{I}}_i^R \cdot \bar{\mathbf{T}}_i^T \tag{2.9}$$

The term $\bar{\mathbf{U}}$ is the unity dyadic, the product $\mathbf{RR}$ is a dyad, and the rigid body angular momentum and rotational inertia are mapped from a body-fixed frame into the inertial frame with the orientation dyadic $\bar{\mathbf{T}}_i$. Then the system total Angular Momentum, Kinetic Energy and Inertia Dyadic are, respectively,

$$\mathbf{H} = \sum_{i=0}^{N}\mathbf{H}_i + M_N\mathbf{R}_N^C \times \mathbf{V}_N^C \tag{2.10}$$

$$T = \sum_{i=0}^{N}T_i + \frac{1}{2}M_N\mathbf{V}_N^C \cdot \mathbf{V}_N^C \tag{2.11}$$

$$\bar{\mathbf{I}} = \sum_{i=0}^{N}\bar{\mathbf{I}}_i + M_N\left[\mathbf{R}_N^C \cdot \mathbf{R}_N^C\ \bar{\mathbf{U}} - \mathbf{R}_N^C\mathbf{R}_N^C\right] \tag{2.12}$$

and the total system energy can be defined as

$$E = \sum_{i=0}^{N}E_i + \frac{1}{2}M_N\mathbf{V}_N^C \cdot \mathbf{V}_N^C \tag{2.13}$$

$$E_i = T_i + \mathcal{U}_i \tag{2.14}$$

where $\mathcal{U}_i = \sum_{j=0}^{i-1}\mathcal{U}_{ji}$. The self-potential of a given body could be included, but is not in this formulation.

2.3. *Lagrangian and Equations of Motion*

The coordinates of the system can be gathered into a standard coordinate and velocity vector

$$\mathbf{Q} = \left[\boldsymbol{\Theta}_0, \ldots, \mathbf{R}_i, \boldsymbol{\Theta}_i, \ldots, \mathbf{R}_N, \boldsymbol{\Theta}_N, \mathbf{R}_N^C\right] \tag{2.15}$$

$$\dot{\mathbf{Q}} = \left[\dot{\boldsymbol{\Theta}}_0, \ldots, \mathbf{V}_i, \dot{\boldsymbol{\Theta}}_i, \ldots, \mathbf{V}_N, \dot{\boldsymbol{\Theta}}_N, \mathbf{V}_N^C\right] \tag{2.16}$$

With this notation, the Lagrangian is defined as

$$\mathcal{L}(\mathbf{Q}, \dot{\mathbf{Q}}) = T(\mathbf{Q}, \dot{\mathbf{Q}}) - \mathcal{U}(\mathbf{Q}) \tag{2.17}$$

The equations of motion for the ith Jacobi coordinate are

$$\frac{M_{i-1}m_i}{M_i}\dot{\mathbf{V}}_i = -\frac{\partial \mathcal{U}}{\partial \mathbf{R}_i} \tag{2.18}$$

The rotational equations of motion for the ith rigid body are

$$\frac{d}{dt}\left(\boldsymbol{\Omega}_i \cdot \bar{\mathbf{I}}_i^R \cdot \frac{\partial \boldsymbol{\Omega}_i}{\partial \dot{\boldsymbol{\Theta}}_i}\right) = \boldsymbol{\Omega}_i \cdot \bar{\mathbf{I}}_i^R \cdot \frac{\partial \boldsymbol{\Omega}_i}{\partial \boldsymbol{\Theta}_i} - \frac{\partial \mathcal{U}}{\partial \boldsymbol{\Theta}_i} \tag{2.19}$$

2.4. *Routh Reduction of the Linear Momentum*

The first step is to remove the linear momentum, which can be formally done via a Routh reduction [Greenwood (1977)]. The Lagrangian is independent of the final center of mass term $\mathbf{R}_N^C$, thus the linear momentum integral is found from $\frac{d}{dt}\frac{\partial \mathcal{L}}{\partial \mathbf{V}_N^C} = 0$, leading to

$$M_N \mathbf{V}_N^C = \mathbf{P} \; = \; \text{Constant} \tag{2.20}$$

Then the center of mass velocity can be solved for as a function of the constant: $\mathbf{V}_N^C = \frac{1}{M_N}\mathbf{P}$

This momentum can be removed from the Lagrangian by first forming the Routhian function

$$\begin{aligned} \mathcal{L}_R &= \mathcal{L} - \mathbf{V}_N^C \cdot \frac{\partial \mathcal{L}}{\partial \mathbf{V}_N^C} \\ &= T - \mathcal{U} - \frac{1}{2M_N}\mathbf{P}\cdot\mathbf{P} \end{aligned} \tag{2.21}$$

The Routhian can be used to recover the center of mass motion after the rest of the solution is found [Greenwood (1977)]

$$\mathbf{V}_N^C = -\frac{\partial \mathcal{L}_R}{\partial \mathbf{P}} \; = \; \frac{1}{M_N}\mathbf{P}$$

which can be solved by quadrature.

This reduction is trivial, but sets the stage for the more consequential reduction of the angular momentum. One can arbitrarily set $\mathbf{P} = \mathbf{0}$, which resets the Lagrangian to the original form, $\mathcal{L}_R = \mathcal{L}$.

2.5. *Routh Reduction of the Angular Momentum Magnitude*

Instead of removing the entire angular momentum, as in [Simo et al. (1991)], one can focus on the simpler reduction of the angular momentum magnitude, which is key to the AMR amended potential definition.

The system has a constant angular momentum which can be stated as $\mathbf{H} = H\hat{\mathbf{H}}$. A simple rotation of the system about the angular momentum direction, which is fixed in inertial space, yields a well defined angular coordinate and its time derivative, θ and $\dot{\theta}$. Note the simple relationship between the system moment of inertia, spin rate and angular momentum magnitude

$$\dot{\theta} = \frac{H}{I_H} \tag{2.22}$$

$$I_H = \hat{\mathbf{H}} \cdot \bar{\mathbf{I}} \cdot \hat{\mathbf{H}} \tag{2.23}$$

The I_H term is only a function of the Jacobi coordinates and Euler angles and is the instantaneous system moment of inertia about the angular momentum direction. It is important to note that the angular rate $\dot{\theta}$ is not constant and is a function of the coordinates.

The Lagrangian system can be rewritten in a rotating coordinate frame defined by the angular velocity $\boldsymbol{\Omega} = \dot{\theta}\hat{\mathbf{H}}$. This is easily done in by rewriting the Jacobi velocities using the transport rule:

$$\mathbf{V}_i = \mathbf{R}_i' + \boldsymbol{\Omega} \times \mathbf{R}_i \tag{2.24}$$

where the $()'$ signifies a time derivative in the rotating frame. Similarly, the angular velocities of the rigid bodies must be redefined relative to this overall rotation, yielding

$$\boldsymbol{\Omega}_i = \omega_i + \boldsymbol{\Omega} \tag{2.25}$$

where ω_i represents the new rotation of these bodies relative to the rotating frame. The Euler angles and their time derivatives are also redefined relative to this rotating frame.

Rewriting the kinetic energy of the system one finds

$$T = \sum_{i=1}^{N} \frac{M_{i-1} m_i}{2\ M_i} \left[\mathbf{R}_i' \cdot \mathbf{R}_i' + 2\boldsymbol{\Omega} \cdot \mathbf{R}_i \times \mathbf{R}_i' - \boldsymbol{\Omega} \cdot \tilde{\mathbf{R}}_i \cdot \tilde{\mathbf{R}}_i \cdot \boldsymbol{\Omega}\right]$$
$$+ \frac{1}{2} \sum_{i=0}^{N} \left[\omega_i \cdot \bar{\mathbf{I}}_i^R \cdot \omega_i + 2\boldsymbol{\Omega} \cdot \bar{\mathbf{I}}_i^R \cdot \omega_i + \boldsymbol{\Omega} \cdot \bar{\mathbf{I}}_i^R \cdot \boldsymbol{\Omega}\right] \tag{2.26}$$

which can be rewritten as

$$T = T_R + \boldsymbol{\Omega} \cdot \mathbf{H}_R + \frac{1}{2} \boldsymbol{\Omega} \cdot \bar{\mathbf{I}} \cdot \boldsymbol{\Omega} \tag{2.27}$$

$$\mathbf{H}_R = \sum_{i=1}^{N} \frac{M_{i-1} m_i}{2\ M_i} \mathbf{R}_i \times \mathbf{R}_i' + \sum_{i=0}^{N} \bar{\mathbf{I}}_i^R \cdot \omega_i \tag{2.28}$$

The quantity $\mathbf{H}_R$ is the total angular momentum of the system relative to the rotating frame, which must equal zero and is a constant. However, it cannot be eliminated from the kinetic energy as the partials of this term are still required to properly state the equations of motion. Also note that the kinetic energy term T_R is the system kinetic energy relative to the rotating frame, and the velocities are the time derivatives of the coordinates relative to the rotating frame, discussed in more detail in [Scheeres (2019)].

Make the explicit substitution $\boldsymbol{\Omega} = \dot{\theta}\hat{\mathbf{H}}$ yielding the transformed Lagrangian

$$\mathcal{L} = T_R + \dot{\theta}\hat{\mathbf{H}} \cdot \mathbf{H}_R + \frac{1}{2}\dot{\theta}^2 \hat{\mathbf{H}} \cdot \bar{\mathbf{I}} \cdot \hat{\mathbf{H}} - \mathcal{U} \tag{2.29}$$

With the velocity $\dot{\theta}$ exposed as such, and the overall orientation angle about the angular momentum vector θ well defined but absent from the Lagrangian, the angular momentum integral can be identified as $\frac{d}{dt}\frac{\partial \mathcal{L}}{\partial \dot{\theta}} = 0$, leading to

$$\dot{\theta}\hat{\mathbf{H}} \cdot \bar{\mathbf{I}} \cdot \hat{\mathbf{H}} + \hat{\mathbf{H}} \cdot \mathbf{H}_R = H = \text{constant} \tag{2.30}$$

Solving for the angular rate results in

$$\dot{\theta} = \frac{H}{I_H}\left[1 - \frac{\hat{\mathbf{H}} \cdot \mathbf{H}_R}{H}\right] \tag{2.31}$$

where I_H is, as before, the total moment of inertia of the system about the angular momentum vector. Again note that $\mathbf{H}_R = \mathbf{0}$, but this equality is not enforced until after the equations of motion and related quantities are defined.

Applying the Routh reduction to the Lagrangian yields the Routhian

$$\begin{aligned}\mathcal{L}_R &= \mathcal{L} - \dot{\theta}\frac{\partial L}{\partial \dot{\theta}} \\ &= \mathcal{L} - \dot{\theta}\left[\dot{\theta} I_H + \hat{\mathbf{H}} \cdot \mathbf{H}_R\right] \\ &= T_R - \mathcal{U} - \frac{1}{2}\dot{\theta}^2 I_H\end{aligned} \tag{2.32}$$

To complete the transformation substitute $\dot{\theta}$ into the expression to find

$$\mathcal{L}_R = T_R - \mathcal{U} - \frac{H^2}{2I_H}\left[1 - \frac{\hat{\mathbf{H}} \cdot \mathbf{H}_R}{H}\right]^2 \tag{2.33}$$

Note that the amended potential has the additional term $\left[1 - \frac{\hat{\mathbf{H}}\cdot\mathbf{H}_R}{H}\right]^2$ which contributes to the equations of motion as the partials of this term are not generally equal to zero.

Once the system is solved, the rotation term can be reconstructed following the Routhian approach

$$\begin{aligned}
\dot{\theta} &= -\frac{\partial \mathcal{L}_R}{\partial H} \\
&= \frac{H}{I_H}\left[1 - \frac{\hat{\mathbf{H}}\cdot\mathbf{H}_R}{H}\right]^2 + \frac{H^2}{2I_H}2\left[1 - \frac{\hat{\mathbf{H}}\cdot\mathbf{H}_R}{H}\right]\frac{\hat{\mathbf{H}}\cdot\mathbf{H}_R}{H^2} \\
&= \frac{H}{I_H}
\end{aligned} \tag{2.34}$$

where, in the last step $\mathbf{H}_R = \mathbf{0}$.

Finally, as the system is time invariant, the Jacobi integral can be found, equal to $\dot{\mathbf{Q}}\cdot\frac{\partial \mathcal{L}_R}{\partial \dot{\mathbf{Q}}} - \mathcal{L}_R$, which yields the classical energy

$$E = T_R + \mathcal{E} \tag{2.35}$$

$$\mathcal{E} = \frac{H^2}{2I_H} + \mathcal{U} \tag{2.36}$$

Note that in this derivation, $\dot{\mathbf{Q}}\cdot\partial\mathbf{H}_R/\partial\dot{\mathbf{Q}} = \mathbf{H}_R = 0$ as the angular momentum is linear in the velocities. From this last statement, along with the fact that $T_R \geqslant 0$, the useful inequality is found

$$\mathcal{E} \leqslant E \tag{2.37}$$

This inequality is sharp, meaning that equality can occur when the internal kinetic energy is zero. This happens, in particular, when the system is at an equilibrium. It can also be verified that when the system is at an equilibrium the condition $\delta\mathcal{E} = \partial\mathcal{E}/\partial\mathbf{Q}\cdot\delta\mathbf{Q} \geqslant 0$ holds for all feasible $\delta\mathbf{Q}$. If the bodies are not in contact, then the variation in each coordinate is zero, whereas if any bodies are in contact then the allowable variations will only increase energy [Scheeres (2019)].

3. Alternate Derivation

The AMR potential can also be derived using a different approach related to Cauchy's inequality and resulting in a modified version of the Sundman inequality. The total angular momentum of the system can be defined as an integral over the entire mass distribution, which contains all of the rigid bodies [Scheeres (2019)].

$$\mathbf{H} = \int_{\mathcal{B}} \mathbf{r}\times\mathbf{v}\ dm \tag{3.1}$$

The magnitude is then H and the unit vector is $\hat{\mathbf{H}} = \mathbf{H}/H$.

Also recall the similarly general definitions for kinetic energy, polar moment of inertia and inertia tensor.

$$T = \frac{1}{2}\int_{\mathcal{B}} \mathbf{v}\cdot\mathbf{v}\ dm \tag{3.2}$$

$$I_P = \int_{\mathcal{B}} \mathbf{r}\cdot\mathbf{r}\ dm \tag{3.3}$$

$$\bar{\mathbf{I}} = -\int_{\mathcal{B}} \tilde{\mathbf{r}}\cdot\tilde{\mathbf{r}}\ dm \tag{3.4}$$

The usual Sundman Inequality derivation is made using the polar moment of inertia, as found in [Pollard (1966)]. Starting with the angular momentum definition, $H^2 = \left|\int_{\mathcal{B}} \mathbf{r} \times \mathbf{v}\, dm\right|^2 \leqslant \left(\int_{\mathcal{B}} |\mathbf{r} \times \mathbf{v}|\; dm\right)^2$. Then the Cauchy-Schwartz inequality can be applied to find $\left(\int_{\mathcal{B}} |\mathbf{r} \times \mathbf{v}|\; dm\right)^2 \leqslant \left(\int_{\mathcal{B}} v^2\, dm\right)\left(\int_{\mathcal{B}} r^2\, dm\right)$. From above these two integrals are the kinetic energy and polar moment of inertia, leading to $H^2 \leqslant 2T\, I_P$. Using the relationship $E = T + \mathcal{U}$, then $H^2 \leqslant 2(E - \mathcal{U}) I_P$, the polar amended potential and related Sundman inequality is found as

$$\mathcal{E}_P = \frac{H^2}{2I_P} + \mathcal{U} \leqslant E \tag{3.5}$$

Here $\mathcal{E}_P$ is the polar moment of inertia version of the amended potential.

A sharper alternate approach to Sundman's inequality can be followed that utilizes the system moment of inertia about $\hat{\mathbf{H}}$, and results directly in the AMR amended potential. At the start, the change is trivial and just recognizes that the angular momentum vector is oriented along the $\hat{\mathbf{H}}$ vector. Thus, $H = \hat{\mathbf{H}} \cdot \mathbf{H} = \int_{\mathcal{B}} \hat{\mathbf{H}} \cdot \mathbf{r} \times \mathbf{v}\, dm$, where since the angular momentum direction is a constant it can be brought inside the integral. Note that $\hat{\mathbf{H}} \cdot \mathbf{r} \times \mathbf{v} = \mathbf{v} \cdot (\hat{\mathbf{H}} \times \mathbf{r})$ and apply the Cauchy-Schwartz inequality to H^2 to find

$$H^2 = \left(\int_{\mathcal{B}} \mathbf{v} \cdot \hat{\mathbf{H}} \times \mathbf{r}\; dm\right)^2 \leqslant \left(\int_{\mathcal{B}} v^2\; dm\right)\left(-\hat{\mathbf{H}} \cdot \int_{\mathcal{B}} \tilde{\mathbf{r}} \cdot \tilde{\mathbf{r}}\; dm \cdot \hat{\mathbf{H}}\right)$$

which immediately leads to $H^2 \leqslant 2T\, I_H$. Again using the relationship $E = T + \mathcal{U}$, then $H^2 \leqslant 2(E - \mathcal{U}) I_H$ which leads to

$$\mathcal{E} = \frac{H^2}{2I_H} + \mathcal{U} \leqslant E \tag{3.6}$$

Note that the defined lower quantity is the AMR amended potential $\mathcal{E}$. Substituting the Jacobi integral from the rotating, reduced dynamical system, $E = T_R + \mathcal{E}$, it is clear that the AMR amended potential equals the energy when the system is at an equilibrium.

4. Comparisons with Other Amended Potentials

There are several versions of the amended potential that can be used to define different variations of the Sundman inequality, relating angular momentum, moment of inertia and energy. It is instructive to compare these to gain insight into the relations between them. These are succinctly summarized with a few theorems and proofs.

4.1. *Comparison with $\mathcal{E}_P$*

THEOREM 4.1.

$$\mathcal{E}_P \leqslant \mathcal{E}$$

If there are any three-dimensional distributions of material then $\mathcal{E}_P < \mathcal{E}$, while if the entire system is planar, including the rigid bodies, then $\mathcal{E}_P = \mathcal{E}$.

Proof. To prove this it is sufficient to show that $I_H \leqslant I_P$ in general, and is either strictly less than or equal for the different cases considered. Note that $I_P = \frac{1}{2}\mathrm{Trace}(\bar{\mathbf{I}}) = \frac{1}{2}\left(I_1 + I_2 + I_3\right)$ where I_i are the principal moments of inertia of the system, assumed ordered as $I_1 \leqslant I_2 \leqslant I_3$. Then $I_H \leqslant I_3$ and the most extreme case to consider is

$$I_3 \leqslant \frac{1}{2}\left(I_1 + I_2 + I_3\right), \tag{4.1}$$

however this can be reduced to $I_3 \leqslant I_1 + I_2$ which is a fundamental result. It can easily be shown that equality only holds for fully planar mass distributions. Thus, if the system has

3-dimensional rigid bodies the result becomes a strict inequality, $I_3 < \frac{1}{2}(I_1 + I_2 + I_3)$. If the systems are fully planar then $I_H = I_3 = I_P$. □

Thus the polar moment of inertia compresses three dimensional mass distributions, including the moments of inertias for rigid bodies, into a lumped value that does not provide a sharp inequality. This implies that at equilibrium, the polar amended potential is not equal to the total energy for systems with three dimensional mass distributions.

4.2. *Smale Amended Potential*

A version of the amended potential that is used in geometric mechanics analysis of these problems is the Smale Amended Potential [Simo et al. (1991)]:

$$\mathcal{E}_{GM} = \frac{1}{2}\mathbf{H} \cdot \bar{\mathbf{I}}^{-1} \cdot \mathbf{H} + \mathcal{U} \tag{4.2}$$

This potential bounds the AMR potential.

THEOREM 4.2.

$$\mathcal{E} \leqslant \mathcal{E}_{GM} \leqslant E$$

The two potentials are equal when the system is in a relative equilibrium.

Proof. One only needs to prove $H^2/I_H^2 \leqslant \mathbf{H} \cdot \bar{\mathbf{I}}^{-1} \cdot \mathbf{H}$. Normalize by H^2, choose a diagonal coordinate representation for $\bar{\mathbf{I}}$ denoted as $\hat{\mathbf{e}}_i$, $i = 1, 2, 3$, and note that $\hat{\mathbf{H}} = \sum_{i=1}^{3} h_i \hat{\mathbf{e}}_i$. Then the inequality can be reduced to

$$\begin{aligned}
1 &\leqslant \left(h_1^2 I_1 + h_2^2 I_2 + h_3^2 I_3\right)\left(h_1^2 \frac{1}{I_1} + h_2^2 \frac{1}{I_2} + h_3^2 \frac{1}{I_3}\right) \\
&= h_1^4 + h_2^4 + h_3^4 + h_1^2 h_2^2 \left(\frac{I_2}{I_1} + \frac{I_1}{I_2}\right) + h_2^2 h_3^2 \left(\frac{I_3}{I_2} + \frac{I_2}{I_3}\right) + h_3^2 h_1^2 \left(\frac{I_1}{I_3} + \frac{I_3}{I_1}\right) \\
&\quad + 2\left(h_1^2 h_2^2 + h_2^2 h_3^2 + h_3^2 h_1^2\right) - 2\left(h_1^2 h_2^2 + h_2^2 h_3^2 + h_3^2 h_1^2\right) \\
&= \left(h_1^2 + h_2^2 + h_3^2\right)^2 + h_1^2 h_2^2 \frac{(I_1 - I_2)^2}{I_1 I_2} + h_2^2 h_3^2 \frac{(I_2 - I_3)^2}{I_2 I_3} + h_3^2 h_1^2 \frac{(I_3 - I_1)^2}{I_3 I_1} \\
&= 1 + h_1^2 h_2^2 \frac{(I_1 - I_2)^2}{I_1 I_2} + h_2^2 h_3^2 \frac{(I_2 - I_3)^2}{I_2 I_3} + h_3^2 h_1^2 \frac{(I_3 - I_1)^2}{I_3 I_1}
\end{aligned}$$

or

$$0 \leqslant h_1^2 h_2^2 \frac{(I_1 - I_2)^2}{I_1 I_2} + h_2^2 h_3^2 \frac{(I_2 - I_3)^2}{I_2 I_3} + h_3^2 h_1^2 \frac{(I_3 - I_1)^2}{I_3 I_1} \tag{4.3}$$

which is true by inspection. When the system is at an equilibrium, it must rotate uniformly about only one of the principle moments of inertia of the system, meaning $h_i h_j = 0$ for $i \neq j$. In this case the bound becomes an equality, meaning that the two potentials are equal when evaluated at an equilibrium. □

Despite the fact that the Smale potential bounds the AMR potential, they both yield conditions for equilibria and by extension for energetic stability of an equilibria.

THEOREM 4.3. *If the system is in a non-contact equilibrium configuration, then*

$$\delta \mathcal{E}_{GM} = \delta \mathcal{E} = 0$$

Proof. Note that, based on the Lagrangian derivation, when the system is at such an equilibrium then $\delta\mathcal{E}=0$. To prove the other equality first express the first variation of both quantities:

$$\delta\mathcal{E}_{GM}=-\frac{1}{2}\mathbf{H}\cdot\bar{\mathbf{I}}^{-1}\cdot\delta\bar{\mathbf{I}}\cdot\bar{\mathbf{I}}^{-1}\cdot\mathbf{H}+\delta\mathcal{U} \tag{4.4}$$

$$\delta\mathcal{E}=-\frac{1}{2}\left(\frac{H}{I_H}\right)^2\delta I_H+\delta\mathcal{U} \tag{4.5}$$

Cancelling like terms reduces the proof to showing the following result

$$\left(\frac{H}{I_H}\right)^2\delta I_H=\mathbf{H}\cdot\bar{\mathbf{I}}^{-1}\cdot\delta\bar{\mathbf{I}}\cdot\bar{\mathbf{I}}^{-1}\cdot\mathbf{H} \tag{4.6}$$

When in an equilibrium $\mathbf{H}\cdot\bar{\mathbf{I}}^{-1}=\bar{\mathbf{I}}^{-1}\cdot\mathbf{H}=\boldsymbol{\Omega}=\dot{\theta}\hat{\mathbf{H}}$ by definition, as it will be rotating about a principal moment of inertia. Thus, substituting for the definition of I_H and noting that $H/I_H=\dot{\theta}$, equality is shown to be trivially true. □

This is an interesting result, given that the one potential in general bounds the other, and implies that either potential can be used to evaluate equilibria and stability.

4.3. *Disaggregated AMR Amended Potential*

A different potential version that bounds the AMR amended potential can be defined. This form disaggregates the potential into a series of component parts. This approach is specialized to the Jacobi coordinates, and was derived and proven in [Scheeres & Brown (2023)]. Define a disaggregated version of the potential as

$$\mathcal{E}_i=\frac{H_i^2}{2I_{H_i}}+\mathcal{U}_i \tag{4.7}$$

where H_i is the angular momentum arising from the ith Jacobi coordinate, $I_{H_i}=\hat{\mathbf{H}}\cdot\bar{\mathbf{I}}_i\cdot\hat{\mathbf{H}}$, and $\mathcal{U}_i=\sum_{j=0}^{i}\mathcal{U}_{ij}$. Note that $\sum_{i=1}^{N}H_i=H$.

Then in [Scheeres & Brown (2023)] it was proven that:

$$\mathcal{E}\leqslant\sum_{i=1}^{N}\mathcal{E}_i\leqslant E \tag{4.8}$$

and that at an equilibrium the two potentials are equal. The individual $\mathcal{E}_i$ components do not provide information on the dynamics. One should note, specifically, that the angular momentum terms H_i are not necessarily conserved, and thus can individually vary when taking variations. Only when they are summed is that quantity conserved.

5. Conclusions

In this contribution the Angular Momentum Relative amended potential is derived and compared with other amended potential versions. When the dynamical system is composed of rigid bodies, it is shown that the AMR potential dominates the potential defined using the polar moment of inertia. It is also shown that the Smale amended potential bounds the AMR potential, but that they are equal at an equilibrium and that they can both yield conditions for evaluating such equilibria and their stability.

References

D.T. Greenwood. Classical Dynamics. Prentice-Hall, 1977.
H. Pollard. Mathematical Introduction to Celestial Mechanics Prentice-Hall, 1966.

D.J. Scheeres. Minimum Energy Configurations in the N-Body Problem and the Celestial Mechanics of Granular Systems. *Celestial Mechanics and Dynamical Astronomy*, 113(3):291–320, 2012.

Scheeres, D.J. (2019). Celestial Mechanics of Rubble Pile Bodies. In: Baù, G., Celletti, A., Gales, C., Gronchi, G. (eds) Satellite Dynamics and Space Missions. Springer INdAM Series, vol 34. Springer, Cham. https://doi.org/10.1007/978-3-030-20633-8_5

D.J. Scheeres. Relative equilibria in the spherical, finite density three-body problem. *Journal of Nonlinear Science*, 26(5):1445–1482, 2016.

D.J. Scheeres and G.M. Brown. Bounds on energy and angular momentum loss in the full n-body problem *Celestial Mechanics and Dynamical Astronomy*, 135:35, 2023.

Simo JC, Lewis D, Marsden JE (1991) Stability of relative equilibria. Part I: the reduced energy-momentum method. Arch Ration Mech Anal 115(1):15–59

Complex Planetary Systems II: Latest Methods for an Interdisciplinary Approach
Proceedings IAU Symposium No. 382, 2024
A. Lemaitre & A.-S. Libert, eds.
doi:10.1017/S1743921323004180

Progress report on the BEBOP search for circumbinary planets with radial velocities

Thomas A. Baycroft[1], Amaury H. M. J. Triaud[1] and Dong Lai[2]

[1]University of Birmingham (UK)
email: txb187@bham.ac.uk

[2] Cornell University (US)

Abstract. The BEBOP (Binaries Escorted By Orbiting Planets) survey is a search for circumbinary planets using the radial velocity spectrographs HARPS and SOPHIE, currently focusing on single-lined binaries with a mass ratio < 0.3. Circumbinary systems are an important testing ground for planet formation theories as the dynamically complex influence of the binary makes planet formation and survival more difficult. Here we present the results of the survey so far including: confirmed planets such as BEBOP-1c the first circumbinary planet detected in radial velocity; the status of our observations; and preliminary occurrence rates. We compare the early results of the radial velocity survey to the population of circumbinary planets discovered in transit, and suggest that there may be a population of inflated planets close to the inner binary which are detectable in transit but more difficult in radial velocity. Using time-lag tidal theory, we show that this inflation is unlikely caused by tides.

Keywords. Circumbinary planets, Eclipsing binaries, ocurrence rates, tidal interactions

1. Introduction

Over 5000 planets orbiting other stars have been confirmed to date. The majority of these have been detected around single stars. This is widely assumed to be an observation bias rather a reflection of the underlying population.

Planets in binary systems come in two flavours: planetary-type (p-type), in a wider orbit around the centre of mass of the two stars; and satellite-type (s-type), orbiting in a close orbit around one of the two stars. The planetary-type are also known as circumbinary planets, and these are the rarer of the two flavours with only 15 such planets known to orbit main-sequence binaries (Standing et al. 2023). Again, this is most likely due to lack of observations rather than a true dearth of such planets.

Around the inner binary, there is an instability zone, in which planets would not be dynamically stable. The extent of this zone has been numerically determined (e.g. Dvorak et al. 1989; Holman and Wiegert 1999), in general it is between $4\times$ and $6\times$ the period of the inner binary. Around the inner binary there is also a zone in which, according to current theories, planet formation is inhibited (Moriwaki and Nakagawa 2004; Paardekooper et al. 2012), this extends to $\sim 50\times$ the binary period. This zone is much more extended than the instability zone, meaning that a planet can live on an orbit in a region where it cannot have been born. Hence, circumbinary planets that are found to be near the instability zone must have formed further out in the protoplanetary disc and migrated inwards to their current location (Kley and Haghighipour 2014). Circumbinary planets are therefore a great probe of migration pathways as they can have populations that are *known* to have migrated, which cannot be said for most exoplanets hosted by single stars.

Circumbinary exoplanets have been detected in transit by both the *Kepler* (e.g. Doyle et al. 2011) and *TESS* (e.g. Kostov et al. 2020) space telescopes. There are 14 of them in 12 systems. Their orbital periods put almost all of them very close to the instability region, and therefore means they most likely have had to migrate to reach their current locations. The fact that they are all so close to the instability limit has been interpreted as a sign that there is a pile-up of circumbinary planets near the instability limit at around $6\times$ the binary period, and that this is not due to an observational bias (Martin and Triaud 2014).

The radial-velocity method discovered planets around single stars before the transit method. For circumbinary planets it is the other way around, as the binary itself has hampered efforts to detect them. Double-lined binaries (which account for 92% of binaries; Kovaleva et al. 2016) are those where spectral lines from both components' are combined into the observed spectrum. The spectral lines' interference with each other in the standard cross-correlation technique for radial-velocity extraction adds extra scatter to the data which makes it difficult to detect planets. One way to circumvent this is to observe single-lined binaries. With a mass ratio < 0.3, the secondary star in these cases is not bright enough to cause a significant effect on the primary star's measured radial velocities. This is the method that has been used by the BEBOP survey whose results so far are presented here.

2. The BEBOP survey and published results

The BEBOP (Binaries Escorted By Orbiting Planets) survey, is a blind, radial velocity survey on a sample of single-lined eclipsing binaries (Martin et al. 2019). These were identified from the EBLM (Eclipsing Binaries with Low Mass companions; Triaud et al. 2013) survey and from *TESS*. Data has been collected with the HARPS spectrograph (Mayor et al. 2003) and the SOPHIE spectrograph (Perruchot et al. 2008) for up to 5 years. There is a main sample of around 100 stars, and a secondary sample also numbering around 100.

The first detection from the BEBOP survey was the confirmation of Kepler-16 b (Triaud et al. 2022). This had been the first circumbinary planet discovered in transit (Doyle et al. 2011), and BEBOP independently detected it using the radial velocity method. This provided the proof-of-concept that the detection of circumbinary planets in radial velocity is attainable, as well as confirmed the mass of the planet which had been determined from eclipse timing variations.

The second detected planet from the BEBOP survey is a new discovery, BEBOP-1/ TOI-1338 c (Standing et al. 2023). This is a multi-planet system with an inner transiting circumbinary planet discovered by *TESS* (Kostov et al. 2020) and an outer planet discovered in radial velocity by BEBOP (Standing et al. 2023). The outer planet has not yet been seen in transit, this is likely due to the longer period as well as the fact that circumbinary planets alternate being in and out of *transitability* (Martin and Triaud 2015). The inner planet is not yet detected with the radial velocity data, this allows us to put an upper limit on the mass of planet b of $21.8 \pm 0.9\,\mathrm{M}_{\oplus}$. This along with the planetary radius being $6.85 \pm 0.19\,\mathrm{R}_{\oplus}$ (Kostov et al. 2020) implies a bulk density $\rho_b \lesssim 0.36\,\mathrm{g\,cm^{-3}}$. This means that planet b is under-dense, perhaps meaning that it is inflated in some way.

3. Results and Occurrence rates

We now present very preliminary occurrence rates of the BEBOP survey based on detected and candidate planets.

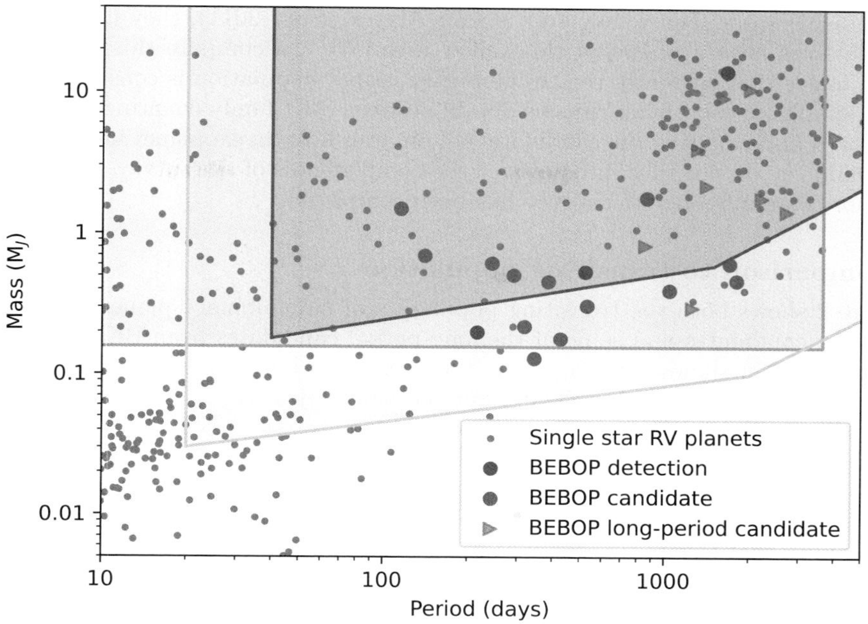

Figure 1. Preliminary occurrence plot showing the BEBOP planets and candidates in colour with the radial velocity population of planets around single stars show in grey (`https://exoplanetarchive.ipac.caltech.edu/`). Dark green the median sensitivity of BEBOP. Light green the best sensitivity of BEBOP. Yellow the area inside which Mayor et al. (2011) find an occurrence rate of 14-15%.

The radial velocity analysis undertaken by BEBOP uses the analysis package `kima` (Faria et al. 2018), this utilises diffusive nested sampling (DNEST4; Brewer and Foreman-Mackey 2018) to efficiently explore parameter space. The trans-dimensional nested sampling allows to explore different numbers of planetary signals in the data and to compute a Bayes' Factor comparing them. We claim strong evidence for a planet when the Bayes' Factor for the model with a planet to the one without is greater than 150 (as in Kass and Raftery 1995). `kima` has a BINARIES radial-velocity model tailored to fitting binary orbits by including apsidal precession of the binary and accounting for GR and tidal effects (Baycroft et al. 2023). The binary orbit can also be fit with independent priors to the planets meaning this is very suited to the detection of circumbinary planets in radial velocity.

From the most recent analysis of the BEBOP sample we have three categories of *planets*: detected, candidate, long-period candidate. The detected planets consist of BEBOP-1c as well as two currently unpublished planets with Bayes' Factors exceeding 150 putting them in the confident detection band. There are 13 candidate planets, these have a Bayes' Factor less than 150 so are not (yet) formally classed as detections. The long-period candidates consists of signals which have a period longer than the baseline of observations, these may have Bayes Factors exceeding 150 but the nature and parameters of the signal are uncertain. There are 9 of these signals for which we take a lower-bound on the orbital period and mass based on the shortest period circular orbit within the 1σ confidence interval for the orbital period.

The 3 detections, 13 candidates and 9 long-period candidates are shown on a mass-period plot in Figure 1 with the colours and shapes of the points differentiating between detections, candidates and long-period candidates. The green areas show the best and median sensitivity of the BEBOP survey, the grey points are radial velocity planets around single stars. The yellow area is from Mayor et al. (2011), they find that 14-15% of single stars have a planet in this yellow area. We can compare this to the BEBOP planets to see that currently the circumbinary planet population is consistent with this. From a simplistic occurrence rate we obtain between 3% (if only counting the detections) and 23% (if also counting the candidates) of binaries host an exoplanet in the yellow box. These numbers do not take into account the completeness of the survey, and this will be done more carefully once the survey has been completed.

4. Comparison to transiting population

Figure 2 shows both the transiting population of circumbinary planets as well as the detections, candidates and some of the long-period candidates from BEBOP. The transiting population shows a pile-up near the instability limit, this pile-up is not seen in the radial velocity population. The lower plot motivates why this is not solely due to the transit method's geometric observation bias towards short-period orbits. This bias is solely dependent on the distance of the planet from the star, this means that a vertical strip in the bottom figure should show planets that are all equally as likely to transit (Martin and Triaud 2014). The fact that all the transiting planets are near the *top* of such strips implies that this pattern cannot solely be an observation bias due to the distance of such planets from the central star.

To explain why the pile-up is seen with transiting planets, but not with radial-velocity planets, a possible solution is that circumbinary planets near the instability limit are inflated, making them both easier to detect in transit (increasing their radius), and harder with radial-velocity (reducing their mass). One example of this is TOI-1338/BEBOP-1 b which is inflated as described above.

5. Tidal energy dissipation

One possible mechanism for the inflation of circumbinary planets close to the stability limit is from tidal energy dissipation. From the any of the two stars' reference frames, the planet occupies a constantly eccentric orbit. Tides raised on the planet will attempt to circularise its orbit and synchronise its rotation (e.g. Guillot et al. 1996). However, the planet's motion about the binary already being circular (or close to it), tidal equilibrium is prevented, and therefore more energy could be dissipated over the lifetime of the system. We do a back of the envelope calculation for TOI-1338/BEBOP-1 b using time-lag tidal theory (Hut 1981).

We consider a tide raised on the planet of mass $m_{\rm p}$ by the secondary of mass m_2 ($\lesssim m_1$). Relative to the barycentre, the position of the planet is $\vec{r_{\rm p}}(t)$ and that of the secondary is $\vec{r_2}(t)$, and we write $\vec{d_{\rm p}}(t) = \vec{r_{\rm p}}(t) - \vec{r_2}(t)$. Now $d_{\rm p}(t)$ varies with mean frequency $\Omega_2 - \Omega_{\rm p} = \Delta\Omega$, with $\Omega_{\rm p}$ and Ω_2 the orbital angular velocities of the planet and secondary. We can give an approximate expression for the tidal kinetic energy:

$$E_{\rm K} \sim k_2 \frac{m_2^2}{m_{\rm p}} \frac{R_{\rm p}^8}{d_{\rm p}^6} (\Delta\Omega)^2, \tag{1}$$

where k_2 is the Love number and $R_{\rm p}$ is the planet's radius. Then the tidal energy dissipation can be taken as $\dot{E}_{\rm K} \sim \frac{E_{\rm K}}{t_{\rm dissipation}} \sim E_{\rm K} \frac{\Delta\Omega}{Q}$ where Q is the tidal quality factor. This leads us to an expression for the tidal energy dissipation rate:

$$\dot{E}_{\rm K} \sim \frac{k_2}{Q} m_{\rm p} \left(\frac{m_2}{m_{\rm p}}\right)^2 \left(\frac{R_{\rm p}}{d_{\rm p}}\right)^6 R_{\rm p}^2 (\Delta\Omega)^3. \tag{2}$$

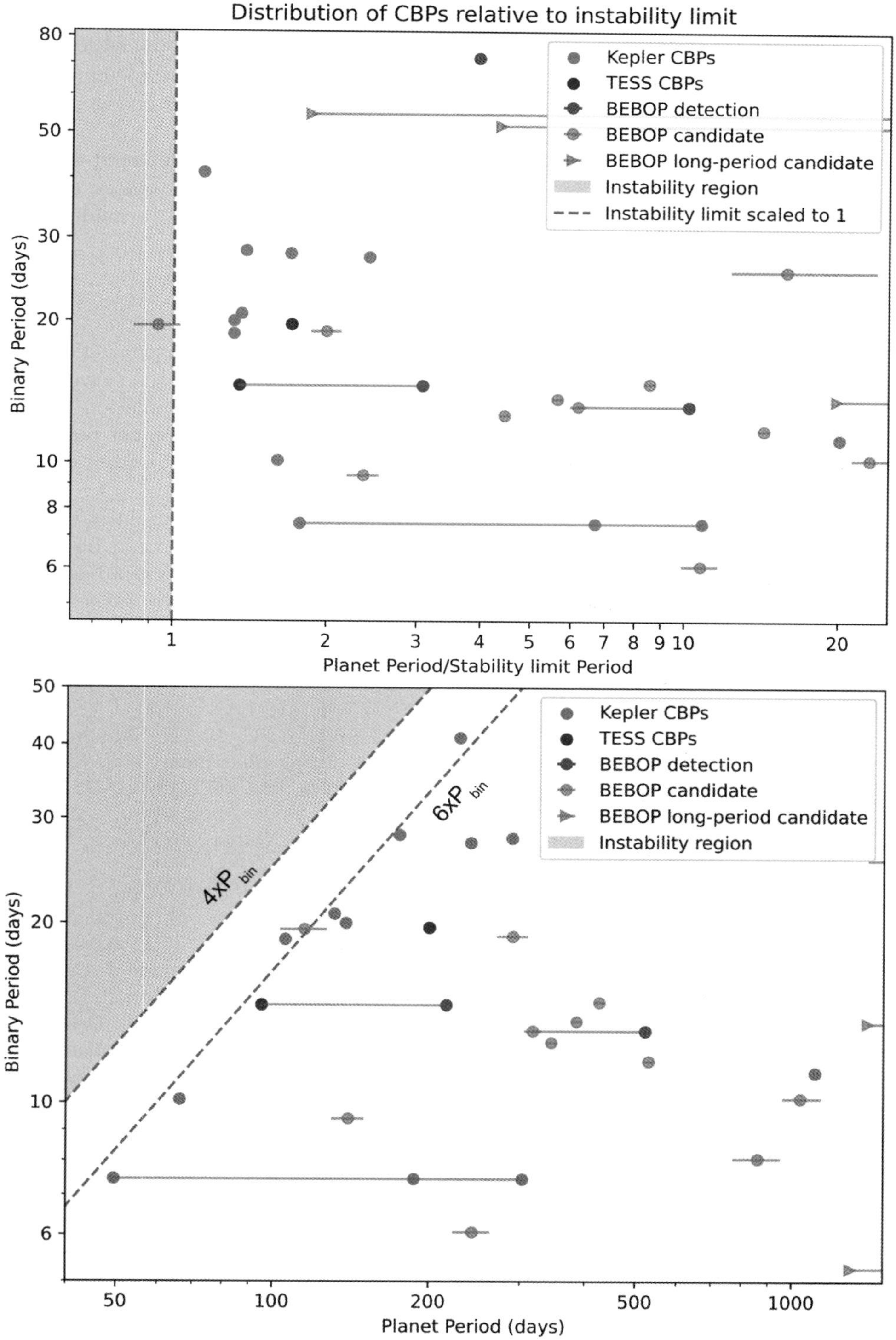

Figure 2. Period distribution of circumbinary planets detected in transit and the BEBOP radial velocity planets and candidates. Top: The periods are scaled by the period that a planet would have if on a circular orbit at the stability boundary. Bottom: The true orbital periods of the planets are shown, with the dashed lines showing four times and six times the binary period.

Substituting in parameters for the TOI-1338/BEBOP-1 system, and assuming $\frac{k_2}{Q} \sim 10^{-5}$, yields an order of magnitude estimate $\dot{E}_{\rm K} \sim 10^{21}$ erg/s. An order-of-magnitude estimate for the energy coming from insolation gives $L_{\rm in} \sim 10^{28}$ erg/s. This implies that energy dissipation from tides is not significant enough to have an impact on the planetary radius, meaning there is likely another explanation.

However, this is still an open question, more circumbinary planets detected with both radial-velocities and with transits will help understand the populations better, and more theoretical models of circumbinary planets, including but not limited to models of the tidal effects, will be beneficial.

6. Conclusions

We have presented the BEBOP radial velocity survey for circumbinary planets orbiting single-lined spectroscopic binaries, as well as preliminary results from the survey.

We have compared the detections and candidates to both the population of radial velocity planets orbiting single stars and the transiting circumbinary planet population. The current results do not show a significant difference in occurrence rate from the single star population.

The pile-up of planets seen in the transiting population is not as apparent in the radial velocity population. We suggest that a possible explanation for this could be that the planets that migrate close to the instability limit are inflated. We perform a back-of-the-envelope calculation that casts doubt on this inflation being caused by tides raised by the binary.

References

Baycroft, T. A., Triaud, A. H. M. J., Faria, J., Correia, A. C. M., & Standing, M. R. 2023, Improving circumbinary planet detections by fitting their binary's apsidal precession. *Monthly Notices of the Royal Astronomical Society*, 521, 1871–1879. ADS Bibcode: 2023MNRAS.521.1871B.

Brewer, B. J. & Foreman-Mackey, D. 2018, DNest4: Diffusive Nested Sampling in C++ and Python. *Journal of Statistical Software*, 86, 1–33.

Doyle, L. R., Carter, J. A., Fabrycky, D. C., Slawson, R. W., Howell, S. B., Winn, J. N., Orosz, J. A., Přsa, A., Welsh, W. F., Quinn, S. N., Latham, D., Torres, G., Buchhave, L. A., Marcy, G. W., Fortney, J. J., Shporer, A., Ford, E. B., Lissauer, J. J., Ragozzine, D., Rucker, M., Batalha, N., Jenkins, J. M., Borucki, W. J., Koch, D., Middour, C. K., Hall, J. R., McCauliff, S., Fanelli, M. N., Quintana, E. V., Holman, M. J., Caldwell, D. A., Still, M., Stefanik, R. P., Brown, W. R., Esquerdo, G. A., Tang, S., Furesz, G., Geary, J. C., Berlind, P., Calkins, M. L., Short, D. R., Steffen, J. H., Sasselov, D., Dunham, E. W., Cochran, W. D., Boss, A., Haas, M. R., Buzasi, D., & Fischer, D. 2011, Kepler-16: A Transiting Circumbinary Planet. *Science*, 333, 1602. ADS Bibcode: 2011Sci...333.1602D.

Dvorak, R., Froeschle, C., & Froeschle, C. 1989, Stability of outer planetary orbits (P-types) in binaries. *Astronomy and Astrophysics*, 226, 335–342.

Faria, J. P., Santos, N. C., Figueira, P., & Brewer, B. J. 2018, kima: Exoplanet detection in radial velocities. *The Journal of Open Source Software*, 3, 487. ADS Bibcode: 2018JOSS....3..487F.

Guillot, T., Burrows, A., Hubbard, W. B., Lunine, J. I., & Saumon, D. 1996, Giant Planets at Small Orbital Distances. *ApJL*, 459, L35.

Holman, M. J. & Wiegert, P. A. 1999, Long-Term Stability of Planets in Binary Systems. *The Astronomical Journal*, 117, 621–628. ADS Bibcode: 1999AJ....117..621H.

Hut, P. 1981, Tidal evolution in close binary systems. *A&A*, 99, 126–140.

Kass, R. E. & Raftery, A. E. 1995, Bayes Factors. *Journal of the American Statistical Association*, 90(430), 773–795. Publisher: Taylor & Francis _eprint: `https://www.tandfonline.com/doi/pdf/10.1080/01621459.1995.10476572`.

Kley, W. & Haghighipour, N. 2014, Modeling circumbinary planets: The case of Kepler-38. *Astronomy and Astrophysics*, 564, A72. ADS Bibcode: 2014A&A...564A..72K.

Kostov, V. B., Orosz, J. A., Feinstein, A. D., Welsh, W. F., Cukier, W., Haghighipour, N., Quarles, B., Martin, D. V., Montet, B. T., Torres, G., Triaud, A. H. M. J., Barclay, T., Boyd, P., Briceno, C., Cameron, A. C., Correia, A. C. M., Gilbert, E. A., Gill, S., Gillon, M., Haqq-Misra, J., Hellier, C., Dressing, C., Fabrycky, D. C., Furesz, G., Jenkins, J. M., Kane, S. R., Kopparapu, R., Hodžić, V. K., Latham, D. W., Law, N., Levine, A. M., Li, G., Lintott, C., Lissauer, J. J., Mann, A. W., Mazeh, T., Mardling, R., Maxted, P. F. L., Eisner, N., Pepe, F., Pepper, J., Pollacco, D., Quinn, S. N., Quintana, E. V., Rowe, J. F., Ricker, G., Rose, M. E., Seager, S., Santerne, A., Sgransan, D., Short, D. R., Smith, J. C., Standing, M. R., Tokovinin, A., Trifonov, T., Turner, O., Twicken, J. D., Udry, S., Vanderspek, R., Winn, J. N., Wolf, E. T., Ziegler, C., Ansorge, P., Barnet, F., Bergeron, J., Huten, M., Pappa, G., & van der Straeten, T. 2020, TOI-1338: TESS' First Transiting Circumbinary Planet. *The Astronomical Journal*, 159, 253. ADS Bibcode: 2020AJ....159..253K.

Kovaleva, D., Malkov, O., Yungelson, L., & Chulkov, D. 2016, Visual binary stars: data to investigate formation of binaries. *Baltic Astronomy*, 25, 419–426. ADS Bibcode: 2016BaltA..25..419K.

Martin, D. V. & Triaud, A. H. M. J. 2014, Planets transiting non-eclipsing binaries. *Astronomy & Astrophysics*, 570, A91.

Martin, D. V. & Triaud, A. H. M. J. 2015, Circumbinary planets - why they are so likely to transit. *Monthly Notices of the Royal Astronomical Society*, 449, 781–793. ADS Bibcode: 2015MNRAS.449..781M.

Martin, D. V., Triaud, A. H. M. J., Udry, S., Marmier, M., Maxted, P. F. L., Collier Cameron, A., Hellier, C., Pepe, F., Pollacco, D., Sgransan, D., & West, R. 2019, The BEBOP radial-velocity survey for circumbinary planets. I. Eight years of CORALIE observations of 47 single-line eclipsing binaries and abundance constraints on the masses of circumbinary planets. *Astronomy and Astrophysics*, 624, A68.

Mayor, M., Marmier, M., Lovis, C., Udry, S., Sgransan, D., Pepe, F., Benz, W., Bertaux, J. L., Bouchy, F., Dumusque, X., Lo Curto, G., Mordasini, C., Queloz, D., & Santos, N. C. 2011,. The HARPS search for southern extra-solar planets XXXIV. Occurrence, mass distribution and orbital properties of super-Earths and Neptune-mass planets. Technical report. Publication Title: arXiv e-prints ADS Bibcode: 2011arXiv1109.2497M Type: article.

Mayor, M., Pepe, F., Queloz, D., Bouchy, F., Rupprecht, G., Lo Curto, G., Avila, G., Benz, W., Bertaux, J. L., Bonfils, X., Dall, T., Dekker, H., Delabre, B., Eckert, W., Fleury, M., Gilliotte, A., Gojak, D., Guzman, J. C., Kohler, D., Lizon, J. L., Longinotti, A., Lovis, C., Megevand, D., Pasquini, L., Reyes, J., Sivan, J. P., Sosnowska, D., Soto, R., Udry, S., van Kesteren, A., Weber, L., & Weilenmann, U. 2003, Setting New Standards with HARPS. *The Messenger*, 114, 20–24. ADS Bibcode: 2003Msngr.114...20M.

Moriwaki, K. & Nakagawa, Y. 2004, A Planetesimal Accretion Zone in a Circumbinary Disk. *The Astrophysical Journal*, 609, 1065–1070. ADS Bibcode: 2004ApJ...609.1065M.

Paardekooper, S.-J., Leinhardt, Z. M., Thbault, P., & Baruteau, C. 2012, How Not to Build Tatooine: The Difficulty of In Situ Formation of Circumbinary Planets Kepler 16b, Kepler 34b, and Kepler 35b. *The Astrophysical Journal*, 754, L16. ADS Bibcode: 2012ApJ...754L..16P.

Perruchot, S., Kohler, D., Bouchy, F., Richaud, Y., Richaud, P., Moreaux, G., Merzougui, M., Sottile, R., Hill, L., Knispel, G., Regal, X., Meunier, J. P., Ilovaisky, S., Le Coroller, H., Gillet, D., Schmitt, J., Pepe, F., Fleury, M., Sosnowska, D., Vors, P., Mgevand, D., Blanc, P. E., Carol, C., Point, A., Laloge, A., & Brunel, J. C. 2008, The SOPHIE spectrograph: design and technical key-points for high throughput and high stability, 7014, 70140J. Conference Name: Ground-based and Airborne Instrumentation for Astronomy II ADS Bibcode: 2008SPIE.7014E..0JP.

Standing, M. R., Sairam, L., Martin, D. V., Triaud, A. H. M. J., Correia, A. C. M., Coleman, G. A. L., Baycroft, T. A., Kunovac, V., Boisse, I., Cameron, A. C., Dransfield, G., Faria, J. P., Gillon, M., Hara, N. C., Hellier, C., Howard, J., Lane, E., Mardling, R., Maxted, P. F. L.,

Miller, N. J., Nelson, R. P., Orosz, J. A., Pepe, F., Santerne, A., Sebastian, D., Udry, S., & Welsh, W. F. 2023, Radial-velocity discovery of a second planet in the TOI-1338/BEBOP-1 circumbinary system. *Nature Astronomy*, 7, 702–714. ADS Bibcode: 2023NatAs...7..702S.

Triaud, A. H. M. J., Hebb, L., Anderson, D. R., Cargile, P., Collier Cameron, A., Doyle, A. P., Faedi, F., Gillon, M., Gomez Maqueo Chew, Y., Hellier, C., Jehin, E., Maxted, P., Naef, D., Pepe, F., Pollacco, D., Queloz, D., Ségransan, D., Smalley, B., Stassun, K., Udry, S., & West, R. G. 2013, The EBLM project. I. Physical and orbital parameters, including spin-orbit angles, of two low-mass eclipsing binaries on opposite sides of the brown dwarf limit. *A&A*, 549, A18.

Triaud, A. H. M. J., Standing, M. R., Heidari, N., Martin, D. V., Boisse, I., Santerne, A., Correia, A. C. M., Acua, L., Battley, M., Bonfils, X., Carmona, A., Collier Cameron, A., Corts-Zuleta, P., Dransfield, G., Dalal, S., Deleuil, M., Delfosse, X., Faria, J., Forveille, T., Hara, N. C., Hbrard, G., Hoyer, S., Kiefer, F., Kunovac, V., Maxted, P. F. L., Martioli, E., Miller, N. J., Nelson, R. P., Poveda, M., Rein, H., Sairam, L., Udry, S., & Willett, E. 2022, BEBOP III. Observations and an independent mass measurement of Kepler-16 (AB) b - the first circumbinary planet detected with radial velocities. *Monthly Notices of the Royal Astronomical Society*, 511, 3561–3570. ADS Bibcode: 2022MNRAS.511.3561T.

Complex Planetary Systems II: Latest Methods for an Interdisciplinary Approach
Proceedings IAU Symposium No. 382, 2024
A. Lemaitre & A.-S. Libert, eds.
doi:10.1017/S174392132300399X

Transversal Yarkovsky acceleration for Apophis through jet transport

Luis Benet[1] **and Jorge A. Pérez Hernández**[2]

[1]Instituto de Ciencias Físicas, Universidad Nacional Autónoma de México (UNAM)
Av. Universidad s/n, Col. Chamilpa, 62210 Cuernavaca, México
email: benet@icf.unam.mx

[2]Telespazio Germany GmbH, Darmstadt, Germany
email: perez.hz@gmail.com

Abstract. In this contribution we describe the jet transport techniques that we used in Pérez-Hernández and Benet (2022) for the estimation of the Yarkovsky transversal acceleration for (99942) Apophis, which included optical and radar astrometry observations obtained during 2021 Apophis' fly-by. Our numerical approach exploits automatic differentiation techniques which improve the orbital determination problem. We obtain a non-zero Yarkovsky parameter $A_2 = (-2.899 \pm 0.025) \times 10^{-14}$ au d^{-2} which is consistent with other recent determinations of this parameter. Our results allow to constrain the collision probabilities for the close approaches in 2029, 2036 and 2068.

Keywords. Yarkovsky effect, Apophis, Jet transport

1. Introduction

Accurate predictions of the orbital motion of asteroid (99942) Apophis are driven by the Yarkovsky effect (Giorgini et al. 2002; Chesley 2006). This is a non-gravitational acceleration arising from the anisotropic thermal re-emission of absorbed radiation (Vokrouhlický et al. 2000; Bottke Jr et al. 2006). Since the main dynamical manifestation of the Yarkovsky effect is a drift in the semimajor axis, this acceleration can be modeled as $\dot{\vec{v}}_a = A_2\, g(r_{a,S})\hat{t}_{a,S}$, where $r_{a,S}$ is the radial distance from the Sun to the asteroid, and $\hat{t}_{a,S}$ represents the transversal unit vector along the motion (Farnocchia et al. 2013). The function $g(r)$ describes the dependence on the solar distance, and it is often modelled as $g(r) = (r_0/r)^2$ with $r_0 = 1$ au. The parameter A_2 corresponds to a transversal, non-conservative Yarkovsky acceleration, which is obtained during the orbit determination process. Early attempts to determine A_2 for Apophis resulted in values whose uncertainty could not rule out a pure gravitational interaction; see e.g. Vokrouhlický et al. (2015); Brozović et al. (2018); Del Vigna et al. (2018); Greenberg et al. (2020). Apophis is of interest because, while a collision with Earth has been ruled out for its next close approach in April 13th 2029, more high-quality observations are required to rule out collisions for later close approaches, in particular in 2036 and 2068.

In this contribution, we provide details about the techniques, known as jet transport, which we used to determine the value of A_2 and its uncertainty Pérez-Hernández and Benet 2022, exploiting most optical and radar observations publically available, including observations made during the last close encounter in 2021.

2. Orbit determination

We briefly describe the orbit determination problem, following Milani and Gronchi (2010). First, in order to determine the orbit of the asteroid, a precise dynamical model is required. This is essentially a system of differential equations which includes all important interactions, as well as the initial conditions $\tilde{\mathbf{x}}_0$ for the asteroid defined at a given epoch t_0. The goal is to obtain the nominal orbital elements $\mathbf{x}_0$, together with other unknown parameters of the dynamical model such as A_2, which is the best fit to the observations. The observational quantities are written as $r_k \pm \sigma_k$, where the index k labels the observation, and also refers to the time t_k when the observation was obtained. Here, r_k can be the right ascension and the declination for optical astrometry, or the time delay or the Doppler shift for radar observations. The associated observational uncertainty is σ_k.

In order to link the observations with the variables which are computed from the dynamical model, $\mathbf{x}(t)$, we require the observation function $R(\mathbf{x}(t_k), t_k)$ for each different observed quantity, which depends indirectly on the initial condition $\tilde{\mathbf{x}}_0$. The observation function R is used to construct the observed-minus-computed (O-C) residuals, $\xi_k = r_k - R(\mathbf{x}(t_k), t_k)$, which provide a description of the quality of the dynamical model.

In order to obtain the nominal orbital elements, we introduce the mean-squared residual function

$$Q(\tilde{\mathbf{x}}_0) = \frac{1}{m}\sum_{k=1}^{m} \frac{\xi_k^2}{\sigma_k^2} = \frac{1}{m}\xi^T \mathbf{W} \xi, \tag{1}$$

where m denotes the number of observations, and $\mathbf{W}$ is the diagonal matrix with the inverse squared observational uncertainties. The nominal orbital elements, which we shall denote as $\mathbf{x}_0$, correspond to the minimum of $Q(\tilde{\mathbf{x}}_0)$. The orbit determination problem can thus be interpreted as an optimization problem, with the mean-squared residual function Q as the target function. We note that any (constant) parameters to be determined, such as A_2, can be included as dynamical variables in $\mathbf{x}$, adding new equations of motion to the dynamical model for the NEA, i.e., $\dot{A}_2 = 0$.

2.1. *Differential correction approach*

In order to find the minimum $\mathbf{x}_0$ of the target function, $Q(\tilde{\mathbf{x}}_0)$ is Taylor-expanded around the unknown nominal orbit, up to second order in the deviations with respect to $\mathbf{x}_0$; we note that third- and higher-order corrections are disregarded. Imposing $\frac{\partial Q}{\partial \mathbf{x}_0}(\mathbf{x}_0) = 0$ at the nominal orbit, and using a first guess for the nominal orbit $\mathbf{x}_0^*$, which is assumed to be close enough to $\mathbf{x}_0$, the optimization problem is solved from the (truncated) Taylor expansion for $Q(\tilde{\mathbf{x}}_0)$ by a Newton-like iteration method as

$$\mathbf{x}_{j+1}^* = \mathbf{x}_j^* - \Gamma \frac{\partial Q}{\partial \mathbf{x}_0}(\mathbf{x}_j^*). \tag{2}$$

Here, $\mathbf{x}_j^*$ is the j-th iterate of the Newton method and $\Gamma = \left(\frac{\partial^2 Q}{\partial \mathbf{x}_0^2}\right)^{-1}$, which is evaluated at $\mathbf{x}_j^*$, is the covariance matrix. Denoting by $\mathbf{B} = \frac{\partial \xi}{\partial \mathbf{x}_0}(\mathbf{x}_j^*)$ the derivative of the O-C residuals with respect to the nominal solution, also known as the design matrix, and by $\mathbf{H} = \frac{\partial^2 \xi}{\partial \mathbf{x}_0^2}$ the Hessian, from Eq. (1) we have

$$\frac{\partial Q}{\partial \mathbf{x}_0} = \frac{2}{m}\xi^{\mathbf{T}} \mathbf{W} \mathbf{B}, \tag{3}$$

$$\frac{\partial^2 Q}{\partial \mathbf{x}_0^2} = \frac{2}{m}\left(\mathbf{B}^T \mathbf{W} \mathbf{B} + \xi^T \mathbf{W} \mathbf{H}\right). \tag{4}$$

These quantities are used in (2) to obtain an approximation of the nominal solution.

The design matrix is constructed by solving the first variational equations associated to the dynamical model. Denoting by N the dimensionality of $\mathbf{x}(t)$ including any model parameters, there are N first-order equations of motion associated to the dynamical model, and N^2 for the first variational equations. The three-index Hessian is clearly even more expensive to determine in terms of the number of differential equations to be solved, and it is often neglected. The variant of Newton's method that neglects the higher order corrections in $(\mathbf{x}_k^* - \mathbf{x}_0)$ of $Q(\mathbf{x}_k^*)$ and disregards the Hessian is known as the differential corrections' method (Milani and Gronchi 2010).

2.2. *Jet transport*

Jet transport is a technique used to solve initial value problems which, by exploiting automatic differentiation, allows obtaining insight into the dynamics in the neighborhood of the initial condition. Instead of solving the initial value problem for a specific initial condition $\tilde{\mathbf{x}}_0$, the equations of motion are solved for $\tilde{\mathbf{x}}_0 + \underline{\boldsymbol{\delta}}$, where the *symbol* $\underline{\boldsymbol{\delta}}$ denotes the deviations from the initial condition $\tilde{\mathbf{x}}_0$, which are assumed to be small enough. Jet transport only requires that the specific integration solver used for the differential equations works with a polynomial manipulator.

The solutions of the differential equations at time t are then truncated polynomials in $\underline{\boldsymbol{\delta}}$ around the initial condition, which we write as

$$\mathbf{x}(t, \underline{\boldsymbol{\delta}}) \simeq \mathbf{x}^{[0]}(t) + \mathbf{x}^{[1]}(\underline{\boldsymbol{\delta}})(t) + \mathbf{x}^{[2]}(\underline{\boldsymbol{\delta}}^2)(t) + \cdots + \mathbf{x}^{[q]}(\underline{\boldsymbol{\delta}}^q)(t). \tag{5}$$

Here, the zero-order solution in $\underline{\boldsymbol{\delta}}$, $\mathbf{x}^{[0]}(t)$, is just the solution at time t for the initial condition $\tilde{\mathbf{x}}_0$. In turn, $\mathbf{x}^{[k]}(\underline{\boldsymbol{\delta}}^k)(t)$ represents the k-order correction at time t in $\underline{\boldsymbol{\delta}}$. This k-order correction is a homogeneous polynomial in $\underline{\boldsymbol{\delta}}$ of degree k, which we have emphasized by using the notation $\underline{\boldsymbol{\delta}}^k$; it includes the contribution of all monomials of degree k in the independent deviations of $\underline{\boldsymbol{\delta}}$. It can be shown that the q-order solution given by Eq. (5) is a solution of the variational equations of order less or equal to q (Gimeno et al. 2023). Then, jet transport techniques allow including higher order approximations of the target function $Q(\tilde{\mathbf{x}}_0)$, and consequently a better approximation of the nominal orbital elements.

From the jet-transport solution (5) at the epochs of the observations t_k, we have the O-C residuals as truncated polynomials up to degree q. From them, we can compute the expansion, up to degree q in the deviations $\underline{\boldsymbol{\delta}}$, of the mean square residual function (1), which we thus write as

$$Q(\underline{\boldsymbol{\delta}}) \simeq Q^{[0]} + Q^{[1]}(\underline{\boldsymbol{\delta}}) + Q^{[2]}(\underline{\boldsymbol{\delta}}^2) + \cdots + Q^{[q]}(\underline{\boldsymbol{\delta}}^q). \tag{6}$$

The minimization problem for the nominal orbit can be stated as above. The subtle difference is that Eq. (6) permits implementing Newton's method (2) directly, since all needed derivatives with respect to $\mathbf{x}_0$ can be computed directly from the polynomial expansion (6), and $Q(\underline{\boldsymbol{\delta}})$ includes all contributions up to order q in the target function. Aside from obtaining the nominal solution $\mathbf{x}_0$, the correlation matrix Γ can also be computed directly from its definition.

3. Results for Apophis

In this section we briefly describe the dynamical model and the results for Apophis; further details can be found in Pérez-Hernández and Benet (2022). We have produced our own planetary ephemeris (Pérez-Hernández et al. 2021), integrating a model essentially equivalent to JPL DE430 (Folkner et al. 2014). The dynamical model for Apophis includes

Table 1. International Celestial Reference Frame barycentric initial conditions (position, velocity and nominal Yarkovsky parameter) for the gravity-only (OR6) and non-gravitational (OR7) solutions. The epoch is December 17, 2020 00:00:00.0 (TDB). The symbols x_0, y_0, z_0 represent initial positions in au, and v_{x0}, v_{y0}, v_{z0} represent initial velocities in au/d. A_2 is the Yarkovsky transversal parameter in au/d^2, multiplied by 10^{14}. Numbers in parentheses represent the order of magnitude of the formal 1-sigma uncertainty.

Init. cond./sol.	OR6	OR7
x_0 [au]	$-0.18034816(472) \pm 7.05 \times 10^{-9}$	$-0.18034828(526) \pm 7.12 \times 10^{-9}$
y_0 [au]	$0.94069116(764) \pm 1.71 \times 10^{-9}$	$0.94069105(951) \pm 1.94 \times 10^{-9}$
z_0 [au]	$0.34573641(385) \pm 4.03 \times 10^{-9}$	$0.34573599(029) \pm 5.41 \times 10^{-9}$
v_{x0} [au/d]	$-0.0162659357(312) \pm 3.31 \times 10^{-11}$	$-0.0162659397(882) \pm 4.79 \times 10^{-11}$
v_{y0} [au/d]	$4.39163(792) \times 10^{-5} \pm 6.68 \times 10^{-11}$	$4.39154(800) \times 10^{-5} \pm 6.72 \times 10^{-11}$
v_{z0} [au/d]	$-0.000395210(949) \pm 1.24 \times 10^{-10}$	$-0.000395204(013) \pm 1.38 \times 10^{-10}$
A_2 [$\times 10^{14}$ au/d^2]	0.0 (fixed)	$-2.8(988) \pm 2.48 \times 10^{-2}$

the Sun, all planets, the Moon and Pluto, 16 main-belt asteroids, post-Newtonian corrections (except for the asteroids), corrections due to J_2 for the Earth, and the transversal Yarkovsky effect A_2; the code is publically available (Pérez-Hernández et al. 2021). The integration of the differential equations has been performed using Taylor's method of degree 25 with an absolute tolerance 10^{-30}.

Jet transport is applied up to degree $q = 5$, and we have produced two best-fit solutions corresponding to slightly different dynamical models of Apophis: First, a purely gravitational model which does not include the Yarkovsky effect ($A_2 = 0$), and thus jet transport is applied using 6 variables; this solution is denoted as OR6 below. Second, the same model now including A_2; in this case jet transport is applied for 7 degrees of freedom, and the corresponding solution is referred as OR7. For the initial guess, we used JPL's #197 gravity-only solution at the epoch December 17, 2020 00:00:00.0 (TDB) epoch (Brozović et al. 2018). The nominal initial conditions obtained for each model are given in Table 1, using the same epoch indicated above. Notice that the quotient of the nominal value of A_2 obtained in OR7, divided by the formal uncertainty of the same parameter, i.e., the signal-to-noise ratio for A_2, is $\mathrm{SNR}_{A_2} \approx 117$, implying a non-zero value.

In Fig. 1 we present the optical astrometry O-C residuals and in Fig. 2 the O-C residuals for the radar observations, obtained from the best orbital fit for the gravity-only model (OR6) with purple symbols, and for the model including non-gravitational interactions (OR7) with green symbols. We note that the distributions of the projected optical O-C residuals for the OR6 solution, plotted in the side and upper panels of Fig. 1, are bi-modal distributions, whereas the OR7 solution yields single bump distributions; the latter are closer to Gaussian distributions, which we expect from the central limit theorem, under the assumption of uncorrelated observations. We also remark that the uncertainties associated to the O-C radar residuals of the OR7 solution in Fig. 2 are all consistent with zero. In more quantitative terms, the normalized root-mean-square error for the combined optical and radar fits is 0.979 and 0.306, respectively for the OR6 and OR7 solutions. Other quantities comparing the quality of the fit provided by each solution appear in Table 1 of Pérez-Hernández and Benet (2022); they show consistently better fit properties for the non-gravitational model OR7. From these observations we conclude that the OR7 solution better describes the combined optical and radar astrometry observations, and that the fit obtained by this solution yields a non-zero transversal Yarkovsky parameter.

Using the OR7 solution, Apophis orbit can be propagated forward in time to analyze the collision probability for the next close encounter in April 13th 2029, as well as in 2036 and 2068. We obtain vanishingly small values of the collision probabilities in all those close approaches (Pérez-Hernández and Benet 2022).

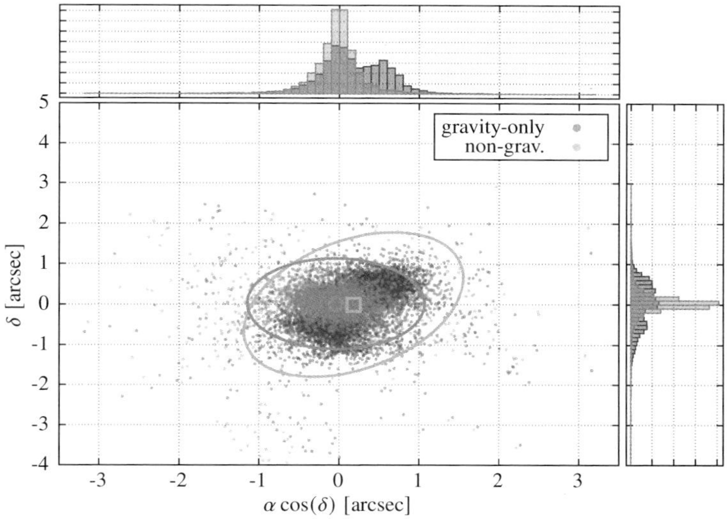

Figure 1. Optical astrometry O-C residuals for the OR6 (purple symbols) and OR7 (green symbols) orbital fits. The x-axis displays the right-ascension residuals, and the y-axis those for the declination. In the top and side panels, the distributions of these residuals are displayed for OR6 and OR7 solutions: for OR6 (in purple) notice that both distributions are bi-modal.

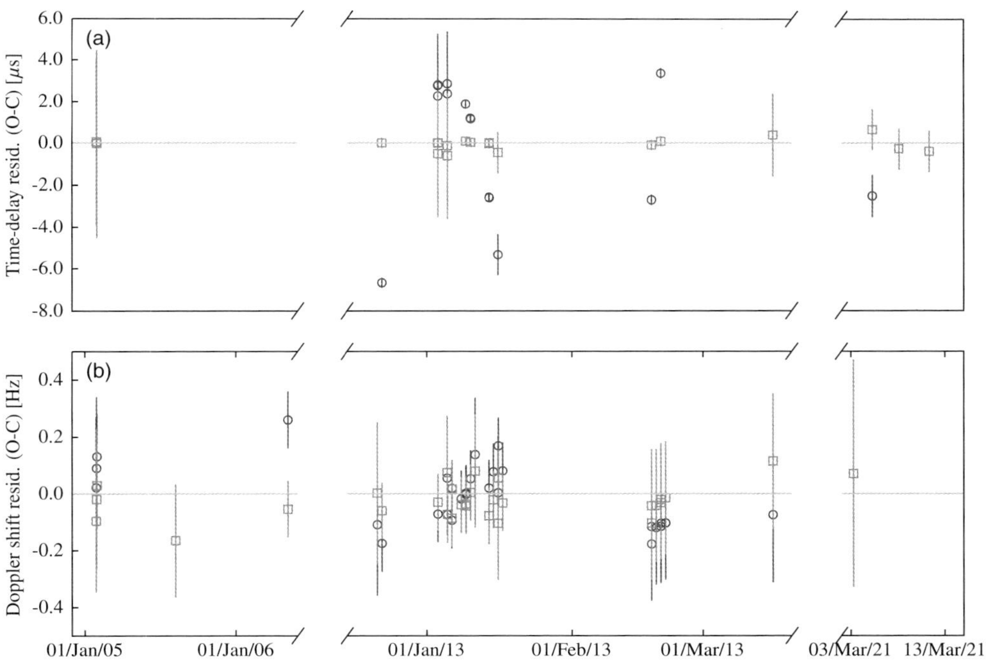

Figure 2. Radar astrometry O-C residuals in time, displaying separately the time delays and the Doppler shifts, for the OR6 (purple symbols) and OR7 (green symbols) orbital fits. We notice that all residuals of the OR7 solution, taking into account the observational error bars, are compatible with zero.

4. Summary and conclusions

In this contribution we have provided further details of our approach using jet transport to determine the transversal Yarkovsky parameter for Apophis. Using jet transport techniques we obtain semi-numeric high-order approximations of the target function, whose minimum we find through a Newton iteration method. Having such a high-order approximation of the target function allows us to compute directly the covariance matrix. With this scheme we obtained two nominal solutions compatible with all public available optical and radar astrometry: a purely gravitational model, the OR6 solution, and a model that incorporates the transversal Yarkovsky acceleration, the OR7 solution. Comparing both solutions, the one that yields a better fit is OR7, and the associated 1-sigma uncertainty associated with A_2 yields conclusively a non-zero value.

The OR7 value obtained for the transversal Yarkovsky parameter is $A_2 = -(2.899 \pm 0.025) \times 10^{-14}$ au d^{-2}, which yields a non-gravitational semi-major axis drift of $\langle \dot{a} \rangle = (-199.0 \pm 1.5)$ m yr^{-1}. This value of A_2 is consistent with the current value reported at JPLs small-body database for Apophis ssd, namely $A_2^{\mathrm{JPL}} = -(2.901 \pm 0.019) \times 10^{-14}$ au d^{-2}. In addition, it is also consistent with the value obtained by combining five very accurate stellar occultations by Apophis with the optical and radar astrometry observations by Souami et al. (2023) and Desmars et al. (2023), who obtained $A_2^{\mathrm{occ}} = -(2.8992 \pm 0.0161) \times 10^{-14}$ au d^{-2}. The consistency of these values, in particular with the last one, shows the efficiency of jet transport methods in the orbit determination problem.

Acknowledgments: We acknowledge financial support from the PAPIIT-UNAM project IG101122, and computer time provided through the project LANCAD-UNAM-DGTIC-284. We thank Bruno Sicardi for bringing to our attention the results obtained by Souami et al. (2023) and Desmars et al. (2023) through stellar occultations.

References

JPL Small-Body Database (data for Apophis). Accessed: 2023-08-23.

Bottke Jr, W. F., Vokrouhlický, D., Rubincam, D. P., & Nesvorný, D. 2006, The Yarkovsky and YORP effects: Implications for asteroid dynamics. *ARAA*, 34, 157–191.

Brozović, M., Benner, L., McMichael, J., Giorgini, J., Pravec, P., Scheirich, P., Magri, C., Busch, M., Jao, J., Lee, C., *et al.* 2018, Goldstone and Arecibo radar observations of (99942) Apophis in 2012–2013. *Icarus*, 300, 115–128.

Chesley, S. R. Potential impact detection for Near-Earth asteroids: the case of 99942 Apophis (2004 MN 4). In Lazzaro, D., Ferraz-Mello, S., & Fernández, J. A., (eds.), *Asteroids, Comets, Meteors,* Proc. IAU Symposium No. 229, 2006, pp. 215–228.

Del Vigna, A., Faggioli, L., Milani, A., Spoto, F., Farnocchia, D., & Carry, B. 2018, Detecting the Yarkovsky effect among near-Earth asteroids from astrometric data. *A & A*, 617, A61.

Desmars, J., Souami, D., Vavilov, D., Hsu, H. M., De Pater, I., & Hestroffer, D. Apophis orbit with stellar occultations. In *Asteroids, Comets, Meteors Conference*, 2023, Abstract #2376, Houston. Lunar and Planetary Institute.

Farnocchia, D., Chesley, S., Vokrouhlický, D., Milani, A., Spoto, F., & Bottke, W. 2013, Near Earth asteroids with measurable Yarkovsky effect. *Icarus*, 224(1), 1–13.

Folkner, W., Williams, J., Boggs, D., Park, R., & Kuchynka, P. 2014, The planetary and lunar ephemerides DE430 and DE431. *Interplanet. Netw. Prog. Rep*, 196, 1–81.

Gimeno, J., Jorba, À., Jorba-Cuscó, M., Miguel, N., & Zou, M. 2023, Numerical integration of high-order variational equations of ODEs. *App Maths Comp*, 442, 127743.

Giorgini, J., Ostro, S., Benner, L., Chodas, P., Chesley, S., Hudson, R., Nolan, M., Klemola, A., Standish, E., Jurgens, R., *et al.* 2002, Asteroid 1950 DA's encounter with Earth in 2880: Physical limits of collision probability prediction. *Science*, 296(5565), 132–136.

Greenberg, A. H., Margot, J.-L., Verma, A. K., Taylor, P. A., & Hodge, S. E. 2020, Yarkovsky drift detections for 247 near-Earth asteroids. *AJ*, 159(3), 92.

Milani, A. & Gronchi, G. 2010,. *Theory of orbit determination.* Cambridge University Press.

Pérez-Hernández, J. A. & Benet, L. 2022, Non-zero Yarkovsky acceleration for near-Earth asteroid (99942) Apophis. *Comm Earth & Environm*, 3(1).

Pérez-Hernández, J. A. Ramírez-Montoya, L. E., & Benet, L. 2021,. NEOs.jl: Jet Transport-based Near-Earth Object orbital propagator and fitter in Julia. `https://github.com/PerezHz/NEOs.jl`.

Pérez-Hernández, J. A., Ramírez-Montoya, L. E., & Benet, L. 2021,. PlanetaryEphemeris.jl: A planetary and lunar ephemerides integrator based on JPL DE430 dynamical model. `https://github.com/PerezHz/PlanetaryEphemeris.jl`.

Souami, D., Desmars, J., Tanga, P., Tsiganis, K., de Pater, I., Hsu, Y. M., Ferreira, J., Siakas, A., Chesley, S., Dunham, D., Venable, R., Irwin, J., Watanabe, H., Bouquillon, S., Herald, D., & Preston, S. Stellar occultations by sub-km sized near Earth asteroids: Apophis and Didymos. In *Asteroids, Comets, Meteors Conference* 2023,, Abstract #2099, Houston. Lunar and Planetary Institute.

Vokrouhlický, D., Farnocchia, D., Čapek, D., Chesley, S., P.Pravec, P.Scheirich, & Müller, T. 2015, The Yarkovsky effect for 99942 Apophis. *Icarus*, 252, 277–283.

Vokrouhlický, D., Milani, A., & Chesley, S. 2000, Yarkovsky effect on small near-Earth asteroids: Mathematical formulation and examples. *Icarus*, 148(1), 118–138.

Complex Planetary Systems II: Latest Methods for an Interdisciplinary Approach
Proceedings IAU Symposium No. 382, 2024
A. Lemaitre & A.-S. Libert, eds.
doi:10.1017/S1743921323004970

Tidal effects in resonant chains of close-in planets: TTV analysis of Kepler-80

Carolina Charalambous[1,2] **and Anne-Sophie Libert**[3]

[1]Instituto de Astrofésica, Pontificia Universidad Católica de Chile,
Av. Vicuña Mackenna 4860, 782-0436 Macul, Santiago, Chile.
email: `ccharalambous@unc.edu.ar`

[2]Millennium Institute for Astrophysics, Chile

[3]naXys, Department of Mathematics, University of Namur, 61 Rue de Bruxelles,
5000 Namur, Belgium

Abstract. Systems near mean-motion resonances (MMRs) are subject to large transit-timing variations (TTVs). The amplitude and period of the TTVs strongly depend on the distance to exact MMR and the planetary eccentricities which are shaped during the formation and long-term evolution of the system. For close-in planets, the tides raised by the star provide a source of dissipation, placing the planets further away from the MMR. In this work, we will discuss how the tidal interactions with the central star play an important role in shaping the period ratios and resonant angles in resonant chains. Moreover, we will show how they can impact the TTVs and therefore how the TTVs could serve as a means to put constraints on the tidal history of planetary systems. The study will focus on the four-planet resonant chain of Kepler-80.

Keywords. Planet-disk interactions – Long-term evolution – Transit timing variations

1. Introduction

Studying Systems with Tightly Packed Inner Planets (STIPs) is particularly important for understanding the evolution of planetary systems. During the migration phase in the protoplanetary disk, planet pairs get trapped in MMRs which excite their eccentricities, while the tides raised by the host star contribute to dissipation on long timescales. The effect of stellar tidal forces is more pronounced for planets closer to the star (e.g. Baruteau and Papaloizou 2013). Regarding STIPs harboring chains of resonances, Charalambous et al. (2023) showed that 3-planet MMRs among the planets can transport this effect outward and thus explain the observed departures from exact MMR for adjacent planet pairs.

Planets near MMRs exhibit prominent TTVs (i.e., deviations of the transit times with respect to the strictly periodic Keplerian case). The short orbital periods of the STIPS allow for multiple cycles of the variations to be observed. TTVs are a powerful technique for identifying previously undetected planets by measuring the non-periodicity of the times of transit (Agol et al. 2005; Holman and Murray 2005). Beyond merely revealing or confirming the presence of non-transiting planets, TTVs offer valuable insights into inferring and better constraining the masses and eccentricities of planets in multiple transiting systems. The amplitude and period of the TTV signal strongly depend on the proximity to exact MMR and the eccentricities of the planets. As previously said, these quantities are shaped during planet-disk interactions and further modified during the long-term evolution of the system.

In the present work, we will focus on the Kepler-80 system, a well-known resonant chain of four close-in planets. Our objective is twofold. Firstly, we will analyze the role of the tidal interactions between the planets and the central star in shaping the period ratios and the resonant angles in Kepler-80, extending the previous study of Charalambous et al. (2023) by considering different tidal parameters (i.e., different internal structure) among the planets. Secondly, we will present a preliminary result on how the TTVs vary with different tidal parameters, suggesting that the TTVs could serve as a means to put constraints on the tidal history of planetary systems.

2. Formation and tidal evolution of resonant chains

During the late stage of planetary formation, interactions between the planets and the gas disk in which they are embedded produce a transfer of angular momentum, changing the radial distance of the planets. During their migration, planets are expected to get trapped in MMRs. Resonant chains are particularly common for the STIPs, whose planets are found in (or close to) 2-planet and 3-planet MMRs (similar to the Laplace resonance).

Once the disk has dispersed, dissipative mechanisms such as tidal interactions with the central star become dominant. A classical formulation of the tidal damping timescale on the eccentricity is (Goldreich and Soter 1966; Papaloizou 2021)

$$\tau_{e_i}^{\mathrm{tid}} = 7.63 \times 10^5 \left(\frac{a_i}{0.05\mathrm{au}}\right)^{13/2} \left(\frac{M_\odot}{M_\star}\right)^{3/2} \left(\frac{m_\oplus}{m_i}\right)^{2/3} \left(\frac{\rho_i}{\rho_\oplus}\right)^{5/3} Q'\mathrm{yr}, \tag{1}$$

with a_i the planetary semi-major axis (in au), $M_\star$ the mass of the star (in $M_\odot$), m_i the planetary mass (in $m_\oplus$), ρ_i the planetary mean density (in $\rho_\oplus$), and $Q' = 3Q/(2k_2)$, where Q represents the tidal quality factor and k_2 the Love number. The main effect of tides raised by the central star on resonant chains is to drive the planets out of resonance. The deviation from exact MMR, also known as *resonance offset*, for two planets near a $(p+q)/p$ MMR, is defined by

$$\Delta_{(p+q)/q} = P_2/P_1 - (p+q)/p, \tag{2}$$

where P_i is the orbital period of planet i.

TTVs are enhanced for planets near MMR (Agol et al. 2005). The characteristic timescale of the TTVs (i.e., the period of the sinusoidal TTV signal) for two planets near a $(p+1)/p$ MMR is the sinusoidal super-period (Lithwick et al. 2012; Teyssandier et al. 2022)

$$P_{\mathrm{TTV}} = \frac{P_2}{p|\Delta_{(p+1)/p}|}. \tag{3}$$

Thus, the closer a pair of planets to a MMR (i.e., the smaller the Δ), the larger their TTVs. When more than two planets evolve in a resonant chain, the typical independent sinusoidal expressions as given in Lithwick et al. (2012) fail to describe the TTV signals. As shown by Libert and Renner (2013), the TTV signals are interlinked with the dynamical evolution of the planetary system: the shorter periods (e.g., the TTVs characteristic timescales) are associated with 2-planet MMRs, while longer periods arise from 3-planet MMRs (see also the frequency analysis of Teyssandier et al. 2022, for the TRAPPIST-1 resonant chain).

3. Study of Kepler-80

Kepler-80 is a 6-planet STIP with all adjacent pairs near a 2-planet MMR (see Table 1) and orbiting a M0/K5 star with $M_\star = 0.73\,\mathrm{M}_\odot$ and star age $t_\star \sim 2$ Gyr (MacDonald et al.

Table 1. Parameters of Kepler-80 from MacDonald et al. (2021). The columns successively show the planetary masses and densities, the orbital periods, the 2-planet MMRs between adjacent planets, and their respective resonance offsets.

		m_i [$m_\oplus$]	ρ_i [$\rho_\oplus$]	P_i [day]	$(p+q)/p$	$\Delta_{(p+q)/p}$
	f	–	–	0.98678	3/1	0.1133
	d	5.95	14.6	3.07222	3/2	0.0118
Kepler-80	e	2.97	6.9	4.64489	3/2	0.0186
	b	3.5	1.45	7.05246	4/3	0.0168
	c	3.49	1.22	9.52355	3/2	0.0385
	g	0.065	–	14.6455		

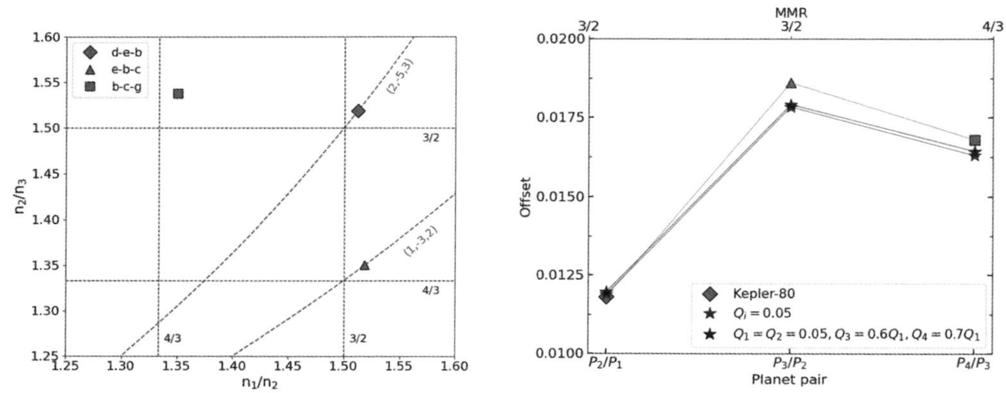

Figure 1. Left: Departures from exact 2-planet MMRs for the Kepler-80 resonant chain in the $(n_1/n_2, n_2/n_3)$ plane. Vertical and horizontal lines indicate 2-planet MMRs, while the dashed purple curves indicate 3-planet MMRs. Right: Resonance offsets for the successive planet pairs within the Kepler-80 d-e-b-c resonant chain. The observed values are represented by cyan symbols.

2016). By looking at the TTVs, MacDonald et al. (2016, 2021) concluded that the four planets Kepler-80 d, e, b, and c dynamically interact in a resonant chain, while the innermost planet Kepler-80 f is decoupled from the rest of the planets and the libration of resonant angles involving the outermost planet Kepler-80 g is not guaranteed. As a result, we only focus here on the 4-planet resonant chain between the middle planets of Kepler-80. Regarding the triplets in Kepler-80, planets d-e-b are in a $(2, -5, 3)$ 3-planet MMR†, while the second triplet e-b-c is in a $(1, -3, 2)$ 3-planet MMR, both being thus in a zeroth-order ($q_{3\text{pl}} = 0$) resonance (see the purple curves in the left panel of Figure 1).

3.1. *Offset analysis*

In Charalambous et al. (2023), we showed that the resonance offsets for the four middle planets of Kepler-80, whose planet pairs are successively close to the 3/2, 3/2, and 4/3 MMRs, do not increase monotonically with the distance to the star, but follow a specific trend. The offset of the second pair increases with respect to the offset of the first pair, while the offset of the last pair decreases with respect to the one of the second pair (see the cyan symbols in the right panel of Figure 1). We computed the period of the

† A 3-planet MMR is characterized by the relation $k_1 n_1 + k_2 n_2 + k_3 n_3 = 0$ for some set of $(k_1, k_2, k_3) \neq (0, 0, 0)$ with $k_i \in \mathbb{N}$ and n_i the mean motion of planet i. The order of the 3-planet MMR is given by $q_{3\text{pl}} = |k_1 + k_2 + k_3|$ (see e.g., Charalambous et al. 2018, for details on the resonant structure).

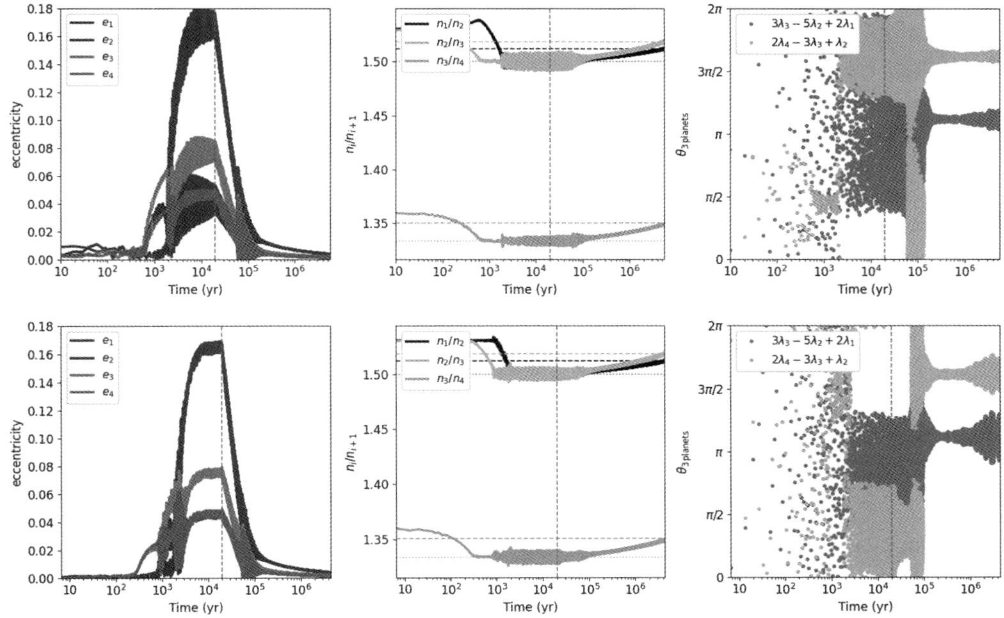

Figure 2. Evolutions of the eccentricities, mean-motion ratios and 3-planet resonant angles of Kepler-80, when Q' is fixed to 0.05 for all the planets (top panels) and to $Q'_1 = Q'_2 = 0.05$, $Q'_3 = 0.6 \times Q'_1$ and $Q'_4 = 0.7 \times Q'_1$ (bottom panels).

third (resp. fourth) planet assuming that the two preceding ones are in an exact 3-planet MMR with the third (resp. fourth) one. We found that the 3-planet MMRs are guiding the dynamics of Kepler-80 since the trend in the observed resonance offsets perfectly fits the analytic estimation.

To analyze whether the formation history of the system is consistent with the observed resonance offsets, we performed simulations of the late-stage formation (planetary migration) and long-term evolution under tidal effects of a Kepler-80-like system. For the tidally-damped N-body simulations we used the REBOUNDx N-body code (Tamayo et al. 2020) with the WHFast symplectic integrator (Rein and Tamayo 2015). Both disk-induced planetary migration and tidal interactions with the host star were included. The planets were assumed to be coplanar with eccentricities initially fixed to 10^{-4} and random orbital angles in $[0, 2\pi]$. Regarding the disk-induced migration which is applied on the outer planet only, we assumed a constant damping timescale of $\tau_a = 10^7$ days and a circularization timescale of $\tau_e = 10^5$ days. Tidal damping effects were included for all the planets, initially by setting Q' to 0.05 for all planets and then adjusting it individually for each planet in a subsequent step.

The time evolution of the 4-planet resonant chain of Kepler-80, with equal Q' for all the planets (top panels of Figure 2), is detailed in Charalambous et al. (2023). During the migration phase induced by the protoplanetary disk, when the planet pairs become trapped in MMRs, their eccentricities gradually increase and eventually stabilize at specific values (left panel). Simultaneously, both the 2- and 3-planet resonant angles undergo libration (right panel). After the disk dispersal, tidal interactions with the central star cause the eccentricities to decrease, reaching values consistent with observational estimates. Meanwhile, the planet pairs cease their oscillations around exact MMR and tend towards the observed departure from MMR (middle panel). The orbital circularization driven by the tides with the host star precisely reproduces the observed values of the

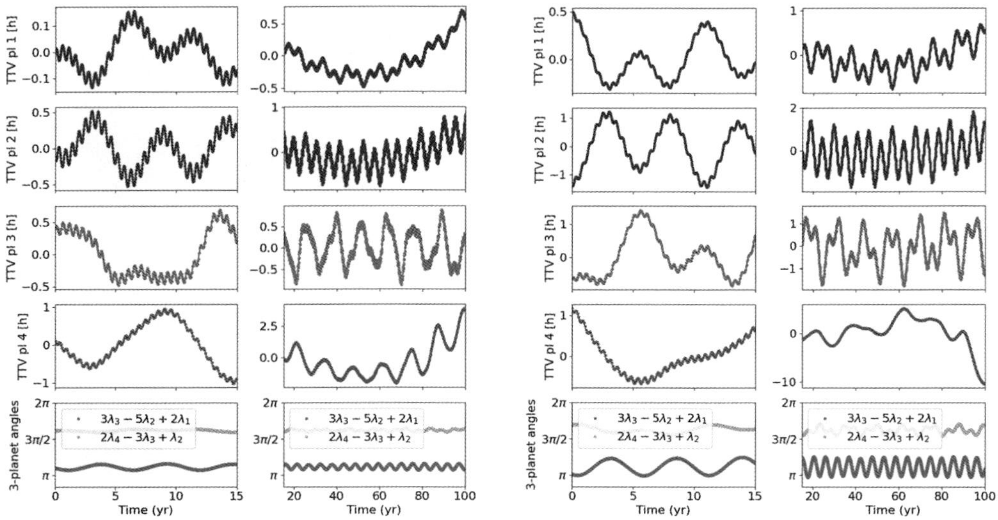

Figure 3. TTVs of the two simulations shown in Figure 2, on 15 yr and 100 yr. In the two left panels, the Q' tidal quality factor is fixed to 0.05, while it is varied individually for each planet in the two right panels.

resonance offsets, as shown in the right panel of Figure 1 by the dark-red stars which represent the simulation outcome when $Q'_i = 0.05$ for all the planets.

A second evolution of the Kepler-80 system is shown in the bottom panels of Figure 2 for different tidal parameter values among the planets, namely $Q'_1 = Q'_2 = 0.05$, $Q'_3 = 0.6 \times Q'_1$ and $Q'_4 = 0.7 \times Q'_1$. As explained in Ferraz-Mello (2013), the Q' value can be understood as a measure of planetary viscosity or elasticity, representing the internal friction inside the body and how fast the planet returns to its equilibrium figure. The fact that different planets have the same Q' is an indication that they dissipate different amounts of energy because their orbital frequencies are different, likely because their internal structure are different as well. The resonance offsets of this second evolution are very similar to the previous ones, as depicted by the blue stars in the right panel of Figure 1. However, during the migration phase, the 3-planet resonant angle $\theta_{234} = \lambda_2 - 3\lambda_3 + 2\lambda_4$ (with λ_i the mean-longitude of planet i) exhibits libration around different centers in the two simulations and after the dispersal of the disk, as tidal forces come into play, the libration center of θ_{234} in the second evolution shifts to finally adopt the same libration center values as in the previous simulation. This second simulation suggests that the migration history with regard to the libration centers of the resonant angles could be erased in the long-term evolution of planetary systems under tidal effects. This observation will be investigated in more detail in future work.

3.2. *TTV analysis*

In line with prior studies that inferred potential formation scenario of resonant extrasolar systems by analyzing the TTVs of the planets (e.g., Teyssandier and Libert (2020) for K2-24 and Teyssandier et al. (2022) for TRAPPIST-1), we computed the TTVs of the two previously described evolutions, at the end of the simulations displayed in Figure 2. While the final system configurations are very similar in terms of resonance offsets and dynamical evolution of the resonant angles (as shown in Figures 1 and 2), the TTV signals of the two evolutions present some significant differences, in both the amplitudes and frequencies (see Figure 3). The latter quantities strongly depend on the proximity to exact MMR and the eccentricity of the planets. As previously said, they are shaped during the formation of the system, in particular the disk-induced migration phase and the tidal dissipation process. According to our preliminary investigation, a change in the tidal parameters among the planets slightly modifies the TTVs and this leads us to think that TTV analysis could be useful to put constraints on the tidal history of planetary systems.

4. Conclusions

We observed distinct patterns in the resonance offsets of detected systems with resonant chains, as for the Kepler-80 system presented here. These resonance offset trends are governed by 3-planet MMRs, align with analytical estimates, and are replicable through tidal damping effects induced by the host star, regardless of the chosen value for the tidal quality factor Q' among the planets. Interestingly, during the tidal phase, a change of libration center was observed for Kepler-80.

We also conducted a preliminary study of the TTVs of Kepler-80 for different tidal parameters. The change in periods and amplitudes of the TTVs observed when considering different tidal parameter values among the planets suggests that resonant chains could prove valuable for uncovering the formation history of planetary systems and eventually help to constrain the Q' values (i.e., the internal structure) of individual planets.

Acknowledgements. C.C. was supported by FONDECYT grant n° 3230283. Computational resources have been provided by the Consortium des Équipements de Calcul Intensif (CÉCI), supported by the FNRS-FRFC, the Walloon Region, and the University of Namur (Conventions No. 2.5020.11, GEQ U.G006.15, 1610468 and RW/GEQ2016).

References

Agol, E., Steffen, J., Sari, R., & Clarkson, W. 2005, On detecting terrestrial planets with timing of giant planet transits. *MNRAS*, 359(2), 567–579.

Baruteau, C. & Papaloizou, J. C. B. 2013, Disk-Planets Interactions and the Diversity of Period Ratios in Kepler's Multi-planetary Systems. *ApJ*, 778(1), 7.

Charalambous, C., Martí, J. G., Beaugé, C., & Ramos, X. S. 2018, Resonance capture and dynamics of three-planet systems. *MNRAS*, 477(1), 1414–1425.

Charalambous, C., Teyssandier, J., & Libert, A. S. 2023, Tidal interactions shape period ratios in planetary systems with three-body resonant chains. *A&A*, 677, A160.

Ferraz-Mello, S. 2013, Tidal synchronization of close-in satellites and exoplanets. A rheophysical approach. *Celestial Mechanics and Dynamical Astronomy*, 116(2), 109–140.

Goldreich, P. & Soter, S. 1966, Q in the Solar System. *Icarus*, 5(1), 375–389.

Holman, M. J. & Murray, N. W. 2005, The Use of Transit Timing to Detect Terrestrial-Mass Extrasolar Planets. *Science*, 307(5713), 1288–1291.

Libert, A. S. & Renner, S. 2013, Detection of Laplace-resonant three-planet systems from transit timing variations. *MNRAS*, 430(2), 1369–1375.

Lithwick, Y., Xie, J., & Wu, Y. 2012, Extracting Planet Mass and Eccentricity from TTV Data. *ApJ*, 761(2).

MacDonald, M. G., Ragozzine, D., Fabrycky, D. C., Ford, E. B., Holman, M. J., Isaacson, H. T., Lissauer, J. J., Lopez, E. D., Mazeh, T., Rogers, L., Rowe, J. F., Steffen, J. H., & Torres, G. 2016, A Dynamical Analysis of the Kepler-80 System of Five Transiting Planets. *AJ*, 152(4), 105.

MacDonald, M. G., Shakespeare, C. J., & Ragozzine, D. 2021, A Five-Planet Resonant Chain: Reevaluation of the Kepler-80 System. *AJ*, 162(3), 114.

Papaloizou, J. C. B. 2021, The orbital evolution of resonant chains of exoplanets incorporating circularisation produced by tidal interaction with the central star with application to the HD 158259 and EPIC 245950175 systems. *Celestial Mechanics and Dynamical Astronomy*, 133(6), 30.

Rein, H. & Tamayo, D. 2015, WHFAST: a fast and unbiased implementation of a symplectic Wisdom-Holman integrator for long-term gravitational simulations. *MNRAS*, 452(1), 376–388.

Tamayo, D., Rein, H., Shi, P., & Hernandez, D. M. 2020, REBOUNDx: a library for adding conservative and dissipative forces to otherwise symplectic N-body integrations. *MNRAS*, 491(2), 2885–2901.

Teyssandier, J. & Libert, A.-S. 2020, Transit timing variation signature of planet migration: the case of K2-24. *A&A*, 643, A11.

Teyssandier, J., Libert, A. S., & Agol, E. 2022, TRAPPIST-1: Dynamical analysis of the transit-timing variations and origin of the resonant chain. *A&A*, 658, A170.

Complex Planetary Systems II: Latest Methods for an Interdisciplinary Approach
Proceedings IAU Symposium No. 382, 2024
A. Lemaitre & A.-S. Libert, eds.
doi:10.1017/S1743921323004118

Revisiting the Cassini States of synchronous satellites with an angular momentum approach

Alexis Coyette[1], Rose-Marie Baland[2] and Tim Van Hoolst[2]

[1]Naxys Institute, Namur University, Namur, Belgium
email: alexis.coyette@unamur.be

[2]Royal Observatory of Belgium, Brussels, Belgium

Abstract. Like our Moon, the large icy satellites of Jupiter are thought to be in a Cassini State, an equilibrium rotation state characterized by a synchronous rotation rate and a precession rate of the rotation axis equal to that of the normal to the orbit. In these equilibrium states (up to four Cassini States are possible for a solid and rigid satellite), the spin axis of the satellite, the normal to its orbit and the normal to the inertial plane remain coplanar with an obliquity that remains theoretically constant. However, as the gravitational torque exerted on the satellite shows small periodic variations, the orientation of the rotation axis will also vary with time and nutations in obliquity will appear.

Here we present a dynamical model for the study of the Cassini States. This model includes the coupling between the polar motion and the spin axis precession/nutation which is neglected in the classical studies. We study the influence of the triaxiality of Ganymede on its four possible Cassini States, use a Toy model of the Moon to illustrate the nutations in obliquity obtained with the dynamical model, and investigate the influence of the presence of a subsurface ocean on the Cassini State I of Europa.

Keywords. Obliquity, Cassini States, Icy Satellites

1. Cassini States - Definition

The large satellites of our Solar System, like our Moon or the Galilean satellites, are in synchronous rotation, with a period of rotation equal to their period of revolution. This synchronous rotation is often referred to as a Cassini state in reference to Cassini (1693) who described this state of equilibrium for the Moon at the end of the XVIIth century. In this rotation state, the precession rates of both the orbital and equatorial planes are assumed to be constant and equal. Moreover, the spin axis of the Moon, the normal to its orbit and the normal to its Laplace plane are assumed to be coplanar, with a constant obliquity ε (defined as the angle between the normal to the orbit and the spin axis). Due to the precession of the orbit, the orbit normal deviates from the spin axis and the obliquity therefore tends to an equilibrium value for which the torque and the satellite response balances each other. Up to four Cassini states are possible for a biaxial body (Peale 1969).

1.1. *Biaxial case*

By assuming the obliquity to be constant over time, the different Cassini states of a solid and rigid biaxial satellite can be obtained from the following classical equation (see e.g. Colombo (1966); Ward and Hamilton (2004))

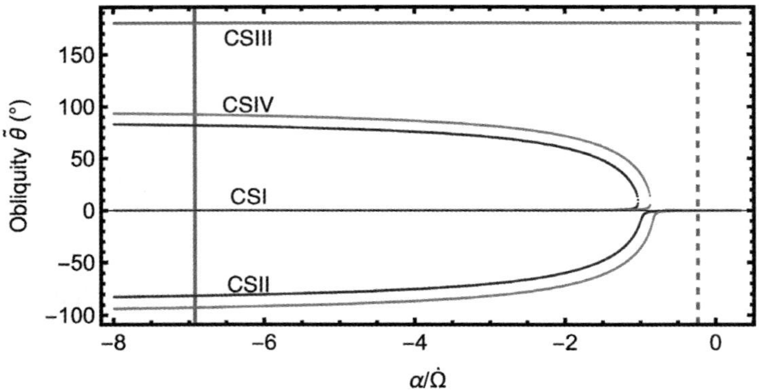

Figure 1. Cassini States of a biaxial (blue) and triaxial (green) Ganymede for different values of $\alpha/\dot{\Omega}$ ranging from -8 to 0.5. For the actual value of $\alpha/\dot{\Omega}$, indicated by the vertical black lines, four Cassini States are possible. If Ganymede were to have a ratio $\alpha/\dot{\Omega}$ similar to that of the Moon (see vertical dashed lines), only two Cassini States would be possible.

$$\alpha_{\rm bi}\sin[2(\tilde{\theta}-i)]+2\dot{\Omega}\sin\tilde{\theta}=0. \tag{1}$$

In this equation, $\alpha_{\rm bi}=\frac{3}{2}\frac{C-\overline{A}}{C}n$ is the frequency of the Free Precession, with n the mean motion of the satellite and $\overline{A}$ and C are the mean equatorial and polar moments of inertia of the biaxial satellite, respectively ($\overline{A}<C$). $\tilde{\theta}$ is the angle between the figure axis and the normal to the inertial plane, or inertial obliquity angle, and $\dot{\Omega}$ the frequency of the precession which is retrograde ($\dot{\Omega}<0$) for the Galilean satellites and the Moon. In these classical studies, the gravitational torque exerted by the central planet is averaged over short period terms. Morevover, the polar motion is neglected so that the rotation axis is assumed to be aligned with the figure axis. We therefore denote this obliquity by $\tilde{\theta}$ to make the distinction between this figure axis inertial obliquity and the spin axis inertial obliquity θ. Obliquities measured from the orbit axis are denoted similarly $\tilde{\varepsilon}=(\tilde{\theta}-i)$ and $\varepsilon=(\theta-i)$, with i the constant orbital inclination with respect to the Laplace plane.

For satellites like the Galilean moons, with a slow orbital precession in comparison with the free precession (i.e. $|\dot{\Omega}|\ll\alpha$), Eq. (1) possesses four solutions characterized by an obliquity close to 0 (Cassini State I, CSI), $-\frac{\pi}{2}$ (CSII), π (CSIII), and $\frac{\pi}{2}$ (CSIV), see Fig. 1. For a satellite characterized by a short orbital precession period (i.e. $|\dot{\Omega}|\gg\alpha$), however, the first term of Eq. (1) can be neglected in front of the second term and we are then left with only two possible Cassini States (CSII and CSIII, with obliquities close to zero and π, respectively, see Fig. 1). This is the case for our Moon that is considered to be in Cassini State II (see e.g. Beletskii 1972; Ward 1975; Henrard and Schwanen 2004).

1.2. *Triaxial case*

The classical equation describing the different Cassini states of a rigid and solid traxial satellite is (see e.g. Ward 1975)

$$\alpha_{\rm tri}\sin[2(\tilde{\theta}-i)]+\frac{3}{4}n\frac{B-A}{C}\sin(\tilde{\theta}-i)\left[1-\cos(\tilde{\theta}-i)\right]+2\dot{\Omega}\sin\tilde{\theta}=0, \tag{2}$$

where $\alpha_{\rm tri}=\frac{3}{2}n\frac{C-A}{C}$ is the frequency of the Free Precession of the triaxial body and $A<B<C$ are the principal moments of inertia of the triaxial satellite. The obliquities obtained for Cassini States I and III are very close to the ones for a biaxial satellite. However, for a large orbital precession period in comparison with the period of the Free

Table 1. Figure axis obliquities $\tilde{\theta}$ for a biaxial (Eq. 1) or triaxial (Eq. 2) Ganymede.

	biaxial	triaxial
$\tilde{\theta}_{\mathrm{I}}^{\mathrm{Cl}}$ (°)	0.238	0.213
$\tilde{\theta}_{\mathrm{II}}^{\mathrm{Cl}}$ (°)	−76.543	−92.827
$\tilde{\theta}_{\mathrm{III}}^{\mathrm{Cl}}$ (°)	180.148	180.148
$\tilde{\theta}_{\mathrm{IV}}^{\mathrm{Cl}}$ (°)	76.908	93.196

Precession, the Cassini States II and IV present a strong deviation due to the triaxiality of the body (deviation of 16° for Ganymede, see Tab. 1).

2. Our dynamical model

The classical equations (Eqs. 1-2) overlook the possibility of time variations (or nutations) in obliquity and in longitude induced by the time-varying gravitational torque acting on the satellite. Here, we use dynamical equations with a non averaged torque to study the behaviour of the obliquity. Our model also couples the spin axis precession/nutation in space with the smaller polar motion, which was neglected in the classical studies. It is described by means of two equations:

(1) The Angular Momentum equation stating that the time derivative of the angular momentum $\vec{H}$ is equal to the applied torque $\vec{\Gamma}$ (Eckhardt 1981)

$$\frac{d\vec{H}}{dt} + \vec{\Omega} \times \vec{H} = \vec{\Gamma}, \tag{3}$$

where $\vec{\Omega} = n(m_X(t), m_Y(t), 1 + m_Z(t))$ is the rotation vector of the satellite and where the angular momentum $\vec{H}$ is given by the product of the inertia tensor and the rotation vector

$$\vec{H} = \begin{pmatrix} A & 0 & 0 \\ 0 & B & 0 \\ 0 & 0 & C \end{pmatrix} \cdot \vec{\Omega}. \tag{4}$$

(2) The equation describing the invariance of the Laplace Pole in inertial space:

$$\frac{d\hat{p}}{dt} + \vec{\Omega} \times \hat{p} = 0, \tag{5}$$

where $\hat{p} = (p_X(t), p_Y(t), p_Z(t))$ is the unit vector along the normal to the Laplace plane.

These equations are written in the Cartesian coordinates of the Body Frame (BF) of the satellite and split into a system of 4 dynamical equations with 4 unknowns $m_X(t)$, $m_Y(t)$, $p_X(t)$ and $p_Y(t)$ (see also Baland et al. (2019)):

$$(C - B)n^2 m_Y(t) + Anm'_X(t) = \Gamma_X(t) \tag{6a}$$
$$(A - C)n^2 m_X(t) + Bnm'_Y(t) = \Gamma_Y(t) \tag{6b}$$
$$-np_Y(t) + nm_Y(t)p_Z(t) + p'_X(t) = 0 \tag{6c}$$
$$np_X(t) - nm_X(t)p_Z(t) + p'_Y(t) = 0. \tag{6d}$$

Note that a fifth variable ($p_Z(t)$) appears in these equations. In Cassini State I (resp. CSIII), the obliquity is close to zero (resp. π) so that the normal to the Laplace plane can be approximated by $\hat{p} \approx (p_X(t), p_Y(t), 1)$ (resp. $\hat{p} \approx (p_X(t), p_Y(t), -1)$). In the following, we analytically solve the system of equations (6a-6d) in the case of the Moon and obtain the time-varying equatorial components of the polar motion $(m_X(t), m_Y(t))$ and of the

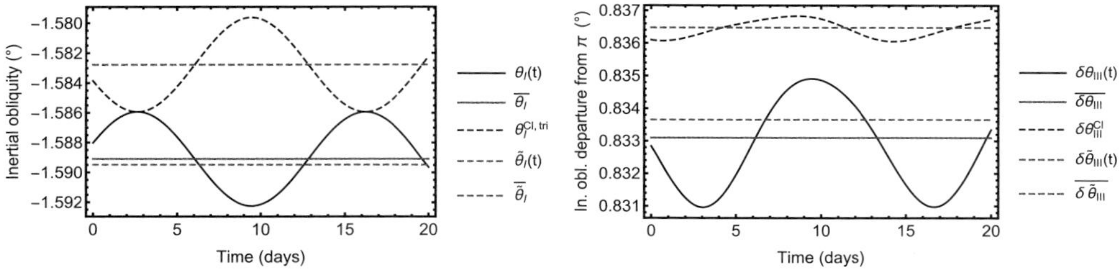

Figure 2. Left panel: CSI inertial obliquities $\theta_\mathrm{I}(t)$ and $\tilde{\theta}_\mathrm{I}(t)$ as a function of time for the Moon (solid and dashed black lines, respectively). Right-panel: CSIII departure from π in inertial obliquities $\theta_\mathrm{III}(t)$ and $\tilde{\theta}_\mathrm{III}(t)$ as a function of time for the Moon (solid and dashed black lines, respectively) The blue solid and dashed lines represent the mean values of the inertial obliquities (CSI) and of the inertial obliquity departure from π (CSIII). The red dashed lines represent the constant-over-time results of the Classical model (Eq. 2).

Laplace pole motion $(p_X(t), p_Y(t))$. The figure axis inertial obliquity $\tilde{\theta}_\mathrm{I}(t)$ (resp. $\tilde{\theta}_\mathrm{III}(t)$) and the spin axis inertial obliquity $\theta_\mathrm{I}(t)$ (resp. $\theta_\mathrm{III}(t)$) are then obtained as byproducts (see Coyette et al. (2023), in prep.).

2.1. *Cassini State I*

Even if the Moon is formally considered to be in the Cassini State II due to its short orbital precession period compared to its free precession period (Ward 1975), its obliquity can be obtained from the small-angle approximation and therefore by the same formulae used here for the Cassini State I. We obtain mean inertial obliquities values of $\overline{\theta}_\mathrm{I} = -1.589°$ and $\overline{\tilde{\theta}}_\mathrm{I} = -1.583°$ (see Fig. 2, left panel, blue lines), close to the value obtained with the classical "triaxial" equation (red line). The slight difference between the "dynamical" $\overline{\tilde{\theta}}_\mathrm{I}$ and the "classical" $\tilde{\theta}_\mathrm{I}^{\mathrm{Cl,tri}}$ values is due to the coupling between the polar motion and the spin precession. The nutations in obliquity of the spin and figure axes have an amplitude of 11 arcsec and are both semi-diurnal and 180° out of phase from each other (black lines). The spin axis and the figure axis are confounded two times per day as the polar motion semi-diurnally vanishes (see Coyette et al. (2016)).

Note that $\overline{\theta}_\mathrm{I}$ and $\overline{\tilde{\theta}}_\mathrm{I}$ differ from the obliquity observed by lunar laser ranging observations (Dickey et al. 1994, −1.543°), likely because of the assumptions made here. In particular we assume a rigid satellite and the small-angle approximation, as in Dumberry and Wieczorek (2016). Note also that neglecting the solar torque acting on the Moon in the Cassini State model leads to a discrepancy with respect to observations, Stys and Dumberry (2018). We do not aim to reproduce the observed value and the results presented here for the Moon should only be considered as those of a Toy Model intended to discuss the semi-diurnal nutations in obliquity and the coupling between the polar motion and the spin precession/nutation in the case of a satellite precessing at constant rate.

2.2. *Cassini State III*

We here hypothetically assume that the Moon is in Cassini State III and model its time-varying spin and figure axes inertial obliquities. The inertial obliquity departures $\delta\theta_\mathrm{III}(t) = \theta_\mathrm{III}(t) - \pi$ and $\delta\tilde{\theta}_\mathrm{III}(t) = \tilde{\theta}_\mathrm{III}(t) - \pi$ are presented on Fig. 2 (right panel, black lines). The mean inertial obliquities are $\overline{\theta}_\mathrm{III} = 180.833°$ and $\overline{\tilde{\theta}}_\mathrm{III} = 180.836°$, respectively (blue lines). Similarly as for CSI, the value of $\overline{\tilde{\theta}}_\mathrm{III}$ is close but not equal to the value $\tilde{\theta}_\mathrm{III}^{\mathrm{Cl,tri}}$ derived with the classical equations (red line), due to the coupling between the polar motion and the spin precession that is here fully considered.

The nutations in obliquities are the superimposition of two semi-diurnal terms with amplitudes of 1.3 and 6.0 arcsec and of one quarter-diurnal term with an amplitude of 0.7 arcsec. In CSIII, the polar motion never vanishes so that the spin axis and the figure axis always point toward different directions.

3. Influence of a subsurface ocean

Many large icy satellites of the Solar System likely possess a global ocean beneath a thin ice shell. Such an ice shell is sometimes assumed to be mechanically decoupled from the solid interior and to precess somewhat independently, with an obliquity larger than that of an entirely solid and rigid body (e.g. Bills and Nimmo 2008, 2009 for the case of Titan, see also Chen et al. (2014); Nimmo (2023)). Baland et al. (2011) drawn a different conclusion and argued that the observed large obliquity of Titan (0.3°, Stiles et al. 2008; Meriggiola et al. 2016, about three times the value expected for an entirely solid Titan) is explained by a resonant amplification of the obliquity of the shell which is partially mechanically coupled to the interior. Baland et al. (2012) also found that the obliquity can be smaller than that of an entirely solid satellite, due to a lack of such a resonance, as could be the case for Europa. We revisit below the obliquity of the shell of Europa, starting from the coupled model of Baland et al. (2019).

3.1. *Equations*

To model the influence of the ocean on the obliquity, we consider one angular momentum equation for each layer k of the satellite (shell s, ocean o and interior i), expressed in the Cartesian coordinates of the BF of that layer if it is solid, or of the BF of the shell in the case of the ocean:

$$\frac{d}{dt}\vec{H}_k + \vec{\Omega}_k \times \vec{H}_k = \vec{\Gamma}_k, \tag{7}$$

where $\vec{\Gamma}_k$ is the total torque applied on layer k that includes the external gravitational torque exerted by the central planet, the internal gravitational and pressure torques exerted by the other layers, including the inertial torque due to the Poincaré flow inside the subsurface ocean. We also use the equations stating the invariance of the Laplace Pole, expressed in the BF of both solid layers (shell s and interior i)

$$\frac{d\hat{p}_k}{dt} + \vec{\Omega}_k \times \hat{p}_k = 0. \tag{8}$$

We then analytically solve Eqs. (7-8) to obtain the polar motion components $(m_{X,k}, m_{Y,k})$ of the different internal layers of the satellite ($k = i$, $k = o$ or $k = s$) as well as the Laplace pole components $(p_{X,k}, p_{Y,k})$ for the two solid layers ($k = i$ or $k = s$) and compute the obliquity of the shell.

3.2. *Numerical results*

The mean orbital obliquity $\bar{\varepsilon}$ of the shell of Europa is presented on Fig. 3 for a large set of possible interior models of Europa compatible with the observed mass, radius and moment of inertia. We first model the obliquity of Europa without considering the inertial torque (similarly as was done by Baland et al. (2012), their Fig. 5). In this case, the obliquity increases from 0.027° to 0.039° with increasing ice shell thickness (left panel of Fig. 3), and is smaller than the expected obliquity of an entirely solid Europa (0.055°). These obliquities are about 0.005° smaller than those obtained by Baland et al. (2012) as the value we use here for the mass of Europa slightly differs from the value used in this paper.

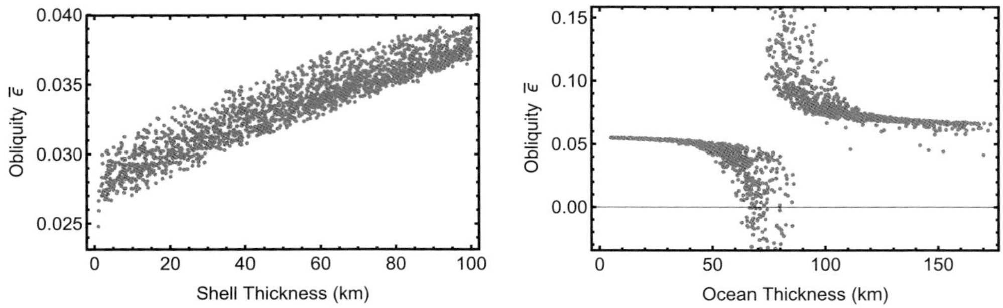

Figure 3. Mean obliquity of Europa's shell for a large set of interior models, without inertial coupling and as a function of the ice shell thickness (left panel), and with inertial coupling and as a function of the ocean thickness (right panel).

We note that the obliquity of Europa's shell is strongly influenced by the inertial torque due to the Poincaré flow inside the subsurface ocean. Due to a resonance with a free mode associated with the presence of the ocean (the Free Ocean Nutation), the mean obliquity of the shell of Europa is either amplified to values as large as 0.15° or decreased to values closer to zero, for models with an ocean thickness ranging from 50 to 120 km. For models far away from this resonance, the obliquity of the shell of Europa is similar to the obliquity of 0.055° expected for an entirely solid satellite.

4. Conclusion

We here introduce a dynamical model based on an angular momentum approach for the study of the Cassini States I and III of large synchronous satellites. Since we do not average the gravitational torque exerted on the satellite, the nutations in obliquity and in longitude can be studied from this model, whereas they were neglected in the Classical Cassini states models (see e.g. Colombo (1966),Ward (1975)). Moreover, the coupling between the spin axis precession and the polar motion is fully included in our model.

Our dynamical model shows good agreement with the classical solutions derived for CSI and CSIII, the slight difference being due to the coupling between the polar motion and the spin axis precession.

With presence of a subsurface ocean at rest, the obliquity of the shell of Europa is smaller than the obliquity obtained for a entirely solid satellite. If we add the inertial torque caused by a Poincaré flow inside the subsurface ocean in our model, depending on the ocean thickness, the obliquity of Europa ranges between 0 and a few tens of degrees. Far from a resonance with the Free Ocean Nutation, the obliquity remains close to the obliquity of an entirely solid satellite. Those results are in contradiction with the assumption that the obliquity of the shell of a satellite with an internal global ocean should necessarily be larger than that of an entirely solid satellite.

Acknowledgment

This work was financially supported by the Belgian PRODEX program managed by the European Space Agency in collaboration with the Belgian Federal Science Policy Office.

References

Baland, R.-M., Coyette, A., & Van Hoolst, T. 2019, *Celestial Mechanics and Dynamical Astronomy*, 131(2), 11.
Baland, R. M., Van Hoolst, T., Yseboodt, M., & Karatekin, Ö. 2011, *A&A*, 530, A141.
Baland, R.-M., Yseboodt, M., & Van Hoolst, T. 2012, *Icarus*, 220, 435–448.

Beletskii, V. V. 1972, *Celestial Mechanics*, 6(3), 356–378.
Bills, B. & Nimmo, F. In *European Planetary Science Congress* 2009, 553.
Bills, B. G. & Nimmo, F. 2008, *Icarus*, 196(1), 293–297.
Cassini, J. 1693,. *De l'Origine et du progrès de l'Astronomie et de son usage dans la Gèographie et dans la navigation.*
Chen, E. M. A., Nimmo, F., & Glatzmaier, G. A. 2014, *Icarus*, 229, 11–30.
Colombo, G. 1966, *The Astronomical Journal*, 71(9), 891–896.
Coyette, A., Baland, R.-M., & Van Hoolst, T. 2023, *in preparation.*
Coyette, A., Hoolst, T. V., Baland, R.-M., & Tokano, T. 2016, *Icarus*, 265, 1 – 28.
Dickey, J. O., Bender, P. L., Faller, J. E., *et al.* 1994, *Science*, 265(5171), 482–490.
Dumberry, M. & Wieczorek, M. A. 2016, *JGR: Planets*, 121(7), 1264–1292.
Eckhardt, D. H. 1981, *Moon and Planets*, 25(1), 3–49.
Henrard, J. & Schwanen, G. 2004, *Celestial Mechanics and Dynamical Astronomy*, 89(2), 181–200.
Meriggiola, R., Iess, L., Stiles, B. W., *et al.* 2016, *Icarus*, 275, 183–192.
Nimmo, F. In *Uranus Flagship: Investigations and Instruments for Cross-Discipline Science Workshop* 2023, 8009.
Peale, S. J. 1969, *The Astronomical Journal*, 74(3), 483–489.
Stiles, B. W., Kirk, R. L., Lorenz, R. D., *et al.* 2008, *The Astronomical Journal*, 135(5), 1669–1680.
Stys, C. & Dumberry, M. 2018, *JGR: Planets*, 123, 2868–2892.
Ward, W. R. 1975, *The Astronomical Journal*, 80(1), 64–70.
Ward, W. R. & Hamilton, D. P. 2004, *The Astronomical Journal*, 128(5), 2501–2509.

Complex Planetary Systems II: Latest Methods for an Interdisciplinary Approach
Proceedings IAU Symposium No. 382, 2024
A. Lemaitre & A.-S. Libert, eds.
doi:10.1017/S1743921323003939

Lagrangian descriptors and their applications to deterministic chaos

Jérôme Daquin

Department of Mathematics and naXys, Namur Institute for Complex Systems,
University of Namur, Rue Grafé 2, B5000 Namur, Belgium
email: jerome.daquin@unamur.be

Abstract. We present our recent contributions to the theory of Lagrangian descriptors for discriminating ordered and deterministic chaotic trajectories. The class of Lagrangian descriptors we are dealing with is based on the Euclidean length of the orbit over a finite time window. The framework is free of tangent vector dynamics and is valid for both discrete and continuous dynamical systems. We review its last advancements and touch on how it illuminated recently Dvorak's quantities based on maximal extent of trajectories' observables, as traditionally computed in planetary dynamics.

Keywords. Chaos indicators, Lagrangian descriptors, Chaotic motions

---

1. Introduction

This present short presentation aims at summarising our recent contributions to the theory of Lagrangian descriptors (LDs) for deterministic chaos detection. It is predominantly based on Daquin et al. (2022) and Daquin and Charalambous (2023) to which we refer for a comprehensive literature review. LDs are scalar quantities rooted in fluid mixing and coined as such by Mendoza and Mancho (2010). There were initially introduced by Madrid and Mancho (2009) for vector fields. Consider the autonomous regular vector field

$$\dot{x} = v(x), \; x \in \mathbb{R}^n, \tag{1}$$

and the orbit associated to Eq. (1) starting at $x_0 \in \mathbb{R}^n$ at $t=0$. Its LD at time $T>0$ is defined as

$$\mathrm{LD}(x_0; T) = \int_0^T \big\| v\big(x(s)\big) \big\|_2 \, \mathrm{ds}, \tag{2}$$

and corresponds to the Euclidean arc-length of the trajectory over the time-window $[0, T]$. LDs have been generalised by considering the integration of others bounded and intrinsic observable associated to an orbit (Mancho et al. 2013). We stick here to the LD based on the arc-length method. Extending LDs to discrete systems is straightforward. Let us denote by $\{z_0, z_1, \dots, z_N\}$, $z_j \in \mathbb{R}^m \; j = 0, \dots, N$, a finite orbit associated to a discrete mapping f, $z_j = f^j(z_0), j = 1, \dots, N$. The LD associated to such orbit is given by

$$\mathrm{LD}(z_0; N) = \sum_{j=0}^{N-1} \sqrt{\sum_{l=1}^{m} (z_{j+1}^{(l)} - z_j^{(l)})^2}, \tag{3}$$

where $z_j^{(l)}$ denotes the l-th component of the state z at time j.

The rest of this contribution is organised as follows. In Sect. 2, we highlight the phenomenology of the LDs for chaos detection by presenting new applications to the paradigmatic one-dimensional logistic map. Sect. 3 presents the finite-time chaos indicator that we have derived from LDs computations and discuss the models on which it has been successfully employed so far. Sect. 4 connects the LD framework with maximal excursions traditionally computed in celestial mechanics in the context of stability maps. Sect. 5 closes the paper with conclusive remarks.

2. Lagrangian descriptors applied to the logistic model

We consider the one-dimensional quadratic logistic map (see *e.g.,* May (1976)) as a testing ground to illustrate the LDs concepts. The mapping is defined by

$$x_{n+1} = \mathscr{L}_\mu(x_n) = \mu x_n(1 - x_n). \tag{4}$$

The state space of the dynamics is $M = [0, 1]$ for $0 < \mu \leq 4$ and $n \in \mathbb{N}$. Dealing with the orbit $\{x_0, x_1, \ldots, x_N\}$, Eq. (5) becomes

$$\mathrm{LD}(x_0; N) = \sum_{i=0}^{N-1} |x_{i+1} - x_i|. \tag{5}$$

The top panel of Fig. 1 shows the bifurcation diagram of the logistic map for $2.9 \leq \mu \leq 4$ with fixed $x_0 = 0.15$. It highlights the presence of periodic orbits, the chain of bifurcations occurring in the dynamics as μ varies, periodic windows immersed within aperiodic range of motions. Such a window is exemplified around $\mu = 3.63$ and will be further discussed. The middle panel of Fig. 1 shows the value of the finite time Lyapunov exponent (FTLE) computed at time $N = 10^6$ as a function of μ. The FTLE is defined as

$$\lambda(x_0; N) = \frac{1}{N} \log_{10}\left(\prod_{i=0}^{N} |D\mathscr{L}_\mu(x_i)|\right) = \frac{1}{N}\sum_{i=0}^{N} \log_{10}(|\mu(1 - 2x_i)|). \tag{6}$$

We have colored the final value of λ according to its sign. Negative FTLEs appear in blue and correspond to regular motions. The values $\lambda = 0$ indicate bifurcations. The positive FTLEs appear in red and correspond to orbits with sensitive dependence upon the initial conditions, *i.e.,* chaotic motions. Interestingly enough, there is a sharp link between the values of the FTLEs and the properties of the LD map. Indeed, the bottom panel of Fig. 1 shows the LD landscape (computed also at the final time $N = 10^6$), colored according to the final value of λ. We observe that negative λ's correspond to smooth parts of the LD curve, whilst positive λ's correspond to domains where the LD metric is irregular. Fig. 2 repeats the computations of FTLEs and LDs at a much thinner scale of the control parameter μ (the shaded area around $\mu = 3.63$ highlighted in Fig. 1) and further confirm our former observations. This demonstrates the sensitivity and robustness of the regularity of the LD metric as diagnostic for chaos detection. For $\mu = 4$, $\mathscr{L}_4$ is chaotic for every x_0 (May 1976; Banks et al. 2003). Fig. 3 shows the LD landscape for varying x_0 in $(0, 1)$ with fixed $\mu = 4$ at $N = 10^4$. The obtained landscape is nowhere smooth, in accordance with the former numerical results.

3. The new non-variational geometrical chaos indicators

As just illustrated on the logistic model, the regularity of the LD metric keeps trace of the possible chaotic nature of the orbit. This property has also been observed on a series of integrable and non-integrable mechanical models supporting interacting resonances (Daquin et al. 2022). For integrable 1-degree-of-freedom (DoF) systems, Pédenon-Orlanducci et al. (2022) showed that the rate of divergence of the derivative of the length metric (using a time-free parametrisation of LDs) scales as a power law when

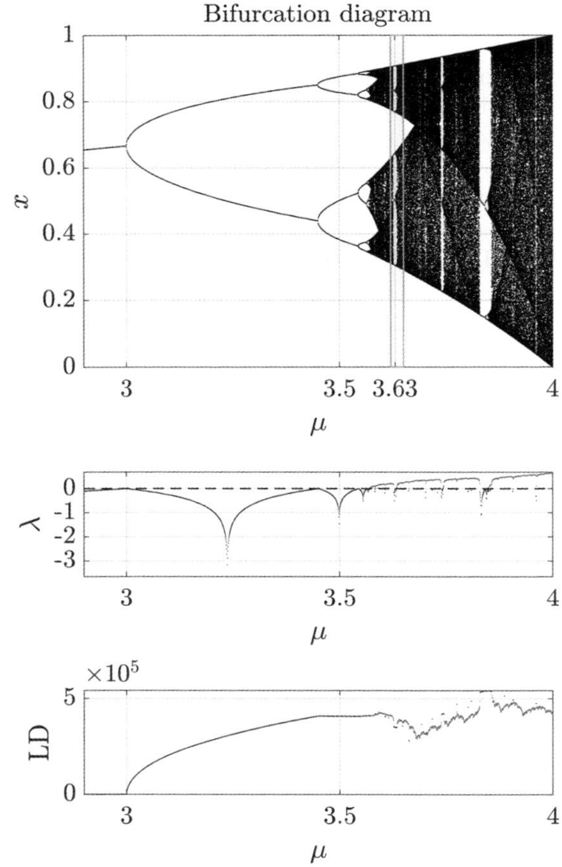

Figure 1. Bifurcation diagram (top panel), Lyapunov exponents (middle panel) and LD lengths (bottom panel) of the logistic equation when μ is varied. There is a strong connection between the regularity of the orbits (sign of Lyapunov exponent) and the regularity of the LD map. See text for further details.

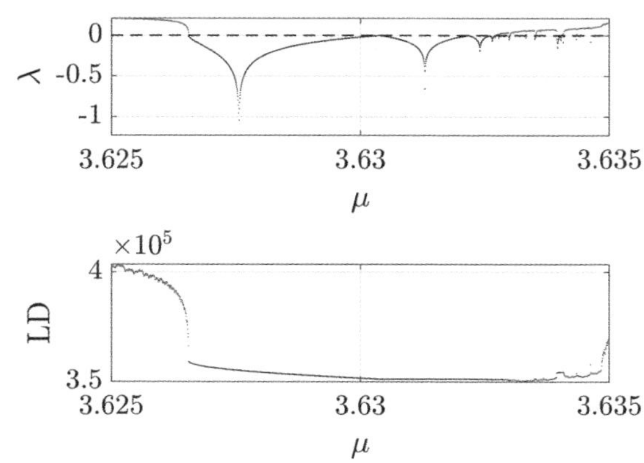

Figure 2. Computation of Lyapunov exponents λ and LD lengths on a small range of the control parameter μ. The regularity of the length application provides a very sensitive and reliable indicator of the ordered and chaotic nature of the motion.

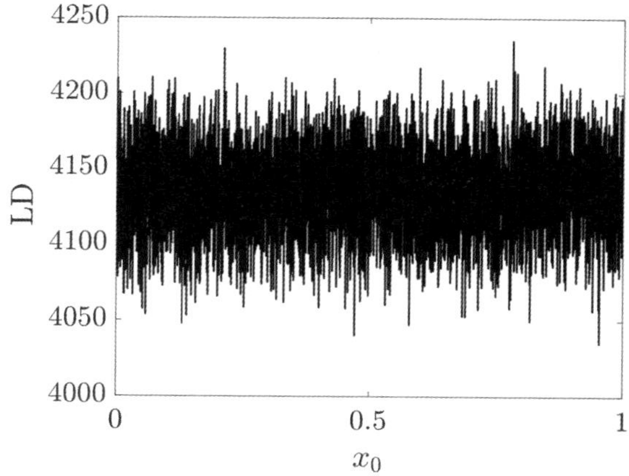

Figure 3. Landscape of LDs as a function of x_0 for the logistic map with $\mu = 4$. For this parameter, the map is known to be chaotic. This is reflected in the absence of regularity of the LD metric.

crossing transversally separatrices. The scaling obeys $\mathscr{O}(1/\sqrt{E})$, where E is the energy labelling the level curve. Those observations led us to assume that the LD map is $\mathscr{C}^\infty$ on the set $\mathscr{R}$ of regular motions. Leveraging on this, Daquin et al. (2022) introduced the $\|\Delta\text{LD}\|$ index measuring the regularity of the LD metric through second-derivatives estimates†, similarly to the frequency analysis method (Laskar 1993). For the dimension $n = 1$ and a point $y \in \mathscr{R}$, $\|\Delta\text{LD}\|$ is defined as

$$\|\Delta\text{LD}\|\,(y;T) = |\text{LD}''(y;T)|. \tag{7}$$

Numerically, the second derivative in Eq. (7) is estimated by finite differences of the type

$$|\text{LD}''(y;T)| \simeq \frac{|\text{LD}(y+h;T) + \text{LD}(y-h;T) - 2\text{LD}(y;T)|}{h^2}, \tag{8}$$

for a small enough discretisation step h. The $\|\Delta\text{LD}\|$ index is proposed as new chaos indicator and undergoes sharp increases on the complement set $\mathscr{R}^C$ of $\mathscr{R}$ when crossing transversally separatrices of integrable model or hyperbolic domains of non-integrable models (Daquin et al. 2022). In the more general case ($n > 1$), $\|\Delta\text{LD}\|$ reads as

$$\|\Delta\text{LD}\|\,(y;T) = \sum_{i=1}^{n} |\partial^2_{y^{(i)}y^{(i)}}\text{LD}(y;T)|. \tag{9}$$

The $\|\Delta\text{LD}\|$ index has been applied and benchmarked on several low-dimensional models to derive stability maps computed on fine domains of initial conditions or parameter space, including: the 2 dimensional standard map and higher 4 dimensional coupled standard maps, non-autonomous pendulum-like models having resonant junctions and interactions supporting resonant webs, as ubiquitous in celestial mechanics. The $\|\Delta\text{LD}\|$ keeps trace of manifolds' oscillations when computed on a short timescale, as observed on the 2-DoF Hénon-Heiles model (Daquin et al. 2022). Qualitative comparison with established variational methods (*e.g.*, the FLI, MEGNO, SALI) have shown excellent agreements in disentangling ordered and chaotic orbits. Recently, quantitative oriented analysis of the performance of LD like diagnostics have been investigated (Hillebrand et al. 2022; Zimper et al. 2023). On $4D$ coupled standard maps in a mixed

† An index based on the first derivatives might miss the geography of resonant webs, see Daquin et al. (2022), Fig. 7.

phase space regime, it has been shown that the probability of agreement P_A in the discrimination of the orbit against the SALI index is on the order of 90%. Thus, the $\|\Delta \mathrm{LD}\|$ index is a cheap, easily implementable and reliable tool for detecting separatrices and chaotic motions.

4. Diameters and Dvorak's amplitudes

The LDs just presented are based on the length of the orbit, see Eq. (3) and (5). For bounded orbits, their amplitudes (or diameters) are another geometrical quantity that might be considered. Consider Eq. (1), a trajectory $x(t; x_0)$ with initial condition x_0 and an observable ϕ. The diameter on the finite time segment $[0, T]$ along the observable ϕ is defined as

$$D(x_0; T) = \max_{0 \le \tau \le T} \phi\big(x(\tau; x_0)\big) - \min_{0 \le \tau \le T} \phi\big(x(\tau; x_0)\big), \tag{10}$$

with similar definitions for the discrete case. The diameter D is called *maximal shift* in Mundel et al. (2014). The study of its cumulative distribution function demonstrated relevance to characterise fluid mixing properties, and is certainly an interesting direction of future research. Daquin and Charalambous (2023) considered diameters of n-DoF Hamiltonian systems by looking more specifically at the diameter of the actions. It turns out that LD and D maps share analogies, and of practical interest is the loss of regularity when crossing separatrices or chaotic layers transversally. This led to introduce the analogue of Eq.(9) for the diameter of Eq. (10), denoted $\|\Delta D\|$. Diameters like quantities have been computed for a while in planetary dynamics in the context of stability maps, by focusing on stretches of some Keplerian elements a, e or i (respectively semi-major axis, eccentricity and inclination), usually denoted by δa, δe or δi, see *e.g.,* Dvorak et al. (2004); Sándor et al. (2007). In the context of 2 and 3-body co-planar simulations triggered towards mean-motion resonances, Daquin and Charalambous (2023) showed in particular how the $\|\Delta D\|$ maps supplements the traditional diameter analysis, allowing to reinflate and recover sharply separatrices, detect thin chaotic lines and the web of resonances, otherwise undetected with the classical diameter metric. As for $\|\Delta \mathrm{LD}\|$, $\|\Delta D\|$ is a cheap non-variational geometrical index allowing sensitive chaos detection.

5. Conclusive remarks

We have presented and summarised our latest contributions to the theory of Lagrangian descriptors for chaos detection. A finite-time non-variational chaos indicator can be easily derived from the study of the regularity of length maps of Eq. (2) and (3). For this, we suggested and introduced the second-derivatives based index of Eq. (9). We have discussed connection of Lagrangian diagnostics with quantities employed to characterise fluid flow mixing and diameter quantities encountered in celestial mechanics. The Lagrangian framework offers a convenient mold for detecting chaos without the need of deriving the variational equations, possibly leading to substantial implementation and computational reductions. Our current efforts focus on the quantitative assessments of their performances.

References

Banks, J., Dragan, V., & Jones, A. 2003,. *Chaos: a mathematical introduction*, volume 18. Cambridge University Press.

Daquin, J. & Charalambous, C. 2023, Detection of separatrices and chaotic seas based on orbit amplitudes. *Celestial Mechanics and Dynamical Astronomy*, 135(3), 1–18.

Daquin, J., Pédenon-Orlanducci, M., Agaoglou, M., Garcia-Sanchez, G., & Mancho, A. M. 2022, Global dynamics visualisation from Lagrangian Descriptors. Applications to discrete and continuous systems. *Physica D: Nonlinear Phenomena*, 442, 133520.

Dvorak, R., Pilat-Lohinger, E., Schwarz, R., & Freistetter, F. 2004, Extrasolar Trojan planets close to habitable zones. *Astronomy & Astrophysics*, 426(2), L37–L40.

Hillebrand, M., Zimper, S., Ngapasare, A., Katsanikas, M., Wiggins, S. R., & Skokos, C. 2022,. Quantifying chaos using Lagrangian descriptors.

Laskar, J. 1993, Frequency analysis for multi-dimensional systems. Global dynamics and diffusion. *Physica D: Nonlinear Phenomena*, 67(1-3), 257–281.

Madrid, J. J. & Mancho, A. M. 2009, Distinguished trajectories in time dependent vector fields. *Chaos: An Interdisciplinary Journal of Nonlinear Science*, 19(1), 013111.

Mancho, A., Wiggins, S., Curbelo, J., & Mendoza, C. 2013, Lagrangian descriptors: A method for revealing phase space structures of general time dependent dynamical systems. *Commun Nonlinear Sci Numer Simulat*, 18, 3530–3557.

May, R. M. 1976, Simple mathematical models with very complicated dynamics. *Nature*, 261(5560), 459–467.

Mendoza, C. & Mancho, A. 2010, Hidden geometry of ocean flows. *Physical review letters*, 105(3), 038501.

Mundel, R., Fredj, E., Gildor, H., & Rom-Kedar, V. 2014, New Lagrangian diagnostics for characterizing fluid flow mixing. *Physics of Fluids*, 26(12), 126602.

Pédenon-Orlanducci, R., Carletti, T., Lemaitre, A., & Daquin, J. 2022,. Geometric parametrisation of Lagrangian Descriptors for 1 degree-of-freedom systems.

Sándor, Z., Süli, Á., Érdi, B., Pilat-Lohinger, E., & Dvorak, R. 2007, A stability catalogue of the habitable zones in extrasolar planetary systems. *Monthly Notices of the Royal Astronomical Society*, 375(4), 1495–1502.

Zimper, S., Ngapasare, A., Hillebrand, M., Katsanikas, M., Wiggins, S. R., & Skokos, C. 2023, Performance of chaos diagnostics based on Lagrangian descriptors. Application to the 4D standard map. *Physica D: Nonlinear Phenomena*, 453, 133833.

Complex Planetary Systems II: Latest Methods for an Interdisciplinary Approach
Proceedings IAU Symposium No. 382, 2024
A. Lemaitre & A.-S. Libert, eds.
doi:10.1017/S1743921323003940

Deconvolving the complex structure of the asteroid belt

Stanley F. Dermott[1], Dan Li[2] and Apostolos A. Christou[3]

[1]Department of Astronomy, University of Florida, Gainesville, FL 32611, US
email: sdermott@ufl.edu

[2]NSF's National Optical-Infrared Astronomy Research Laboratory, Tucson, AZ 85719, US
email: dan.li@noirlab.edu

[3]Armagh Observatory and Planetarium, College Hill, Armagh, BT61 9DG
email: Apostolos.Christou@armagh.ac.uk

Abstract. The asteroid belt is a unique source of information on some of the most important questions facing solar system science. These questions include the sizes, numbers, types and orbital distributions of the planetesimals that formed the planets, and the identification of those asteroids that are the sources of meteorites and near-Earth asteroids. Answering these questions requires an understanding of the dynamical evolution of the asteroid belt, but this evolution is governed by a complex interplay of mechanisms that include catastrophic disruption, orbital evolution driven by Yarkovsky radiation forces, and chaotic orbital evolution driven by gravitational forces. While the timescales of these loss mechanisms have been calculated using estimates of some critical parameters that include the thermal properties, strengths and mean densities of the asteroids, we argue here that the uncertainties in these parameters are so large that deconvolution of the structure of the asteroid belt must be guided primarily by observational constraints. We argue that observations of the inner asteroid belt indicate that the size-frequency distribution is not close to the equilibrium distribution postulated by Dohnanyi (1969). We also discuss the correlations observed between the sizes and the orbital elements of the asteroids. While some of these correlations are significant and informative, others are spurious and may arise from the limitations of the Hierarchical Clustering Method that is currently used to define family membership.

Keywords. Asteroids, Main belt asteroids, Near-Earth objects, Small solar system bodies, Meteorites

1. Introduction

We discuss three classes of family asteroids. (1) Family asteroids that were formed either by catastrophic disruption or crater formation. (2) Halo asteroids that are members of known families. These are family members that, due to the limitations of the Hierarchical Clustering Method (HCM), have not been attached, unambiguously, to a specific family. (3) Asteroids in ghost families. These are older families that were formed in the same way as those in class (1) but are difficult to detect because their orbital elements have dispersed. If we removed all the asteroids in these three classes, then we would be left with the primordial asteroids. We note, however, that each family must have, or have had, a precursor asteroid that is the root source of the other family members. These precursor asteroids are also members of the primordial class. Answering many of the questions facing solar system science requires a determination of the structure of the primordial asteroid belt, that is, the structure of the belt after the transport of the asteroids from their formation locations to the main belt, and after the migration of the major planets that triggered that transport had ceased, but before the time of formation of any of the families. Determining that primordial structure and its transformation into the current structure are the challenges discussed in this paper.

We confine our discussion to the inner main belt (IMB) defined here by 2.1 $au < a < 2.5$ au and $I < 18$ deg, where a is the proper semimajor axis, and I is the proper inclination. Family membership is usually defined using the Hierarchical Clustering Method (HCM) introduced by Zappala *et al.* (1990). This uses the metric

$$d = na\sqrt{d_1^2 + d_2^2 + d_3^2} \tag{1}$$

where

$$na = \sqrt{GM_\odot/a},\ d_1 = \sqrt{5/4}(\Delta a/a),\ d_2 = \sqrt{2}\Delta e,\ d_3 = \sqrt{2}\Delta(\sin I) \tag{2}$$

and na is the average orbital velocity of an asteroid having semi-major axis $a = (a_1 + a_2)/2$, $\Delta a = a_1 - a_2$, $\Delta e = e_1 - e_2$, $\Delta \sin I = \sin I_1 - \sin I_2$, where the indices 1 and 2 denote the two bodies whose mutual distance, d is calculated. Each member of a given family is separated from its nearest neighbor by $d < d_{\text{crit}}$, where d_{crit} is a critical distance chosen such that the members of one family are not linked to the members of another family. Asteroids in the IMB with absolute magnitude $H < 16.5$ are observationally complete (Dermott *et al.* 2018; Hendler & Malhotra 2020). However, while this set of asteroids is devoid of observational selection effects, some correlations between the sizes and the orbital elements of the family asteroids may arise because of the way in which the HCM defines family membership.

2. Orbital element and size correlations in the IMB

Fig. 1 shows the major families in the IMB as defined by Nesvorný (2015) using the HCM. For those asteroids with $H < 16.5$, the IMB is dominated by family and halo asteroids and only an estimated 24% of the asteroids are non-family and possibly primordial (Dermott *et al.* 2022). The largest asteroids in the major families have diameters of 525 km (Vesta), 254 km (Flora), 135 km, 142 km, 495 km (the Nysa, Polana, Eulalia complex), 135 km (Massalia), and 72 km (Erigone), consistent with the argument that the planet-forming planetesimals were large with diameters $D \sim 100$ km (Morbidelli *et al.* 2009). However, the percentage of asteroids in the IMB that are non-family is size-dependent. Fig. 2 shows that the asteroid population with $H < 12$, particularly in the range 2.3 $au < a <$ 2.5 au, is dominated by non-family asteroids that are unlikely to be halo asteroids. For reference, a CC asteroid (defined here as an asteroid with $A < 0.13$) with $H = 12$ and albedo, $A = 0.05$ has a diameter, $D = 24$ km. An NC asteroid (defined here as an asteroid with $A > 0.13$) with $H = 12$ and albedo, $A = 0.24$ has a diameter, $D = 11$ km. These diameters are considerably smaller than those associated with the precursor asteroids of the major families and an outstanding question is what fraction of these non-family, non-halo asteroids are members of ghost families originating from larger primordial asteroids?

The ages of the families have been estimated from the V-shaped spread of the asteroids in $a - 1/D$ space, yielding the result that most families have ages less than half the age of the solar system (Spoto *et al.* 2015). Given that catastrophic disruption was probably greater in the early solar system, this implies that a large fraction of the non-family asteroids could be remnants of ghost families. The spreading rate of the families is determined by the strength of the diurnal Yarkovsky forces and is given by

$$\frac{da}{dt} = \pm\xi \frac{3}{4\pi} \frac{1}{\sqrt{a}(1-e^2)} \frac{L_\odot}{c\sqrt{GM_\odot}} \frac{1}{D\rho} \tag{3}$$

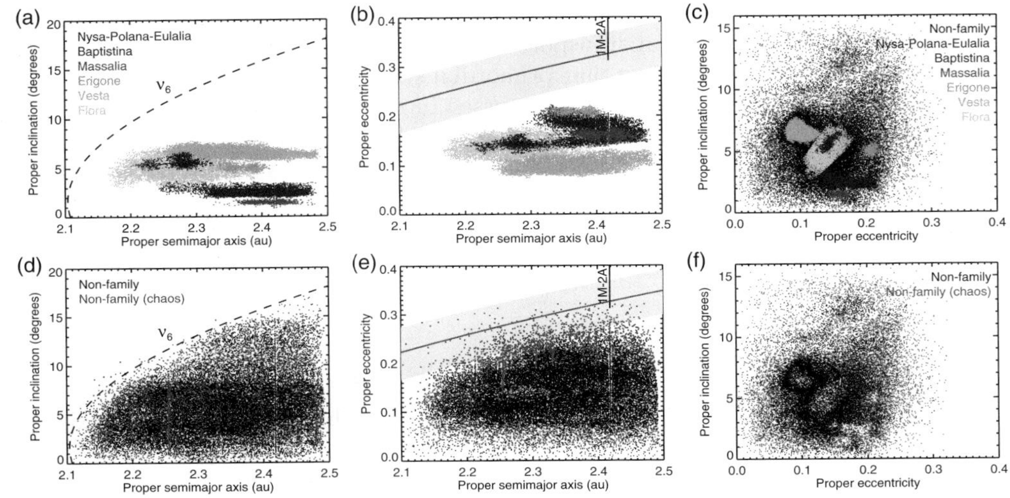

Figure 1. Proper orbital elements of the asteroids in the IMB with $H < 16.5$ (the observational completeness limit). Panels (a)-(c) show the major asteroid families defined by Nesvorný (2015). The green shaded region in panel (b) is the Mars-crossing zone, and the dashed curve in panel (a) is the ν_6 secular resonance. Panels (d)-(f) show the non-family asteroids, color-coded according to the maximum Lyapunov characteristic exponent, $mLCE$ calculated by Knežević & Milani (2000), with red being the most chaotic ($mLCE > 0.00012$ per year) and black the most stable ($mLCE < 0.00012$ per year) (taken from Dermott *et al.* 2021).

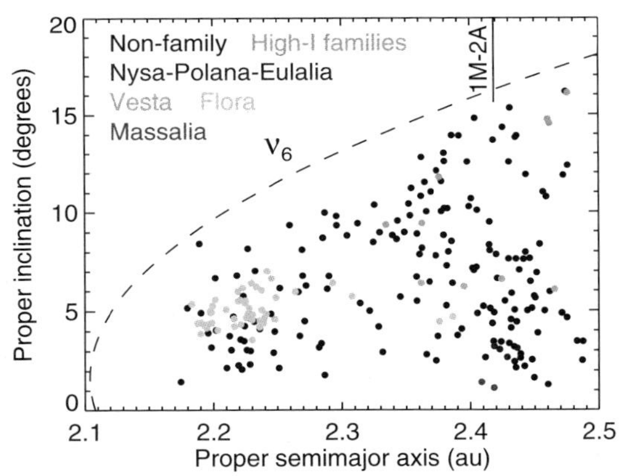

Figure 2. Scatter plot in $a - I$ space of all the asteroids in the IMB with absolute magnitude, $H < 12$. Non-family asteroids are colored black. Asteroids in the prominent Flora family are colored cyan. A CC asteroid with $H = 12$ and albedo, $A = 0.05$ has a diameter, $D = 24\,km$. An NC asteroid with $H = 12$ and albedo, $A = 0.24$ has a diameter, $D = 11\,km$ (taken from Dermott *et al.* 2021)

where $L_\odot$ and $M_\odot$ are the solar luminosity and the solar mass, c is the speed of light, and D and ρ are the diameter and the mean density of the asteroid (Greenberg *et al.* 2020). The Yarkovsky efficiency, ξ, depends on the spin pole obliquity and the thermal properties of the asteroid (Bottke *et al.* 2002). Observations of near-Earth asteroids (NEAs) show that

$$\xi = 0.12^{+0.16}_{-0.06} \tag{4}$$

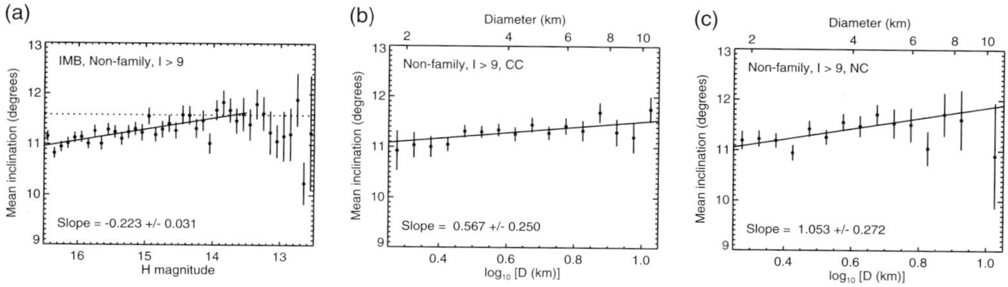

Figure 3. Panel (a): the variation with absolute magnitude, H of the mean proper inclination of the high inclination ($I > 9\ deg$), non-family asteroids. The data is shown binned in H, but the slope has been determined from the individual points in the range $16.5 > H > 13.5$. Panels (b) and (c): in these plots we have divided the data into CC and NC groups as determined by their WISE albedos (taken from Dermott *et al.* 2021)..

(Greenberg *et al.* 2020). Including the uncertainties in the mean densities, the total uncertainties in the evolution rate may be as large as a factor of two and therefore it is possible that the family ages span the age of the solar system. It would be useful to have estimates of the strengths of the Yarkovsky forces for the separate CC and NC main belt asteroids that do not depend on estimates of the thermal parameters and mean densities, etc. Direct observations of the changes in the asteroid orbits are not yet available for main belt asteroids and we must look to other observations. It has been shown that the Martian 1:2 mean motion resonance has a well-defined excess of asteroids (Gallardo *et al.* 2011; Dermott *et al.* 2022). This excess depends on the rate of orbital evolution and modeling could yield an estimate of the strength of the Yarkovsky forces. However, that modeling has yet to be undertaken.

The outer half of the IMB (Fig. 2), the half that is not dominated by the Flora family, has given us one window into the non-family asteroid population. A second significant window is given by the $a - I$ distribution shown in Fig. 1a. Here, we see that the $I > 9\ deg$ region is devoid of major families and the halo asteroids associated with those families. Fig. 3a shows that for the non-family asteroids in this window there is a strong (7σ) correlation between the asteroid magnitude, H and the proper inclination, I (Dermott *et al.* 2021, 2022). This correlation is supported by the correlations of the separate subsets of CC (albedo, $A < 0.13$) and NC (albedo, $A < 0.13$) asteroids (Figs. 3b, c). This correlation arises because the length of the escape route in the IMB, that is, the distance between the ν_6 secular resonance and the 3:1 Jovian mean motion resonance decreases as I increases with the result that small asteroids are preferentially lost from the high inclination orbits (Dermott *et al.* 2021, 2022). To estimate the effectiveness of Yarkovsky forces in the removal of small asteroids, we can evaluate eqn. (2) and write

$$\frac{1}{a}\frac{da}{dt} = \pm 2.0\xi \left(\frac{1\,km}{D}\right)\left(\frac{2000\,kgm^{-3}}{\rho}\right) Gyr^{-1}. \tag{5}$$

Given that

$$D = \frac{1329\,km}{10^{H/5}\sqrt{A}} \tag{6}$$

and assuming that $a = 2.4\ au$, we can write

$$\frac{da}{dt} = 0.00043\,10^{H/5}\sqrt{A}\left(\frac{2000\,kgm^{-3}}{\rho}\right) au\,Gyr^{-1}. \tag{7}$$

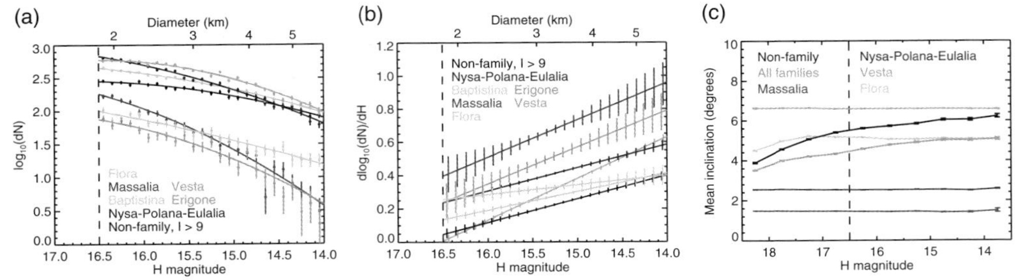

Figure 4. Panel (a): quadratic fits to the SFDs for asteroids in the separate major families, and for the non-family ($I > 9\ deg$) asteroids. Nominal asteroid diameters have been calculated from H assuming an albedo of 0.13. Panel (b): the variation with absolute magnitude, H of the slopes to the quadratic fits shown in panel (a). Panel (c): the dependence of the mean proper inclination on asteroid size for all family asteroids (treated as one group), for all non-family asteroids (also treated as one group), and for the separate major families (taken from Dermott *et al.* 2018, 2021).

For NC asteroids, using $A \approx 0.24$ and $\rho \approx 2000\,kgm^{-3}$ we have

$$\frac{da}{dt} = 0.0002\,10^{H/5}\,au\,Gyr^{-1}. \tag{8}$$

For CC asteroids, more appropriate values would be $A \approx 0.05$ and $\rho \approx 1000\,kgm^{-3}$. However, since $\sqrt{0.24/0.05/2} \approx 1$, either set of parameters results in the same estimate for da/dt and we can usefully apply eqn. (4) to all of the IMB asteroids. If we assume that $da \approx 0.2\,au$ typically results in the loss of an asteroid through one of the escape hatches, then asteroids with $H \gtrsim 15$ will be lost from the IBM in $\lesssim 1\ Gyr$. $H = 15$ corresponds to $D \approx 6\,km$ (CC) or $D \approx 3\,km$ (NC).

Unless the Yarkovsky efficiency, ξ estimated from the NEA observations is an overestimate we must expect a significant loss of small asteroids from the IMB. This loss could be compensated for by the creation of small asteroids through the catastrophic destruction of larger asteroids. However, we have other observations that place constraints on the creation rate. One constraint is provided by the $H - I$ correlation shown in Fig. 3a. Dermott *et al.* (2021, 2022) have argued that the loss of asteroids from the IMB is dominated by Yarkovsky orbital evolution, and if the initial distribution of asteroids in $a - I$ space was uniform, and the loss of asteroids occurred over the age of the solar system, then ξ must be a factor of three smaller than the NEA estimate or the correlation would be three times larger. Alternatively, if the orbital evolution rate is consistent with the NEA observations, then the asteroids must be younger than the age of the solar system by a factor of three, implying that the majority of asteroids in the IMB with high inclination ($I > 9\ deg$) orbits are not primordial but members of families and the products of the catastrophic destruction of a small number of larger asteroids. A diffuse cluster of high-I asteroids, centered on $I = 13.5\ deg$ and $e = 0.18$ is evident in Fig. 1c.

Other evidence for the net loss of small asteroids from the IMB is obtained from the size-frequency distributions (SFDs). The data in Fig. 4 are observationally complete and show that all the major families and the non-family asteroids in the IMB have a depletion of small asteroids. As H increases the observed slopes of all the SFDs tend to zero. Modeling of the variation of the SFD with H has shown that for $H \gtrsim 16.5$ we must expect a marked lack of small asteroids (Dermott *et al.* 2021, 2022). Thus, the expectation that for high H the SFD tends to the equilibrium distribution postulated by Dohnanyi (1969) is not supported by the observations. This is because Dohnanyi (1969) considered a closed system and did not allow for the loss of small asteroids from that system. This has major consequences with respect to calculations of the expected rate of

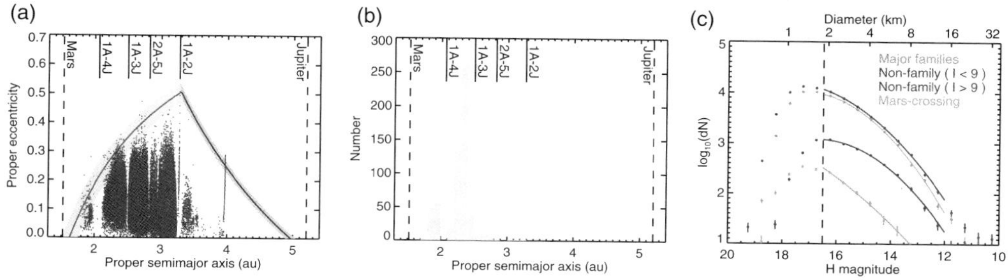

Figure 5. Panel (a): scatter plot of the proper eccentricity e and the semimajor axis, a of the asteroids in the IMB with absolute magnitude $H < 15$. The green shaded zone on the left is the Mars-crossing zone. Panel (b): histogram of the semimajor axes of the asteroids in the Mars-crossing zone. Panel (c): quadratic fits to the SFDs (with $dH = 0.5$) for the asteroids in the separate major families, the non- family ($I < 9$ deg) asteroids, the non-family ($I > 9$ deg) asteroids, and the asteroids in the Mars-crossing zone. Nominal asteroid diameters have been calculated from H assuming an albedo of 0.13 (taken from Dermott *et al.* 2021).

collisional disruption, and the rate of change of asteroid spin directions due to impacts with small asteroids. Given that we do not know the SFD of the small asteroids that impact the larger asteroids, we must accept that these rates are unknown.

3. The Mars-crossing zone

The orbital eccentricities of the main-belt asteroids are largely capped by the Mars-crossing zone (Fig. 5a). Given that the lifetime of the asteroids in this zone is $\lesssim 10^8$ yrs, this indicates that Mars is currently scattering asteroids out of the main belt and into the inner solar system. In Fig. 5b, we observe that most of the asteroids in or above the crossing-zone are in the inner main belt (IMB), suggesting that the IMB is the major source of near-Earth asteroids (NEAs) and meteorites (Dermott *et al.* 2021, 2022). This conclusion is supported by the results of numerical investigations of the likely escape routes from the main belt (Gladman *et al.* 1997; Granvik *et al.* 2017, 2018). The latter studies used estimates of the total number of NEAs after correcting for the incompleteness of the observed NEA population, estimates of the SFD of the main belt asteroids, and estimates of the strengths of the Yarkovsky forces. However, the lifetimes of the asteroids in the Mars-crossing zone are determined solely by the rate of close encounters with Mars and it is possible to determine those lifetimes without making unnecessary assumptions. The calculations for the asteroid set shown in Fig. 5b have yet to be performed, but it is possible, given that corrections for observational selection and any assumptions about SFDs or Yarkovsky evolution rates are not needed, that, if the Mars-crossing population is an equilibrium population, then these calculations could give a robust estimate of the rate of transfer of asteroids from the main belt to near-Earth space.

Asteroids are lost from the IMB through dynamical mechanisms that include chaotic and Yarkovsky driven orbital evolution. These mechanisms result in asteroids diffusing into various weak mean motion resonances (Fig. 1e), and then into the Mars-crossing zone (Morbidelli & Nesvorný 1999; Farinella & Vokrouhlický 1999). A third mechanism depends on Yarkovsky forces driving small asteroids into the ν_6 secular resonance (Farinella *et al.* 1994; Migliorini *et al.* 1998; Farinella, Vokrouhlický & Hartmann 1998). Numerical integrations have shown that all these mechanisms are viable. The questions that remain are which mechanism is dominant and what is the rate of loss. Greenberg *et al.* (2020) have shown from observations of the current orbital evolution of 247 NEAs that the ratio of retrograde to prograde spin of these objects is as high as $2.7^{+0.3}_{-0.6}$. This implies that prior to their escape these objects were evolving towards the

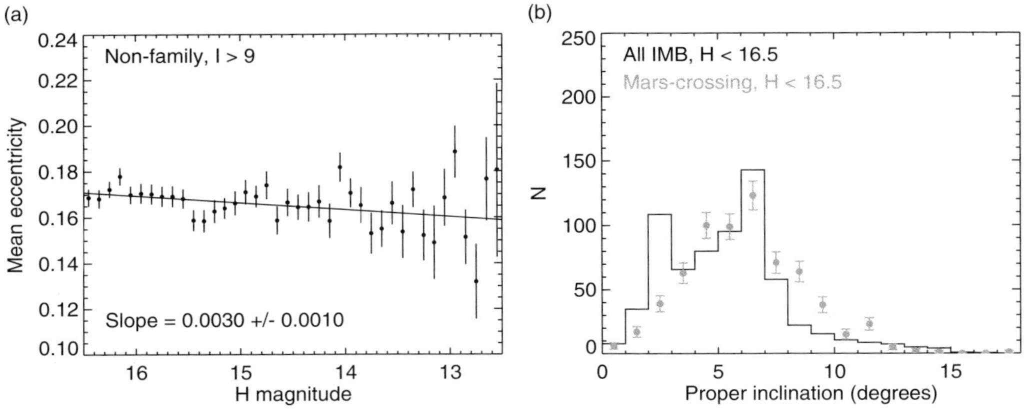

Figure 6. Panel (a): the variation with absolute magnitude, H of the mean proper eccentricity of the high inclination ($I > 9$ deg), non-family asteroids. Panel (b): histograms of the inclination distributions of all the asteroids in the IMB with $H < 16.5$ and the asteroids in the Mars-crossing zone shown in Fig. 1e. These distributions have been normalized (taken from Dermott *et al.* 2021).

Sun and thus that the ν_6 secular resonance is the dominant escape route. There are other observations that are devoid of observational selection effects, that have yet to be analyzed, that could also place constraints on the loss mechanisms. In Fig. 6a we see that the mean proper eccentricity of the high eccentricity asteroids is size-dependent, increasing with decreasing asteroid size, suggesting the action of Yarkovsky forces. Fig. 6b shows the distribution of the inclinations of the asteroids in the Mars crossing zone. This distribution is somewhat similar to the distribution of the major families. There is also an excess of asteroids with high inclinations, but it has not been determined that this excess is strong enough to support the dominant role of the ν_6 secular resonance.

4. Vesta and the origin of the HED meteorites

The action of Yarkovsky forces has resulted in a marked net loss of small asteroids from the IMB, implying that the remnants of many ghost families could now be sparsely distributed among other large, background asteroids. Progress with the detection of these dispersed families has been made by Delbó *et al.* (2017, 2019). Here, we show that further evidence for the loss of small asteroids from old families is provided by an analysis of the SFD of the Vesta family.

One third of the ~29,000 asteroids in the IMB with $H < 16.5$ originate from Vesta. This asteroid is the source of the HED meteorites and this is the only firm asteroid-meteorite link (McCord, Adams & Johnson 1970; McSween & Binzel 2022). Vesta is the largest asteroid in the IMB and has remained intact since its formation, thus the asteroids in this large family must originate from one or more craters. Over 25 years ago, Hubble observations revealed an enormous crater, Rheasilvia, and this crater is thought to be the source of the meteorites (Thomas *et al.* 1997). But later the Dawn spacecraft revealed a second enormous crater, Veneneia, partly underneath Rheasilvia (Schenk *et al.* 2012). Given that Veneneia has diameter of ~ 400 km, its formation would probably have resulted in the formation of a large family. Is it possible that Vesta is currently associated with two families of asteroids, of different ages, that have sampled different surface locations and mantle depths?

We have previously argued (Dermott *et al.* 2018) that the merging of two families with different mean proper orbital elements and different SFDs will give rise to correlations

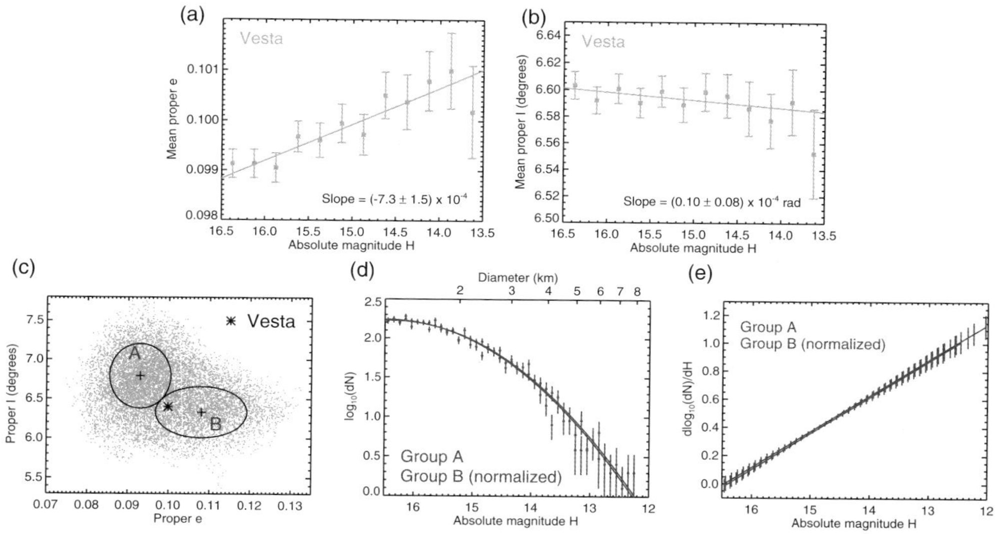

Figure 7. Panels (a) and (b) show the variations with absolute magnitude, H of the mean proper e and the mean proper I of the asteroids in the Vesta family. Panel (c) defines two independent regions of the Vesta family in $e - I$ space. Panel (d) shows the SFDs of the two regions shown in panel (c) after normalization. Panel (e) shows the variation with absolute magnitude, H of the slopes to the quadratic fits of the SFDs shown in panel (d).

between the sizes and the mean proper orbital elements of the family members (treated as a whole) – see Fig. 4c. In Fig. 7a we observe a strong correlation (4.7σ) between the eccentricities and sizes of the Vesta family members. This observation, plus the asymmetric distribution of the asteroids in $e - I$ space (Fig. 7c), suggests the current existence of two families. However, if we isolate two different regions of the Vesta family in $e - I$ space, as shown in Fig. 7c, we observe, after normalization, that while the two regions have different mean orbital elements, the SFDs of the two regions are identical. This implies that the Vesta family must originate from one large crater and, while we believe that the Veneneia crater must have resulted in the formation of a large asteroid family, we can only account for the similarity of the SFDs if a majority of the small asteroids in the older Veneneia family have been lost from the system due to the action of Yarkovsky forces. This argument allows us to place bounds on the age of the Rheasilvia crater and on the strength of the Yarkovsky forces. Full details of these arguments will be published elsewhere.

5. Defects of the HCM and their consequences

What is the origin of the strong correlation observed between the mean eccentricities and sizes of the Vesta family members? Here, we suggest that this correlation may not have a physical explanation, but may be due to a fundamental flaw in the way that the HCM defines family membership. The total distance of the separation, d used in eqn. (2) to define family membership gives approximately equal weights to the $\Delta a/a$, Δe and ΔI components represented by d_1, d_2 and d_3. On the initial formation of a family, these three components are determined by the structure of the debris cloud released close to the crater. It is possible that the distribution of kinetic energy between the particles in this cloud is size-dependent (Leinhardt *et al.* 2015). However, the second step in family formation requires these particles to reaccumulate into rubble-pile asteroids and it is not obvious that any size-energy correlation will be carried over to the resultant rubble-piles.

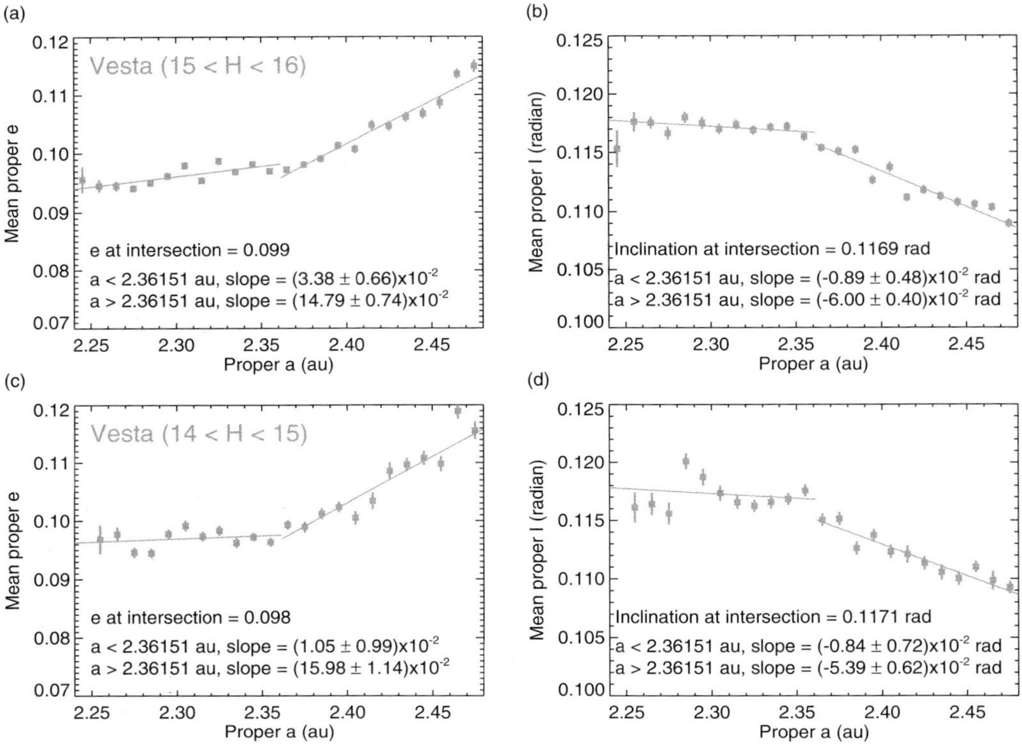

Figure 8. Variations of the mean proper e and the mean proper I with semimajor axis for asteroids in the Vesta family defined by Nesvorný (2015) using the HCM. These variations are shown for two independent size ranges.

The two components of d that depend on Δe and ΔI are comparatively stable: the changes due to Yarkovsky forces are negligible, but we must expect some small changes due to chaotic orbital evolution. If the changes in Δe and ΔI are dominated by the changes that occur while the asteroids are trapped in weak resonances, as shown by Christou, Dermott & Li (2022), then the changes in Δe and ΔI may be size-dependent. However, in contrast, we can be certain that the changes in the semimajor axes, due to Yarkovsky forces, are large and definitely size-dependent. Thus, the final separations of the asteroids as determined by $\Delta a/a$ are decoupled from the family formation mechanism and it is questionable that they should be used in any criterion of family membership. The inclusion of $\Delta a/a$ in the definition of d in eqn. (2) could lead to the criterion that separates family from non-family asteroids being size-dependent. Evidence that this is the case is shown in Fig. 8, where we see that there are strong correlations between the mean proper e and mean proper I and the semimajor axes of the Vesta family asteroids.

In Fig. 9, we show how the mean separations of the asteroids originating from Vesta at $a = 2.36$ au increase with increasing distance from Vesta. We conclude from this that if the criterion for family membership, $d < d_{\text{crit}}$, is fixed but the mean separation of the asteroids in $a - e - I$ space increases with increasing separation from Vesta, then the number of asteroids classified as family members will decrease with increasing separation from Vesta, while the number of non-family members that are halo asteroids will increase. This is shown to be the case in Fig. 10. Panel (a) of that figure shows the structure of the major families in $a - I$ space as defined by Nesvorný (2015) using the HCM. We observe that the width of the family, ΔI decreases as the semimajor axis increases to 2.5 au.

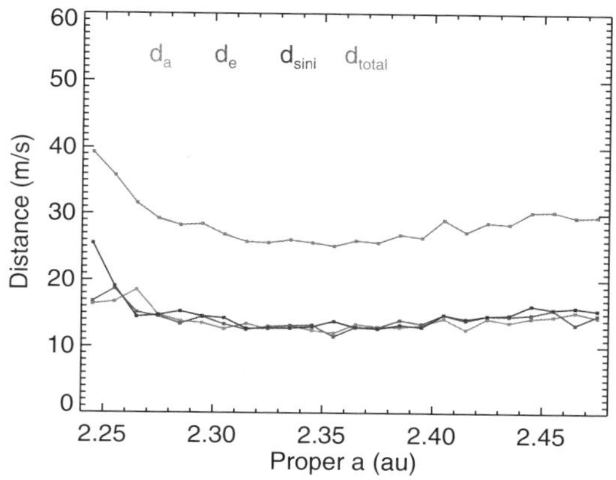

Figure 9. The variation with semimajor axis of the total and the three components of the mean minimum separation of the Vesta family asteroids in $a - e - I$ space determined using the HCM

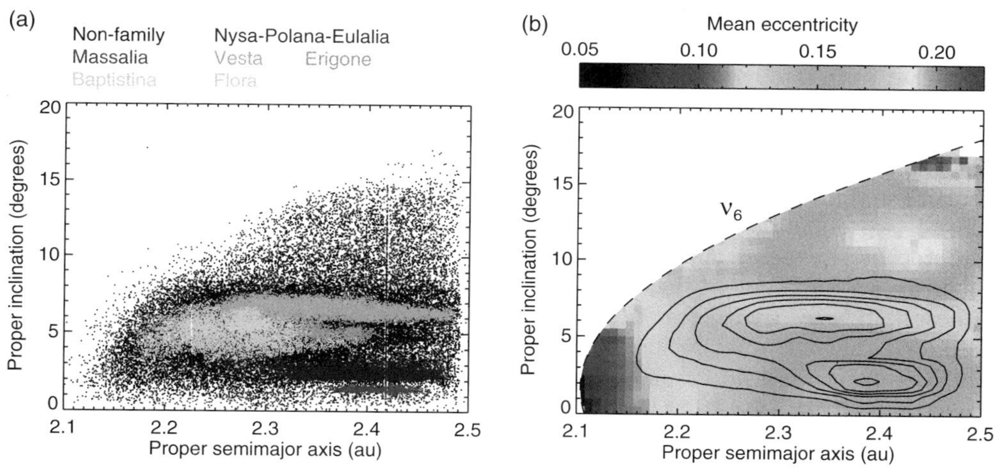

Figure 10. Comparison of the structure of the Vesta family as revealed by two different methods. Panel (a) shows the major families as defined by Nesvorný (2015) using the HCM. The contours in panel (b) show the variation of number density, while the small cells show the variation of the mean proper eccentricity, e in $a - I$ space.

Panel (b) shows the variation of the mean eccentricity of the IMB asteroids in $a - I$ space. In this plot, we have not defined the families, but we show those features that allow us to recognize the families. The contours in panel (b) show the variation of the number density, and by comparing the two panels we can see that the highest number densities are associated with the Vesta family and the Nysa-Polana-Eulalia complex. Fig. 10b also shows the variation of the mean proper eccentricity in $a - I$ space.

Inspection of Fig. 1 shows a distinct contrast between these two major families. While the Vesta family has a high mean inclination and a low mean eccentricity, the opposite is the case for the Nysa-Polana-Eulalia complex. In Fig. 10b, the band of low eccentricity asteroids with mean inclination close to the inclination of Vesta is clearly associated with the Vesta family. However, while the width of the Vesta family, ΔI shown in panel (a), as determined by Nesvorný (2015) using the HCM, decreases as the semimajor axis increases to 2.5 au, the opposite is true for the width of the band of low eccentricity asteroids (i.e., the cyan-blue band between $I \sim 4.5$ *deg* and $I \sim 8$ *deg*) shown in panel (b). The width of this low eccentricity band increases markedly as the 3:1 Jovian resonance is approached, suggesting that some of the asteroids in panel (a) close to $a = 2.5$ *au* that have been classified as non-family by Nesvorný (2015) should be classified as halo asteroids of the Vesta family. This shows a major deficiency in the ability of the HCM to separate family asteroids from the asteroids in the associated halo, and it is this deficiency that leads to the size, orbital element correlations shown in Figs. 7 and 8.

As an aside, the distribution of the mean proper eccentricity in $a - I$ space shown in Fig. 10b allows us to recognize a yellow region of high eccentricity that could be associated with the cluster centered on $I = 13.5$ *deg* and $e = 0.18$ evident in Fig. 1c. Because this cluster does not extend the full extent of the distance between the ν_6 secular resonance and the 3:1 Jovian mean motion resonance, this cluster may be a recent family rather than a ghost family.

6. Conclusions

The SFDs of the asteroids in the IMB are inconsistent with the equilibrium distribution postulated by Dohnanyi (1969). The separation of the family and halo asteroids achieved using the HCM is somewhat inadequate and this could lead to an overestimate of the depletion of small family asteroids. However, this is not a large concern because Fig. 5c shows that both family and non-family are depleted in small asteroids. If this depletion applies to the main belt as a whole, then while the SFD of the target asteroids may be well constrained, the SFD of the smaller bullets is uncertain, and therefore we must accept that the collisional lifetimes of the asteroids are uncertain, that the rate of delivery of small asteroids from the inner belt to the inner solar system is also uncertain, and that the rate of production of small asteroids and the associated NEAs and meteorites may be a stochastic process and time variable. This conclusion is consistent with previous arguments on the origin of the solar system dust bands. Immediately after the discovery of those bands by IRAS, two theories of their origin were proposed. The first theory postulated that the dust bands were an equilibrium feature associated with the three most prominent asteroid families: Eos. Themis and Koronis (Dermott *et al.* 1984). The second theory postulated that they were random features associated with single collisions and not necessarily related to the major families (Sykes & Greenberg 1986). Dynamical modeling of the structure of the "ten-degree band" showed that the initial mean proper inclination of the dust particle orbits needed to account for that band is 9.3 *deg* and equal to the mean proper inclination of the Veritas family asteroids rather than that of the Eos family (Grogan, Dermott & Durda 2001; Dermott *et al.* 2002), thus favoring the stochastic model.

References

Bottke, W. F., Jr., Vokrouhlický, D., Rubincam, D. P., & Broz, M. 2002, *Asteroids III*, 395.
Christou, A.A., Dermott, S.F., & Li, D. 2022, *MNRAS*, 516, 1428.
Delbó, M., Walsh, K., Bolin, B., Avdellidou, C., & Morbidelli, A. 2017, *Science*, 357, 1026.
Delbó, M., Avdellidou, C., & Morbidelli, A. 2019, *A&A*, 624, A69.
Dermott, S. F., Nicholson, P. D., Burns, J. A., & Houck, J. R. 1984, *Nature*, 312, 505.
Dermott, S. F., Kehoe, T. J. J., Durda, D. D., Grogan, K., & Nesvorný, D. 2002, in: B. Warmbein (eds.), *Asteroids, Comets, and Meteors: ACM 2002*, ESA Special Publication, p. 319
Dermott, S.F., Christou, A.A., Li, D., Kehoe, T. J.J., & Robinson, J.M. 2018, *Nature Astronomy*, 2, 549.
Dermott, S.F., Li, D., Christou, A.A., Kehoe, T. J.J., Murray, C.D., & Robinson, J.M. 2021, *MNRAS*, 505, 1917.
Dermott, S.F., Li, D., & Christou, A.A. 2022, *IAU Symposium*, 364, 1.

Dohnanyi, J.S. 1969, *Journal of Geophysical Research*, 74, 2531.
Farinella, P., Froeschlé, C., Froeschlé, C., Gonczi, R., Hahn, G., Morbidelli, A., & Valsecchi, G.B. 1994, *Nature*, 371, 314.
Farinella, P., Vokrouhlický, D., & Hartmann, W. K. 1998, *Icarus*, 132, 378.
Farinella, P. & Vokrouhlický, D. 1999, *Science*, 283, 1507.
Gallardo, T., Venturini, J., Roig, F., & Gil-Hutton, R. 2011, *Icarus*, 214, 632.
Gladman, B.J., Migliorini, F., Morbidelli, A., Zappala, V., Michel, P., Cellino, A., Froeschle, C., Levison, H.F., Bailey, M., & Duncan, M. 1997, *Science*, 277, 197.
Granvik, M., Morbidelli, A., Vokrouhlický, D., Bottke, W.F., Nesvorný, D., & Jedicke, R. 2017, *A&A*, 598, A52.
Granvik, M., Morbidelli, A., Jedicke, R., Bolin, B., Bottke, W.F., Beshore, E., Vokrouhlický, D., Nesvorný, D., & Michel, P. 2018, *Icarus*, 312, 181.
Greenberg, A.H., Margot, J.L., Verma, A.K., Taylor, P.A., & Hodge, S.E. 2020, *AJ*, 159, 92.
Grogan, K., Dermott, S. F., & Durda, D. D. 2001, *Icarus*, 152, 251.
Hendler, N.P. & Malhotra, R. 2020, *The Planetary Science Journal*, 1, 75.
Knežević, Z., & Milani, A. 2000, *Cel. Mech. Dyn. Astr.*, 78, 17
Leinhardt, Z., Dobinson, J., Carter, P. J., & Lines, S. 2015, *ApJ*, 806, 23.
McCord, T. B., Adams, J. B., & Johnson, T. V. 1970, *Science*, 168, 1445.
McSween, Harry Y. and Binzel, Richard P. 2022, *Vesta and Ceres. Insights from the Dawn Mission for the Origin of the Solar System*, 41.
Migliorini, F., Michel, P., Morbidelli, A., Nesvorný, D., & Zappala, V. 1998, *Science*, 281, 2022.
Morbidelli, A. & Nesvorný, D. 1999, *Icarus*, 139, 295.
Morbidelli, A., Bottke, W. F., Nesvorný, D. & Levison, H. F. 2009, *Icarus*, 204, 558.
Nesvorný, D. 2015, HCM Asteroid Families V3.0. *NASA Planetary Data System.*
Schenk, P., O'Brien, D. P., Marchi, S., Gaskell, R., Preusker, F., Roatsch, T., Jaumann, R., Buczkowski, D., McCord, T., McSween, H. Y., Williams, D., Yingst, A., Raymond, C., & Russell, C. 2012, *Science*, 336, 694.
Spoto, F., Milani, A., & Knežević, Z. 2015, *Icarus*, 257, 275.
Sykes, M. V., & Greenberg, R. 1986, *Icarus*, 65, 51.
Thomas, P. C., Binzel, R. P., Gaffey, M. J., Storrs, A. D., Wells, E. N., & Zellner, B. H. 1997, *Science*, 277, 1492.
Zappala, V., Cellino, A., Farinella, P., & Knežević, Z. 1990, *AJ*, 100, 2030.

Complex Planetary Systems II: Latest Methods for an Interdisciplinary Approach
Proceedings IAU Symposium No. 382, 2024
A. Lemaitre & A.-S. Libert, eds.
doi:10.1017/S1743921323004817

Towards a complete picture of the evolution of planetary systems around evolved stars

Mats Esseldeurs[1], **Stéphane Mathis**[2] **and Leen Decin**[1]

[1]Instituut voor Sterrenkunde, KU Leuven, Celestijnenlaan 200D, 3001 Leuven, Belgium
email: mats.esseldeurs@kuleuven.be

[2]Université Paris-Saclay, Université Paris Cité, CEA, CNRS, AIM,
91191 Gif-sur-Yvette, France

Abstract. Solar-like stars evolve through the Asymptotic Giant Branch (AGB) phase. This phase is characterized by increased radii, high luminosities, and significant mass loss. In order to understand the survival of companions during this phase, and explain the presence of planets orbiting white dwarfs, it is essential to examine the orbital evolution of these systems. Several physical mechanisms come into play for AGB stars, including stellar mass loss and tidal interactions between the star and its companion. Assessing mass-loss rates and accretion to the companion requires complex radiation-hydro-chemical simulations. Furthermore, comprehending the full history of tidal dissipation in low-mass stars during their late evolutionary stages, which strongly depends on their internal structure, requires dedicated analytical and numerical studies.

Keywords. AGB stars, stellar winds and mass-loss, tides, star-planet interactions

---

1. Introduction

About 95% of all stars in the galaxy have an initial mass lower than 8 $M_\odot$. When these stars evolve off the main sequence, they will go through the Asymptotic Giant Branch (AGB) phase, just before turning into a white dwarf (WD). This phase is characterized by increased radii, high luminosities, intense pulsations, and significant mass loss (Höfner & Olofsson 2018). In order to get a complete picture of the evolution of planetary systems from the birth to the final evolutionary state of their host star, to understand the survival of planetary or stellar companions during the AGB phase and to explain the presence of planets orbiting WDs, it is essential to examine the orbital evolution of these systems (Mustill & Villaver 2012; Madappatt *et al.* 2016). Several key physical mechanisms come into play for studying orbital evolution around AGB stars, such as the stars significant mass-loss rate, the efficiency of mass accretion onto the companion, and the tidal interactions between the star and its companion.

2. 3D radiation-hydro-chemical simulations of AGB stellar outflows

AGB stars lose a significant amount of mass through a pulsation-enhanced dust-driven wind. Pulsations at the surface of AGB stars push material into sufficiently high layers where it is cool enough for molecules to condensate into dust. These dust particles can be captured by the star's radiation pressure and pushed outwards. Finally, the dust particles collide with the remaining gas particles and create an outflow of material (Höfner & Olofsson 2018). AGB outflows exhibit complex structures, such as arcs,

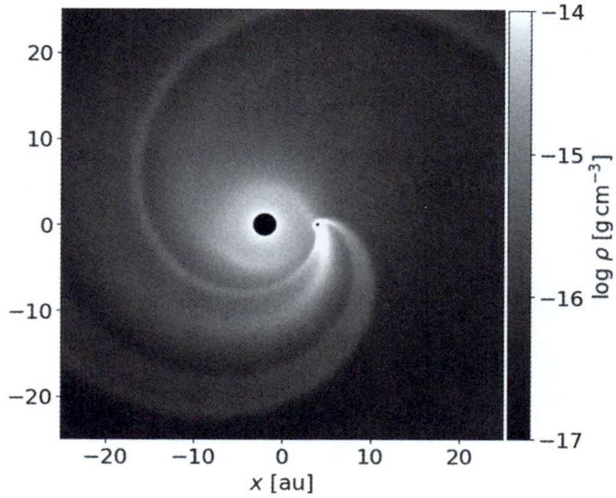

Figure 1. Density structure of a companion-perturbed AGB outflow, shown in a slice through the orbital plane with the AGB star on the left and the companion on the right; simulation from Esseldeurs *et al.* (2023).

spirals, disks, and bipolarity where the current theory suggests that these 3D structures are caused by an unseen companion (Decin *et al.* 2020). To understand this phenomenon, complex 3D radiation-hydro-chemical simulations (see for example Fig. 1; depicting the density structure of a companion-perturbed AGB outflow with the AGB star on the left and the companion on the right) are necessary to understand the influence a companion on the morphological structures of AGB outflows. However, existing simulations are computationally demanding and ongoing efforts are focused on enhancing computational speed (Siess *et al.* 2022; Esseldeurs *et al.* 2023). These simulations will allow us to investigate the impact of the companion on the star's mass-loss rate and the efficiency of accretion onto the companion, currently unaccounted for.

3. Identifying tidal dissipation mechanisms around evolved stars

Tidal dissipation encompasses two components: the equilibrium and dynamical tides. The former occurs due to the hydrostatic displacement induced by the ellipsoidal deformation triggered by the companion. Its energy is dissipated because of turbulent friction in convective layers, generating transfer of angular momentum between the spin and the orbit (e.g. Zahn 1966; Remus *et al.* 2012). This equilibrium tide has been used in studies of AGB stars (e.g. Mustill & Villaver 2012; Madappatt *et al.* 2016), but the dynamical tide has never been formally evaluated during this phase despite its importance for subgiants and RGB stars (Weinberg *et al.* 2017; Ahuir *et al.* 2021). This dynamical tide involves stellar oscillation modes excited by the tidal potential. Depending on the internal structure of the star, as well as the Brunt-Väisälä and Lamb frequency compared to the tidal frequency, different oscillation modes can be exited. In the deep convective envelope of an AGB star, inertial modes in the convective layer can be excited only for stellar companions, as planetary companions don't spin up the star sufficiently. In the radiative core low-frequency progressive gravity waves are triggered by tides, both for stellar and planetary companions (Esseldeurs *et al.* in prep.). Fig. 2 shows how tidal gravity waves compete with the equilibrium tide as a function of the orbital distance of a planetary companion and stellar age for a 1.2 $M_\odot$ host star. In this model, the equilibrium tide is the dominant dissipation mechanism during the AGB phase, while both are important at the white dwarf stage.

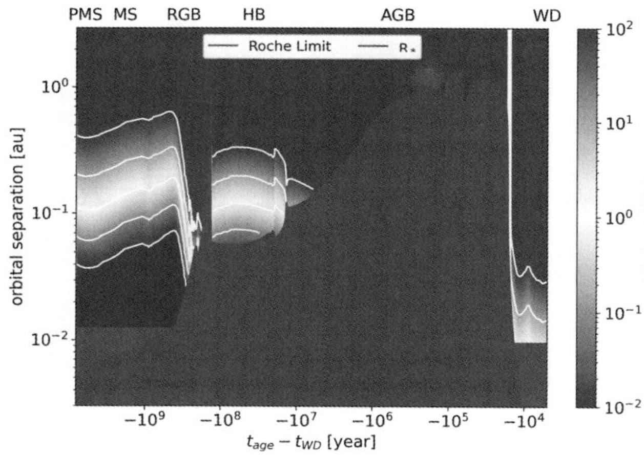

Figure 2. Competing dissipation strengths of the equilibrium ($\Im(k_2)_{eq}$) and dynamical ($\Im(k_2)_{IGW}$) tide as a function of the orbital separation and stellar age (t_{age} upto the age where the star becomes a white dwarf t_{WD}), for a 1.2 $M_\odot$ host star; simulation from Esseldeurs *et al.* (in prep.).

4. The interplay between winds, pulsations, and tides

Tidal dissipation and mass loss in AGB stars is currently always treated separately, but these problems are inherently coupled. On the one hand, the AGB's dust-driven winds are initiated by pulsations, which can be enhanced by additional tidally induced pulsations. On the other hand, tidal energy stored in the form of waves can be transported through the stellar wind, enhancing the dissipation. Therefore, it will be necessary to simultaneously treat mass loss and tidal dissipation.

References

Ahuir J., Mathis S., Amard L. 2021, *A&A*, 651, A3
Decin L., Montargès M. *et al.* 2020, *Science*, 369, 1497
Esseldeurs M., Siess L. *et al.* 2023, *A&A*, 674, A122
Höfner S. & Olofsson H. 2018, *A&AR*, 26, 1
Madappatt N., De Marco O. *et al.* 2016, *MNRAS*, 463, 1040
Mustill A. J. & Villaver E. 2012, *ApJ*, 761, 121
Remus F., Mathis S., Zahn J. P. 2012, *A&A*, 544, A132
Siess L., Homan W. *et al.* 2022, *A&A*, 667, A75
Weinberg, N. N., Sun, M. *et al.* 2017, *ApJ*, 849, L11
Zahn J. P. 1966, *Ann. Astrophys.*, 29, 313

Complex Planetary Systems II: Latest Methods for an Interdisciplinary Approach
Proceedings IAU Symposium No. 382, 2024
A. Lemaitre & A.-S. Libert, eds.
doi:10.1017/S1743921323004805

Multi-planet systems explored: the case of the HD34445 system revisited

Nikolaos Georgakarakos[1,2] and **Ian Dobbs-Dixon**[1,2]

[1]Division of Science, New York University Abu Dhabi,
PO Box 129188, Abu Dhabi, United Arab Emirates
email: ng53@nyu.edu

[2]Center for Astrophysics and Space Science (CASS) New York University Abu Dhabi,
PO Box 129188, Abu Dhabi, United Arab Emirates
email: idd1@nyu.edu

Abstract. HD34445 is a system that consists of a star and six planets. In some previous work, we investigated the dynamical stability of the system by means of numerical simulations. Here, we explore the system further by carrying out additional numerical experiments. A total of 100000 simulations confirm previous findings of the stability status of the system at the 1σ and 99% c.i. level. We find that only 2.7% of the systems with parameters varied within 1σ from the mean were unstable, while that percentage rose to about 28% for systems with parameter values taken within the 99% c.i. These preliminary results are presented and discussed herein.

Keywords. chaos, methods: numerical, techniques: radial velocities, celestial mechanics, planets and satellites: dynamical evolution and stability

1. Introduction

HD34445 is a system that hosts six planets. The first planet, a sub-Jupiter mass body, was discovered a while ago by Howard *et al.* (2010). A few years later, Vogt *et al.* (2017) revised the orbital solution for HD34445b and announced the detection of another five planets in the system. In some previous work Georgakarakos & Dobbs-Dixon (2019) tested the validity of the solution proposed in Vogt *et al.* (2017) by investigating the dynamical stability of the system. Here, we extend that work by exploring further aspects of the problem.

2. Method

The dynamical stability of the system is investigated by means of numerical simulations. Contrary to what was done in Georgakarakos & Dobbs-Dixon (2019), where the stability of the system was mainly investigated by varying only one parameter at a time and for a rather limited number of cases, here, we create a pool of initial conditions for the masses, eccentricities, semi-major axes, mean anomalies and longitudes of pericentre of all planets drawn from normal distributions with means and standard deviations given in Vogt *et al.* (2017). The parameter ranges are within 1σ or within the 99% credibility interval. Then, we simulate 50000 systems per case. As previously, stability means no hyperbolic orbits with respect to the star and no orbit crossings. We first investigate the system assuming that all bodies lie on the same plane of motion. For the simulations we used the Gauss - Radau integrator in Eggl & Dvorak (2010). We have set the code to stop the integration when any planet becomes hyperbolic.

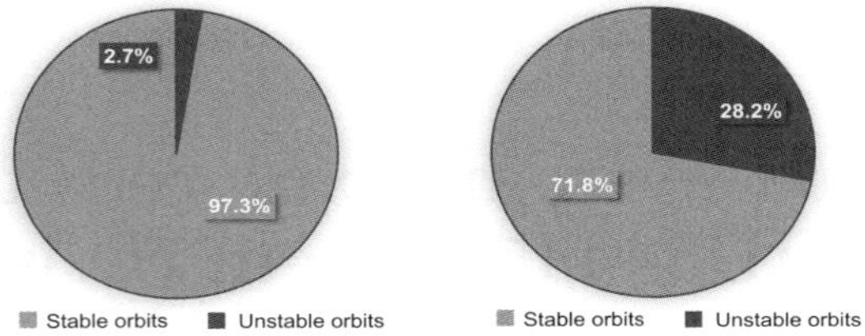

Figure 1. Stability pie charts for the 1σ (left) and 99% c.i. (right) cases.

3. Preliminary Results

Here, we present some results from the 1σ and 99% c.i. simulations. The 1σ results appear pretty stable with only 2.7% unstable orbits (1334 out of 50000), while the 99% c.i. group demonstrates a rather significant percentage of unstable systems (28.2%; 14102 out of 50000). Figure 1 is a graphical representation of these preliminary findings. In most of the unstable cases, it is the smallest planet, HD34445e, that gets ejected from the system. Regarding the initial conditions leading to instability, a quick inspection of them did not reveal any clustering around specific values or any significant difference compared to the initial conditions of the stable orbits. That was the conclusion when inspecting the parameters one by one or in pairs of two, e.g. mass against eccentricity. The only difference we noticed was when we looked at the eccentricities of HD3445d, HD3445c and HD3445f (i.e. the second, third and fourth planet as we move away from the star). It appears that there were more values near the high end of the range that led to unstable orbits than to stable ones. The difference, however, was not significant. It is likely though that as we proceed to grouping three or more parameters at a time, more evident patterns may emerge.

4. Outlook

The analysis on our current results will be extended and additional aspects of the dynamical stability of the system are going to be investigated such as for example the effect of General Relativity, three dimensional orbits and the importance of the presence of the outermost planet HD34445g.

References

Eggl, S., Dvorak, R. 2010, in: J. Souchay & R. Dvorak (eds.), *Lecture Notes in Physics: Dynamics of Small Solar System Bodies and Exoplanets*, No. 790 (Springer Verlag, Berlin) p. 431

Georgakarakos, N., Dobbs-Dixon, I. 2019, *MNRAS*, 488, 3818–3825

Howard, A. W., Johnson, J. A., Marcy, G. W., *et al.* 2010, *ApJ*, 721, 1467–1481

Vogt, S. S., Butler, R. P., Burt, J., *et al.* 2017, *AJ*, 154, 181

Complex Planetary Systems II: Latest Methods for an Interdisciplinary Approach
Proceedings IAU Symposium No. 382, 2024
A. Lemaitre & A.-S. Libert, eds.
doi:10.1017/S1743921323004386

Stability maps for the 5/3 mean motion resonance between Ariel and Umbriel with inclination

Sérgio R. A. Gomes[1] and **Alexandre C. M. Correia**[1,2]

[1]CFisUC, Departamento de Física, Universidade de Coimbra, 3004-516 Coimbra, Portugal
email: sergio.ra.gomes@outlook.com

[2]IMCCE, Observatoire de Paris, PSL Université, 77 Av. Denfert-Rochereau, 75014 Paris, France

Abstract. The evolution of the five largest satellites of Uranus during the crossing of the 5/3 mean motion resonance between Ariel and Umbriel is strongly affected by chaotic motion. Studies with numerical integrations of the equations of motion and analysis of Poincaré surface sections provided helpful insights to the role of chaos on the system. However, they lack of a quantification of this chaos in the phase-space. Here, we construct stability maps using the frequency analysis method. We determine that for low energies (small eccentricity and/or inclinations), the phase-space is mainly stable. As the energy increases, the chaotic regions replace the stable motion, until only small, localized libration regions remain stable.

Keywords. Celestial mechanics, dynamical evolution and stability, tides

1. Introduction

The dynamical evolution that led to the current configuration of the regular satellites of Uranus remains an enigma (e.g. Pollack et al. 1991; Szulágyi et al. 2018; Ishizawa et al. 2019; Ida et al. 2020; Rufu and Canup 2022). The current large eccentricities ($\sim 10^{-3}$) cannot be explained by mutual interactions between the satellites (Squyres et al. 1985; Smith et al. 1986; Peale 1988). Additionally, the surfaces of Miranda, Ariel, and Titania display evidences of large scale surface melting (Dermott et al. 1988; Peale 1988; Tittemore 1990). Tidal interactions raised on the planet induce a differential outward motion of the satellites (eg. Peale 1988; Tittemore and Wisdom 1988, 1989; Tittemoreand and Wisdom 1990; Pollack et al. 1991; Ćuk et al. 2020). This migration likely resulted in multiple encounters with mean motion resonances (MMRs) over their evolution. Although currently we do not observe any MMR in the system, tidal models suggests that the 5/3 MMR between Ariel and Umbriel was the latest to be crossed (eg. Peale 1988; Tittemore and Wisdom 1988; Ćuk et al. 2020; Gomes and Correia 2023), possibly exciting the eccentricities and inclination of the moons.

Tittemore and Wisdom (1988) carried a detailed study on the passage through the 5/3 Ariel-Umbriel MMR in the planar approximation and for small eccentricities. Their results shown that chaotic motion has a significant impact on the dynamical evolution during the resonance crossing. In fact, this chaos can drive the eccentricities of Ariel and Umbriel to much higher values than the initial ones and provides a mechanism to break the resonance.

Using a N-body integrator, Ćuk et al. (2020) studied the passage through the 5/3 Ariel-Umbriel MMR. The authors account for the five regular Uranian moons and assessed the role of the eccentricity, the inclination, and the spin for the outcome of the resonance crossing. Their results show that, during resonant passage, the eccentricities and the inclinations of the five Uranian satellites are excited by chaotic motion, even if they are not in resonance.

Gomes and Correia (2023) also revisited the intricate 5/3 MMR, in order to understand the role of inclination in the outcome of the passage. They performed a similar analysis as the one by Tittemore and Wisdom (1988, 1989), with a two-body secular model in the circular approximation with low inclinations, and a more robust tidal model. Resorting to the Poincaré surface section method, their analysis confirms the results obtained by Tittemore and Wisdom (1988) and Ćuk et al. (2020), as well as the theoretical predictions from Dermott et al. (1988), that chaotic motion rules the dynamics of the MMR between Ariel and Umbriel. In this work, we re-evaluate the stability of the 5/3 MMR, but now applying the frequency analysis method (Laskar 1990, 1993) to access the stability of the system.

2. Model

To easily analyse the stability of the 5/3 MMR, we need to resort to a simplified model, with a reduced number of degrees of freedom. For that, we adopt the secular resonant two-satellite circular model with low inclinations developed in Gomes and Correia (2023).

The model considers an oblate central body of mass m_0 (Uranus) surrounded by two point-mass bodies m_1, $m_2 \ll m_0$ (satellites), where the subscript 1 refers to the inner orbit (Ariel) and the subscript 2 refers to the outer orbit (Umbriel) and departs from a Hamiltonian truncated to the first order in the mass ratios, m_k/m_0, zeroth order in the eccentricities, and second order in the inclinations, I_k (with respect to the equatorial plane of the central body). After the high frequency angles are averaged, the Hamiltonian reads as

$$\begin{aligned}\bar{H} = & (K_a + S_a)\,(y_1\bar{y}_1 + y_2\bar{y}_2) + K_b(y_1\bar{y}_1 + y_2\bar{y}_2)^2 \\ & + (O_a + S_b)\,y_1\bar{y}_1 + (O_b + S_c)\,y_2\bar{y}_2 + \frac{S_d}{2}(y_1\bar{y}_2 + \bar{y}_1 y_2) \\ & + \frac{R_a}{2}(y_1^2 + \bar{y}_1^2) + \frac{R_b}{2}(y_2^2 + \bar{y}_2^2) + \frac{R_c}{2}(y_1 y_2 + \bar{y}_1\bar{y}_2)\ ,\end{aligned} \tag{1}$$

where K stands for the Keplerian coefficients, O for the oblateness coefficients, S for secular the coefficients, and R for the resonant coefficients (see appendix A in Gomes and Correia 2023). The Hamiltonian is written as a function of a set of canonical complex rectangular coordinates $(y_1, \mathrm{i}\,\bar{y}_1, y_2, \mathrm{i}\,\bar{y}_2)$, given by

$$y_k = \sqrt{\beta_k\sqrt{\mu_k a_k}\,(1-\cos I_k)}\,\mathrm{e}^{\mathrm{i}\varphi_k} \approx I_k\sqrt{\frac{\Gamma_k}{2}}\mathrm{e}^{\mathrm{i}\varphi_k}\ , \tag{2}$$

where $\bar{y}_k$ is the complex conjugate of y_k, a_k is the semi-major axis, $\beta_k = m_0 m_k/(m_0 + m_k)$, $\mu_k = G(m_0 + m_k)$, G is the gravitational constant, and $\varphi_k = \frac{5}{2}\lambda_2 - \frac{3}{2}\lambda_1 - \Omega_k$ are the resonant angles, where λ_k is the mean longitude and Ω_k is the longitude of the ascending node.

$$\Gamma_1 = \frac{3}{2}\Gamma\,(1-\Delta) \quad \text{and} \quad \Gamma_2 = -\frac{3}{2}\Gamma\left(1 - \frac{5}{3}\Delta\right) \tag{3}$$

are constant parameters, where $\Delta = (\beta_1\sqrt{\mu_1 a_1}\cos I_1 + \beta_2\sqrt{\mu_2 a_2}\cos I_2)/\Gamma$ is the normalised total orbital angular momentum, with $\Gamma = 2.6684\times 10^{-12}$ $\mathrm{M}_\odot\,\mathrm{au}^2\,\mathrm{yr}^{-1}$.

Table 1. Present physical and orbital properties of the Uranian system (Thomas 1988; Jacobson 2014).

	m ($M_\odot \times 10^{-10}$)	R (km)	$\langle T_{\rm rot}\rangle$ (day)	$J_2(\times 10^{-3})$	$C/(m_0 R_0^2)$
Uranus	436562.8821	25559.	0.7183	3.5107	0.2296
Satellite	m ($M_\odot \times 10^{-10}$)	R (km)	$\langle T_{\rm orb}\rangle$ (day)	$\langle a\rangle$ (R_0)	$\langle I\rangle$ (°)
Ariel	6.291561	578.9	2.479971	7.468180	0.0167
Umbriel	6.412118	584.7	4.133904	10.403550	0.0796

The conservative equations of motion are simply obtained from the Hamilton equations, yielding

$$\dot{y}_1 = \mathrm{i}\left[(K_a + S_a)\, y_1 + 2K_b\, (y_1\overline{y}_1 + y_2\overline{y}_2)\, y_1 + (O_a + S_b)\, y_1 + \frac{S_d}{2} y_2 + R_a \overline{y}_1 + \frac{R_c}{2}\overline{y}_2 \right] \tag{4}$$

and

$$\dot{y}_2 = \mathrm{i}\left[(K_a + S_a)\, y_2 + 2K_b\, (y_1\overline{y}_1 + y_2\overline{y}_2)\, y_2 + (O_b + S_c)\, y_2 + \frac{S_d}{2} y_1 + R_b \overline{y}_2 + \frac{R_c}{2}\overline{y}_1 \right] . \tag{5}$$

The dynamics of the 5/3 MMR essentially depends on Δ. We introduce the quantity

$$\delta = \frac{\Delta}{\Delta_r} - 1 \quad \text{with} \quad \Delta_r = 0.7681 \ , \tag{6}$$

which measures the proximity to the nominal resonance.

3. Stability maps

Gomes and Correia (2023) analysed the behaviour of Ariel and Umbriel at different stages of the crossing of the 5/3 MMR resorting to Poincaré surface sections. Here, we revisit the global dynamics of this resonance using stability maps. To this end, we adopt the frequency analysis method (Laskar 1990, 1993) to map the diffusion of the orbits.

For coherence, we fix $\delta = -2 \times 10^{-6}$ (Eq. (6)) as in Gomes and Correia (2023), for which the most diverse dynamics can be found, and adopt the physical properties of the system from Table 1. Since the Hamiltonian (Eq. (1)) is a four-degree function of y_k, the intersection of the constant energy manifold by a plane may have up to four roots (families). Each family corresponds to a different dynamical behaviour, and so we must plot one of them at a time. However, the families are symmetric, and actually we only need to show two of them. We chose to represent the families with the positive roots (that we dub 1 and 2). Family 1 corresponds to intersections with the plane containing the root with the highest absolute value, while family 2 corresponds to intersections with the plane containing the root with the lowest absolute value.

For each energy value, we build a grid of 200×200 equally distributed initial conditions in the plane $(y_{1,i}, y_{1,r})$, where $y_{k,i}$ and $y_{k,r}$ correspond to the imaginary and real parts of y_k, respectively. We fix $y_{2,r} = 0$ for all initial conditions and compute $y_{2,i}$ for each family from the total energy (Eq. (1)). We then numerically integrate the equations of motion (4) and (5) for a time T. Finally, we perform a frequency analysis of y_1, using the software TRIP (Gastineau and Laskar 2011) over the time intervals $[0, T/2]$ and $[T/2, T]$, and determine the main frequency in each interval, $f_{\rm in}$ and $f_{\rm out}$, respectively. The stability of the orbit is measured by the index

$$D \equiv \left| 1 - \frac{f_{\rm out}}{f_{\rm in}} \right| \ , \tag{7}$$

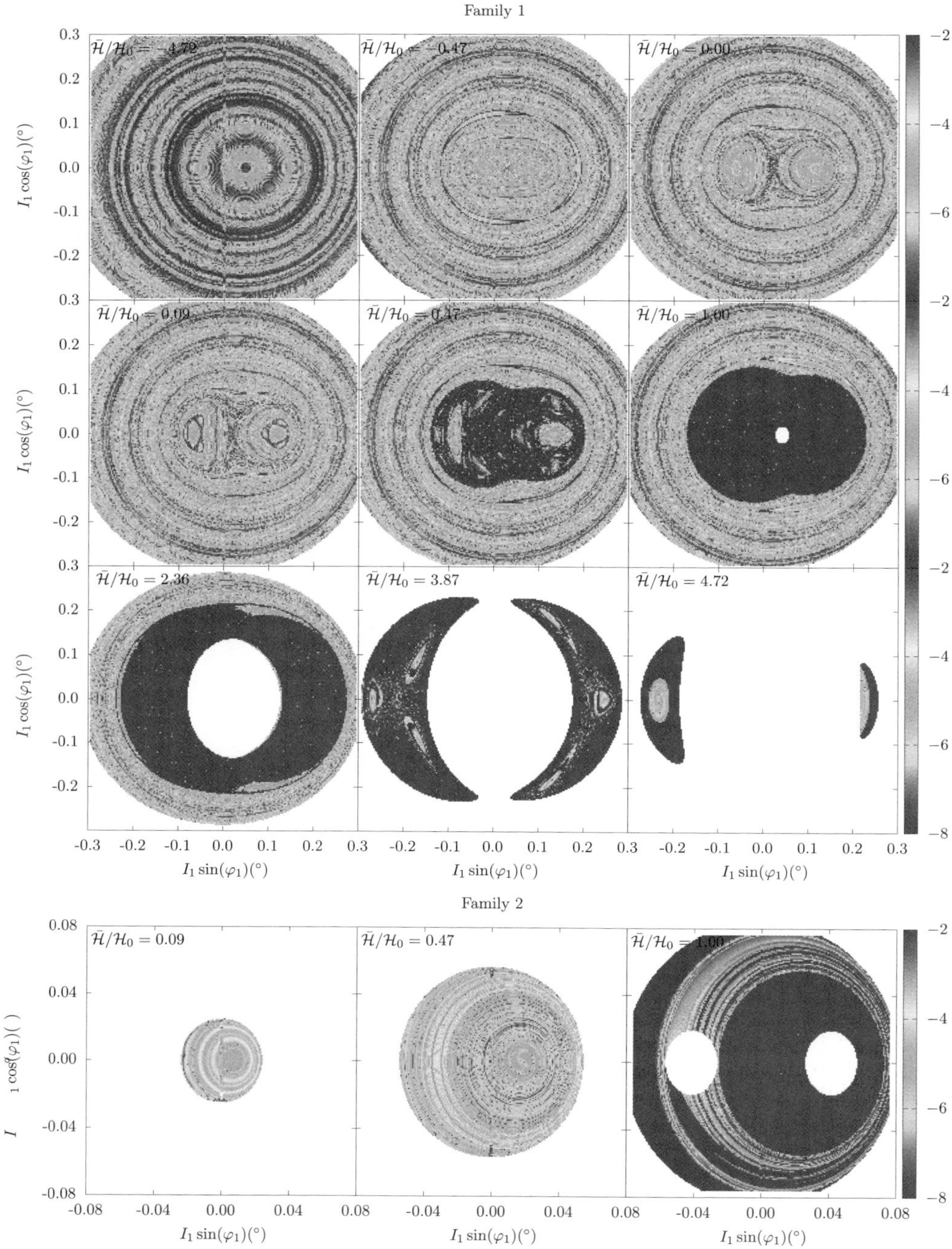

Figure 1. Stability maps for Ariel in the plane $(I_1 \sin \varphi_1, I_1 \cos \varphi_1)$ with $\cos \varphi_2 = 0$ and $\delta = -2 \times 10^{-6}$. The colour scale corresponds to the relative frequency diffusion index in logarithmic scale (Eq. (7)). Each panel was obtained with a different energy value and $H_0 = 1.06 \times 10^{-19}$ $M_\odot$ au^2 yr^{-2}.

which estimates the stability of the orbital long-distance diffusion (Dumas and Laskar 1993). The larger D, the more orbital diffusion exists. For stable motion, we have $D \sim 0$, while $D \ll 1$ if the motion is weakly perturbed, and $D \sim 1$ when the motion is irregular. It is difficult to determine the precise value of D for which the motion is stable or unstable, but a threshold of stability D_s can be estimated such that most of the trajectories with $D < D_s$ are stable (for more details see Couetdic et al. 2010).

The diffusion index depends on the considered time interval. Here, we integrate the equations of motion for $T = 10^4$ yr, because this interval is able to capture the main characteristics of the dynamics regarding the resonant frequency, which lies within the range ~ 60 yr. With this time interval, we estimate that $D_s \sim 10^{-4}$. The diffusion index D is represented by a logarithmic colour scale calibrated such that blue and green correspond to stable trajectories ($D \ll D_s$), while orange and red correspond to chaotic motion ($D \gg D_s$).

In Fig. 6, we show the stability maps for Ariel. We rescale y_k by $\sqrt{\Gamma_k/2}$, and so we actually plot the maps in the plane $(I_1 \sin \varphi_1, I_1 \cos \varphi_1)$ with $\cos \varphi_2 = 0$ (Eq. (2)). Each panel corresponds to a different energy value $\bar{H}/H_0$, where $H_0 = 1.06 \times 10^{-19}\ M_\odot\,\mathrm{au}^2\,\mathrm{yr}^{-2}$ is the energy of the transition between the circulation and libration regions, i.e., the energy of the separatrix. The lowest energies occur in the circulation regions, $\bar{H} < H_0$, while the largest energies occur in the libration region, $\bar{H} > H_0$. The inner circulation region is delimited by $0 < \bar{H} < H_0$, where $\bar{H} = 0$ corresponds to the energy of the equilibrium point with $y_1 = y_2 = 0$. For this energy range, there are four families, while for the remaining energies only two families exist.

For $\bar{H} \ll 0$ (Fig. 6 a), only family 1 exists, and we observe that the system is always stable, corresponding to trajectories in the outer circulation region. As the energy increases, two islands appear, corresponding to trajectories that are in the libration region (in resonance). Initially, the motion in these new regions is also stable, but the sepatrix and some localized concentric regions outside the separatrix are chaotic. As the energy approaches the threshold $\bar{H} = 0$ (Figs. 6 b,c), the chaotic regions expand for the vicinities of the separatrix. For $0 < \bar{H} < H_0$ (family 1), the chaotic regions increase even further, while the resonant islands shrink (Figs. 6 d,e), until they completely disappear for $\bar{H} = H_0$ (Fig. 6 f). Note that up to these energies, outside the chaotic regions, the circulation region remains stable. For this specific energy range, we also need to plot family 2. Close to $\bar{H} = 0$, we observe stable motion in the inner circulation region (Figs. 6 j,k). However, as we approach $\bar{H} = H_0$, this area is replaced by a chaotic region (Fig. 6 l). Finally, for $\bar{H} > H_0$, we observe that the stable region progressively vanishes, and chaotic motion dominates the phase-space, where only small libration regions remain stable (Figs. 6 g,h,i). In this energy range, we only have family 1 and trajectories in the outer circulation region also do not exist. Moreover, there is also a forbidden region at the centre of each panel that grows with the energy value, while the libration areas shrink. The results in Fig. 6 are in perfect agreement with those shown in Fig. 3 from Gomes and Correia (2023) using Poincaré surface sections.

4. Conclusion

As observed by Gomes and Correia (2023) and several previous works (Dermott et al. 1988; Tittemore and Wisdom 1988; Ćuk et al. 2020), chaos has a strong presence on the passage through the 5/3 MMR between Ariel and Umbriel. However, the analysis of this resonance with stability maps allows to quantify the chaos for each region.

The dynamics of the 5/3 MMR between Ariel and Umbriel is very rich and depends on the energy of the system. In fact, the energy depends on the value of the inclinations (Eq. (1)), given by the variables y_1 and y_2 (Eq. (2)). Therefore, the value of the inclinations of Ariel and Umbriel when the system encounters the resonance can trigger completely different behaviours. For $\bar{H} < 0$, the phase-space is dominated by stable orbits. As we approach the sepatrix energy, $\bar{H} \sim H_0$, the chaotic motion engorges the low inclination regions, while the outer circulation regions remain stable. Finally, for $\bar{H} \gg H_0$, only small libration regions remain stable, surrounded by large chaotic areas. This is a new result, since surface sections appeared to have exclusive quasi-periodic motion for energies near the equilibrium resonance points.

The dynamical analysis with stability maps can be extended to several combinations of the variables. Indeed, we do not need to choose a specific projection plane, and so the choice of the phase-space plane is much less restricted than for the Poincaré surface sections (see also Alves-Carmo et al. 2023). Allied to the quantification of chaos, the stability maps method thus provides a more exhaustive analysis of the dynamics of the resonance passage.

Acknowledgements. This work was supported by grant SFRH/BD/143371/2019, and by projects CFisUC (UIDB/04564/2020 and UIDP/04564/2020), GRAVITY (PTDC/FIS-AST/7002/2020), PHOBOS (POCI-01-0145-FEDER-029932), and ENGAGE SKA (POCI-01-0145-FEDER-022217), funded by COMPETE 2020 and FCT, Portugal. We acknowledge the Laboratory for Advanced Computing at University of Coimbra (https://www.uc.pt/lca) for providing the resources to perform the stability maps with high resolution.

References

Alves-Carmo, A. J., Vaillant, T., & Correia, A. C. M. 2023, Dynamics of trans-Neptunian objects near the 3/1 mean-motion resonance with Neptune. *Astronomy and Astrophysics*, 677, A83.

Couetdic, J., Laskar, J., Correia, A. C. M., Mayor, M., & Udry, S. 2010, Dynamical stability analysis of the HD 202206 system and constraints to the planetary orbits. *Astronomy and Astrophysics*, 519, A10.

Dermott, S. F., Malhotra, R., & Murray, C. D. 1988, Dynamics of the Uranian and Saturnian satellite systems - A chaotic route to melting Miranda? *Icarus*, 76, 295–334.

Dumas, H. S. & Laskar, J. 1993, Global dynamics and long-time stability in Hamiltonian systems via numerical frequency analysis. *Physical Review Letters*, 70(20), 2975–2979.

Gastineau, M. & Laskar, J. 2011, Trip: A computer algebra system dedicated to celestial mechanics and perturbation series. *ACM Commun. Comput. Algebra*, 44(3/4), 194–197.

Gomes, S. R. A. & Correia, A. C. M. 2023, Effect of the inclination in the passage through the 5/3 mean motion resonance between Ariel and Umbriel. *Astronomy and Astrophysics*, 674, A111.

Ida, S., Ueta, S., Sasaki, T., & Ishizawa, Y. 2020, Uranian satellite formation by evolution of a water vapour disk generated by a giant impact. *Nature Astronomy*, 4, 880–885.

Ishizawa, Y., Sasaki, T., & Hosono, N. 2019, Can the Uranian Satellites Form from a Debris Disk Generated by a Giant Impact? *The Astrophysical Journal*, 885(2), 132.

Jacobson, R. A. 2014, The Orbits of the Uranian Satellites and Rings, the Gravity Field of the Uranian System, and the Orientation of the Pole of Uranus. *The Astronomical Journal*, 148(5), 76.

Laskar, J. 1990, The chaotic motion of the solar system - A numerical estimate of the size of the chaotic zones. *Icarus*, 88, 266–291.

Laskar, J. 1993, Frequency analysis for multi-dimensional systems. Global dynamics and diffusion. *Physica D Nonlinear Phenomena*, 67, 257–281.

Peale, S. J. 1988, Speculative histories of the Uranian satellite system. *Icarus*, 74(2), 153–171.

Pollack, J. B., Lunine, J. I., & Tittemore, W. C. Origin of the Uranian satellites. In Bergstralh, J. T., Miner, E. D., & Matthews, M. S., editors, *Uranus* 1991,, pp. 469–512.

Rufu, R. & Canup, R. M. 2022, Coaccretion + Giant-impact Origin of the Uranus System: Tilting Impact. *The Astrophysical Journal*, 928(2), 123.

Smith, B. A., Soderblom, L. A., Beebe, R., Bliss, D., Boyce, J. M., Brahic, A., Briggs, G. A., Brown, R. H., Collins, S. A., Cook, A. F., Croft, S. K., Cuzzi, J. N., Danielson, G. E., Davies,

M. E., Dowling, T. E., Godfrey, D., Hansen, C. J., Harris, C., Hunt, G. E., Ingersoll, A. P., Johnson, T. V., Krauss, R. J., Masursky, H., Morrison, D., Owen, T., Plescia, J. B., Pollack, J. B., Porco, C. C., Rages, K., Sagan, C., Shoemaker, E. M., Sromovsky, L. A., Stoker, C., Strom, R. G., Suomi, V. E., Synnott, S. P., Terrile, R. J., Thomas, P., Thompson, W. R., & Veverka, J. 1986, Voyager 2 in the Uranian System: Imaging Science Results. *Science*, 233(4759), 43–64.

Squyres, S. W., Reynolds, R. T., & Lissauer, J. J. 1985, The enigma of the Uranian satellites' orbital eccentricities. *Icarus*, 61(2), 218–223.

Szulágyi, J., Cilibrasi, M., & Mayer, L. 2018, In Situ Formation of Icy Moons of Uranus and Neptune. *The Astrophysical Journal Letters*, 868(1), L13.

Thomas, P. C. 1988, Radii, shapes, and topography of the satellites of Uranus from limb coordinates. *Icarus*, 73(3), 427–441.

Tittemore, W. C. 1990, Tidal heating of Ariel. *Icarus*, 87(1), 110–139.

Tittemore, W. C. & Wisdom, J. 1988, Tidal evolution of the Uranian satellites I. Passage of Ariel and Umbriel through the 5:3 mean-motion commensurability. *Icarus*, 74(2), 172–230.

Tittemore, W. C. & Wisdom, J. 1989, Tidal evolution of the Uranian satellites II. An explanation of the anomalously high orbital inclination of Miranda. *Icarus*, 78(1), 63–89.

Tittemore, W. C. & Wisdom, J. 1990, Tidal evolution of the Uranian satellites III. Evolution through the Miranda-Umbriel 3:1, Miranda-Ariel 5:3, and Ariel-Umbriel 2:1 mean-motion commensurabilities. *Icarus*, 85(2), 394–443.

Ćuk, M., Moutamid, M. E., & Tiscareno, M. S. 2020, Dynamical history of the uranian system. *The Planetary Science Journal*, 1(1), 22.

Complex Planetary Systems II: Latest Methods for an Interdisciplinary Approach
Proceedings IAU Symposium No. 382, 2024
A. Lemaitre & A.-S. Libert, eds.
doi:10.1017/S1743921323004179

Migration of giant planets in low viscosity discs and consequences on the Nice model

P. Griveaud, A. Crida, E. Lega, A. C. Petit and A. Morbidelli

Université Côte d'Azur, Observatoire de la Côte d'Azur, CNRS, Laboratoire Lagrange, France
email: philippine.griveaud@oca.eu

Abstract. Using 2D hydrodynamical simulations, we show that in a low viscosity protoplanetary disc, Jupiter and Saturn get locked in the 2:1 mean motion resonance and migrate slowly inwards, unlike cases at higher viscosities. We conclude that in such discs the scenario of the Grand-Tack is not possible. Additionally, we investigate how the migration of the four (potentially five) giant planets in low viscous discs may affect the initial conditions of another important model for the formation history of our Solar System: the Nice Model. Adding ice giants in our hydrodynamical simulations, we find different possible resonant chains induced by migration. We then let the disc evolve until the gas phase dissipates and study the dynamical stability of the system. We find it possible to recreate the Solar System from such resonant chains, however the likelihood of this outcome remains low.

Keywords. Planets-disc interactions, protoplanetary discs, planetary systems

1. Introduction

Planets form in protoplanetary discs and their interactions with the gas give rise to migration. In most cases, if multiple planets form in a disc, they undergo convergent migration and get locked in resonance with one another. In the case of our Solar System, several works have studied the migration of Jupiter and Saturn (e.g. Masset & Snellgrove 2001; Morbidelli & Crida 2007). The resonance chains planets form strongly depends on how the planets migrate and therefore on the disc's parameters.

At the time of these cited studies, discs were believed to evolve under the effect of a significant viscosity (attributed to turbulence), characterised by a dimensionless α parameter (Shakura & Sunyaev 1973) in the interval $[10^{-3}, 10^{-2}]$. However, recent observations of dust settling have allowed Pinte et al. (2016) and Villenave et al. (2022) to derive values of α of the order of 10^{-4} and 10^{-5} respectively. Such low turbulence and viscosity levels are also supported by recent theoretical developments (e.g. Turner et al. 2014, for a review). In our study, we therefore revisit the migration of multiple giant planets in low viscosity discs.

2. Protoplanetary disc phase: hydrodynamical study

We run 2D global hydrodynamical simulations with the grid-based code FARGOCA†. We set the Shakura & Sunyaev (1973) viscosity parameter to $\alpha = 10^{-4}$.

† A recently re-factorised version of the code that can be found at: https://gitlab.oca.eu/DISC/fargOCA

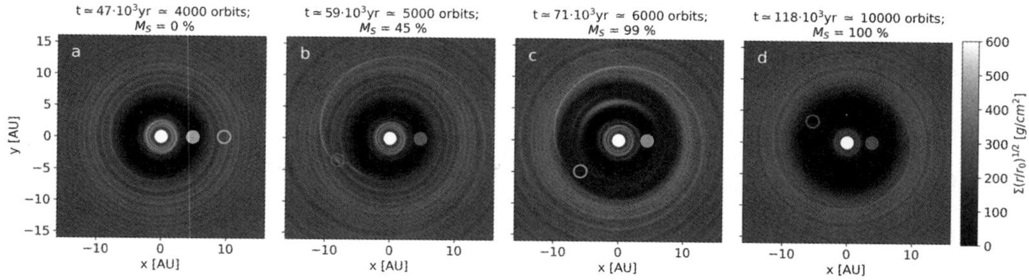

Figure 1. Surface density of the disc at different times of a simulation (as seen in Griveaud et al. 2023). The filled and empty circles mark the positions of Jupiter and Saturn, respectively. Panel a shows the disc at the moment of introduction of Saturn, while the outer planet still has zero mass. In panels b and c, the planet is at 45% and 99% of its total mass, respectively. Panel d shows the disc and the planets closer to the end of the simulation, when both planets are in a large common gap and locked in the 2:1 MMR.

2.1. *Migration of Jupiter and Saturn*

The simulation starts with a planet on a fixed circular orbit whose mass is increased analytically from zero up to Jupiter's mass in 800 local orbits. We wait an additional 400 orbits to let the disc stabilise before allowing Jupiter to migrate. This process avoids triggering instabilities that could arise from introducing a Jupiter mass planet in the disc directly (Hammer et al. 2017; Hallam & Paardekooper 2020). Once Jupiter is migrating smoothly, we add Saturn into the system. The second planet's mass grows similarly to the first one, but it is allowed to migrate as it grows. Figure 1 shows the different stages of a simulation. Panel b shows that Saturn's migration is dominated by its gravitational wake, so the planet is in type I regime at this stage. Panel c on the other hand shows that Saturn starts opening its gap, therefore transitioning from type I to type II migration.

While migrating, Saturn gets locked into the 2:1 mean motion resonance (MMR) with Jupiter. This increases Jupiter's eccentricity to ~ 0.1. Figure 2 shows up to $150 \cdot 10^3$ years the evolution of the semi-major axis of both planets, where the shaded areas represent the radial extent between peri- and apo-center. Both planets continue to migrate slowly inwards (see section 2.3). This result is in contrast with the one of classically viscous discs ($\alpha \geq 10^{-3}$), in which the planets are found most often in the 3:2 MMR.

We extended our study to a wide range of disc parameters to assess the robustness of our result and consistently found capture in the 2:1 MMR. In the case where the aspect ratio is smaller than in the nominal simulation, we find that the planets get locked in the 5:2 MMR. This is explained by the presence of an over density of gas at the edge of Jupiter's gap acting as a planet trap for the growing Saturn. This over density coincides with the location of the 5:2 MMR with Jupiter, allowing Saturn to enter very slowly into this resonance. Nevertheless, we find that in all our parameter exploration, Saturn never crosses the 2:1 MMR with Jupiter, unlike in the classically viscous case.

We explain those results by comparing the migration speed in a simulation of a planet growing and migrating in a low-viscosity disc, with an analytical critical migration rate for resonance crossing (adapted from Batygin 2015). The maximum migration rate reached by the planet is way below that of the critical rate to cross the 2:1 MMR. Therefore we conclude that it is not possible for any outer planet to cross the 2:1 MMR with a Jupiter mass planet in a low viscosity disc. The reader is referred to Griveaud et al. (2023) for more details.

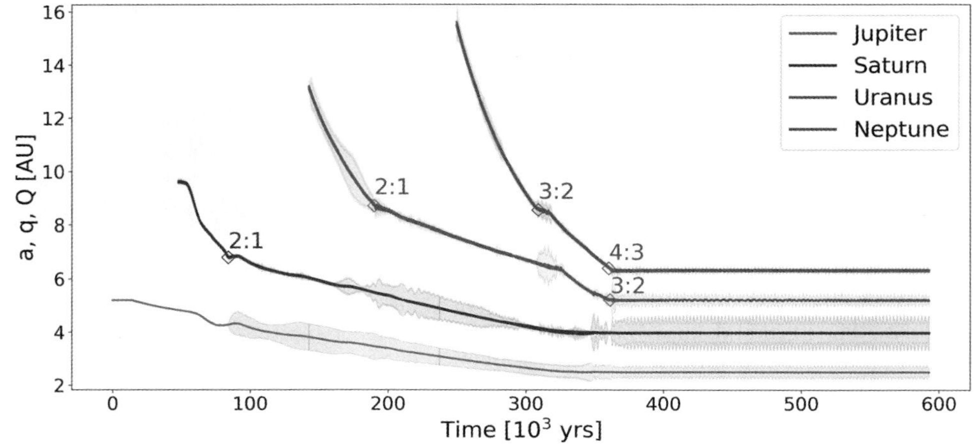

Figure 2. Orbital parameters' evolution of Jupiter, Saturn, Uranus and Neptune in the nominal simulation. The shaded areas mark the positions of the peri- and apo-centre, $q = a(1-e)$ and $Q = a(1+e)$, of the planets, respectively. The beginning of the plot shows Jupiter growing on a fixed orbit, then migrating alone in the disc. Saturn is introduced in the system at about 48 000 years, and gets caught in the 2:1 MMR with Jupiter at about 83 000 years. This phase corresponds to section 2.1. Uranus and Neptune are added consecutively in the simulation and the final resonance chain is (2:1, 3:2, 4:3). Here Neptune's introduction time was $T_{N,0} = 250\,000$ years (see Fig. 3)

2.2. *Building resonance chains*

As a follow-up to Griveaud et al. (2023), aiming at reproducing the Solar System, we now add a third planet with a Uranus mass, in the nominal simulation shown above. In Fig. 2, we show the evolution of the planetary system. Uranus gets initially locked the 2:1 MMR with Saturn. Once the system with three planets is stabilised, we add a fourth planet with the mass of Neptune. When Neptune approaches the 3:2 MMR with Uranus it excites its eccentricity which eventually leads to Uranus being ejected out of both resonances. The two ice giants continue to migrate inwards until Uranus gets locked in the 3:2 MMR with Saturn. At the same time, Neptune locks itself in the 4:3 MMR with Uranus, coincidentally also the 2:1 MMR with Saturn. The system remains stable in a compact chain.

Exploring the stability of such systems, we tried introducing Neptune at a different time. This is shown in Fig. 3. Like previously, Neptune kicks Uranus out of the 2:1 MMR with Saturn however, now Uranus is ejected out of the system after a succession of close encounters with the gas giants (see Fig. 3 at $355 \cdot 10^3$ years). Despite this short instability moment, Neptune is captured in the 2:1 MMR with Saturn and the system of now three planets remains stable. We then add another Uranus mass planet in the disc. This new ice giant migrates inwards, gets locked in the 4:3 MMR with the previous one, and the system remains stable.

These simulations show the sensitivity of the system to the time of introduction of the outer planets. Indeed, in low viscosity discs, gaps carved by giant planets are much deeper and wider than in classically viscous discs. Consequently there is less dissipation around the planets and therefore the resonance chains are weaker. We explored which other resonant chains the planets form by modifying further the time of introduction of the outer planets as well as by considering a colder disc (i.e. smaller aspect ratio). We also added a fifth planet in some systems, motivated by previous studies of the Nice Model. Table 1 summarises the stable systems obtained from our simulations. We note

Table 1. Summary of the resonance chains obtained from hydrodynamical simulations, based on the disc property, the number and order of planets. The last three columns show results of N-body simulations presented in section 3 for the cold disc cases and for planetesimal disc mass M_D. See text for the criteria. The nominal disc's N-body simulations are a work in progress.

Disc	Planet order	Resonant chain	M_D	Crit. A	Crit. B
Nominal	J-S-N-U	(2:1, 2:1, 4:3)			
Nominal	J-S-U-N	(2:1, 3:2, 4:3)			
Nominal	J-S-N-U-U	(2:1, 2:1, 4:3, 5:4)			
Cold	J-S-U-N-P9	(2:1, 3:2, 5:4, 4:3)	$30M_\oplus$	5.6 %	4 %
Cold	J-S-U-N	(2:1, 3:2, 5:4)	$30M_\oplus$	8 %	4 %
			$50M_\oplus$	9.5 %	5.5 %

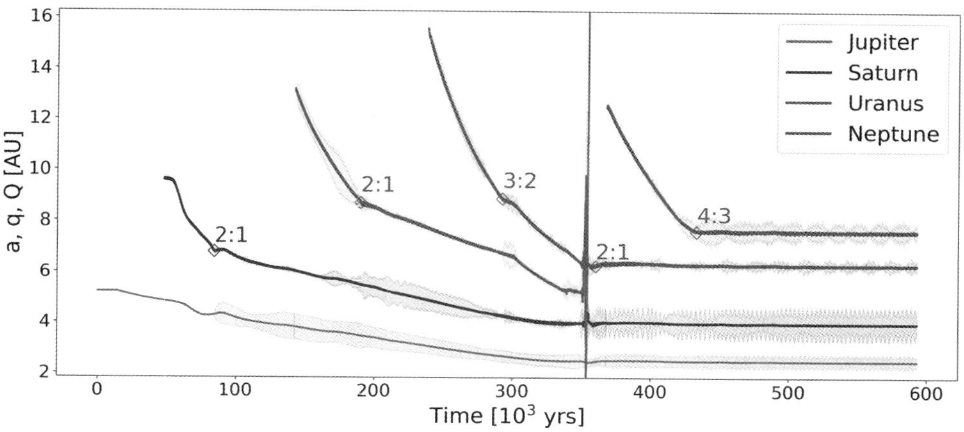

Figure 3. Same as Fig. 2 but with Neptune's introduction time $T_{N,0} = 238\,000$ years. The final resonance chain is (2:1, 2:1, 4:3).

that in the cold disc case, when adding a fifth ice giant, the system is unstable. Instead we were able to add a planet of mass $M_p = 6M_\oplus$, corresponding to the estimated mass of Planet Nine (Brown & Batygin 2021).

2.3. *Migration rate of giant planets in resonance*

In viscous discs, where Jupiter and Saturn are most often in 3:2 MMR, they can reverse their migration and move outwards in the disc (Masset & Snellgrove 2001). This permits the Grand-Tack scenario (Walsh et al. 2011) in which Jupiter first migrates in, then outwards when it reaches 1.5au once Saturn catches up with it. In contrast, here we find that Jupiter and Saturn, in the 2:1 (or 5:2) MMR keep migrating inwards, although more slowly than a single planet. Thus, in Griveaud et al. (2023), we concluded that the Grand Tack scenario is not possible in low viscosity discs.

Interestingly however, we notice that in both cases shown in the previous sub-section, adding two ice giants in the system halts remarkably the inward migration of Jupiter and Saturn. This may help keeping the Solar System giant planets far from the Sun. Their subsequent evolution to their present orbit, after the gas is gone, is the topic of next section.

3. Giant planets instability: N-body study

Once we have obtained some stable resonant chains between 4 or 5 planets, we are interested in studying whether these resonant configurations can recreate the Solar System

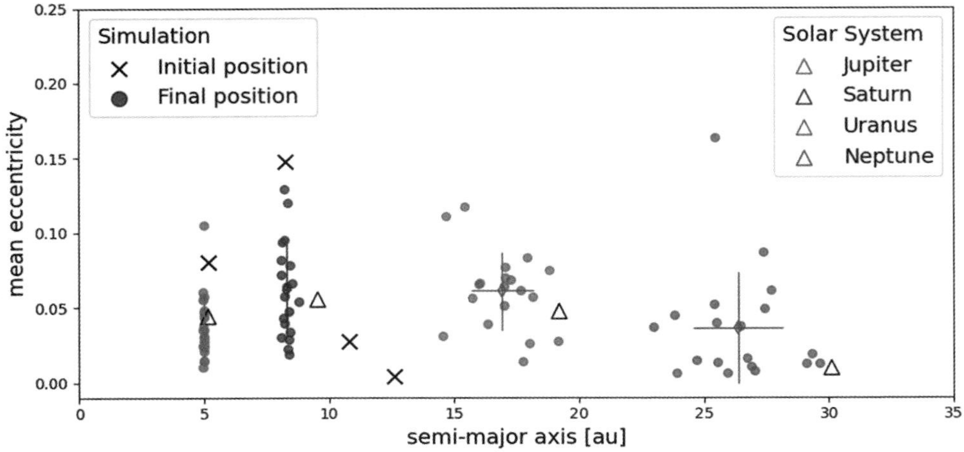

Figure 4. Final orbits of the four giant planets after their interaction with an outer disc of planetesimals (Nice model). The initial positions of the planets are given by the black crosses, corresponding to the (2:1, 3:2, 5:4) resonant chain. Only the cases with four planets at the end of the simulation are shown here ($\sim 10\%$ cases). The bars show the mean and standard deviation of the distribution of orbital parameters for each planet. For reference, the currents position of the Solar System planets are given by the triangles.

after some interactions with a planetesimal disc: the Nice model (Tsiganis et al. 2005). We gradually remove the gas from our hydrodynamical simulations. Using the orbital parameters of the planets outputted by the code FARGOCA, we initialise the same system in the N-body library `rebound` (Rein & Liu 2012). Using the hybrid integrator `Mercurius` (Rein et al. 2019) we integrate the system for a few hundreds of millions of years. In Fig. 4, we show the results of such simulations in the case where we add a planetesimal disc of mass $50M_\oplus$ positioned from 0.5au away from the outer most planet and up to 30au. Starting from the (2:1, 3:2, 5:4) resonant chain, about 10% of the simulations maintain four planets in an order comparable to the current Solar System† (criterion A of Nesvorný & Morbidelli 2012). In about half of them, the semi-major axes of the planets are within 20% of the solar system values and their eccentricities less than 0.11 (criterion B of Nesvorný & Morbidelli 2012, without inclinations).

The analysis of the other resonance chains is in progress, but our preliminary results shown in Table 1 are less favourable. An exploration of the parameters of the disc of planetesimals may be necessary and is the subject of on-going work (Griveaud et al. *in prep.*).

4. Conclusion and perspectives

In low viscosity discs, we find that Jupiter and Saturn are never closer than the 2:1 MMR and both planets migrate inwards. We firstly conclude that the Grand Tack scenario is not possible in such discs. When ice giants are added into these discs, systems are often unstable due to the lack of dissipation around the planets. Nevertheless, we find some resonant chains possible, all of which maintaining Jupiter and Saturn in the 2:1 MMR and with significant eccentricities (similar to Clement et al. 2021). Additionally, adding the ice giants stops the inward migration of the four giant planets, which can help keeping them from migrating to close to the Sun. From these resonant chains, we aim to reproduce the Solar System after a global instability phase as in the Nice Model. We find

† We consider the two ice giants to be interchangeable and therefore name them according to their final radial order.

that it is possible, but at present the likelihood remains lower than in previous Nice Model studies. The conditions (timing/order) of capture in MMR and the morphology of planetesimal disc are critical and can allow for a wide variety of scenarii. This makes it much harder to connect the present configuration with the original state of the system. Therefore, further exploration in the instability phase is the subject of on-going work.

References

Batygin, K. 2015, Monthly Notices of the Royal Astronomical Society, 451, 2589

Brown, M. E. & Batygin, K. 2021, The Astronomical Journal, 162, 219

Clement, M. S., Raymond, S. N., Kaib, N. A., et al. 2021, Icarus, 355, 114122

Griveaud, P., Crida, A., & Lega, E. 2023, Astronomy & Astrophysics, 672, A190

Hallam, P. D. & Paardekooper, S. J. 2020, Monthly Notices of the Royal Astronomical Society, 491, 5759

Hammer, M., Kratter, K. M., & Lin, M.-K. 2017, Monthly Notices of the Royal Astronomical Society, 466, 3533

Masset, F. & Snellgrove, M. 2001, Monthly Notices of the Royal Astronomical Society, 320, L55

Morbidelli, A. & Crida, A. 2007, Icarus, 191, 158

Nesvorný, D. & Morbidelli, A. 2012, The Astronomical Journal, 144, 117

Pinte, C., Dent, W. R. F., Ménard, F., et al. 2016, The Astrophysical Journal, 816, 25

Rein, H., Hernandez, D. M., Tamayo, D., et al. 2019, Monthly Notices of the Royal Astronomical Society, 485, 5490

Rein, H. & Liu, S. F. 2012, Astronomy and Astrophysics, 537, A128

Shakura, N. I. & Sunyaev, R. A. 1973, Astronomy and Astrophysics, 24, 337

Tsiganis, K., Gomes, R., Morbidelli, A., & Levison, H. F. 2005, Nature, 435, 459

Turner, N. J., Fromang, S., Gammie, C., et al. 2014, Transport and Accretion in Planet-Forming Disks (eprint: arXiv:1401.7306), 411

Villenave, M., Stapelfeldt, K. R., Duchêne, G., et al. 2022, The Astrophysical Journal, 930, 11

Walsh, K. J., Morbidelli, A., Raymond, S. N., O'Brien, D. P., & Mandell, A. M. 2011, Nature, 475, 206

Complex Planetary Systems II: Latest Methods for an Interdisciplinary Approach
Proceedings IAU Symposium No. 382, 2024
A. Lemaitre & A.-S. Libert, eds.
doi:10.1017/S1743921323004830

Secular Resonances and Terrestrial Planet Formation in Planetary Systems with Multiple Stars

Michael Andrews[1] **and Nader Haghighipour**[1,2,3]

[1]Institute for Astronomy, University of Hawaii-Manoa, Honolulu, Hawaii, USA

[2]Planetary Science Institute, Honolulu, Hawaii, USA

[3]Institute for Advanced Planetary Astrophysics, Honolulu, Hawaii, USA
email: nader@ifa.hawaii.edu

Abstract. We present the results of a study of secular resonances in a binary star system and their effects on the formation of terrestrial planets. The systems of our interest are binaries with moderate separations (i.e., smaller than 40 AU) where planets revolve around one of the stars. Using numerical simulations, we demonstrate the appearance and evolution of secular resonances in systems with two giant planets. Results indicate that the perturbation of the binary companion suppresses secular resonances and they do not play a significant role on the formation and orbital architecture of terrestrial planets. Unlike in our solar system where the secular resonance of Saturn confines the formation of terrestrial planets to regions interior to its location, in a binary star, terrestrial planets can form interior and exterior to this resonance. We present details of our simulations and discuss the implications of their results.

Keywords. Secular resonances, Binary systems, Terrestrial planet formation

1. Introduction

The mean-motion and secular resonances of giant planets, especially Jupiter and Saturn, have played crucial roles in the formation, dynamical evolution, and orbital architecture of the inner solar system. While the mean-motion resonances of Jupiter have created the Kirkwood gaps, the secular resonance of Saturn has confined the eccentricities and inclinations of asteroids, and has created an inner edge for the asteroid belt (Milani & Knežević 1992, 1994). Collectively, these resonances are the reason that no planet exists between Jupiter and Mars, and instead, the solar system has an asteroid belt.

In addition to the post-formation evolution of the inner solar systems, secular resonances affect the formation of the terrestrial planets as well (Levison & Agnor 2003; Haghighipour & Winter 2016). In a comprehensive study, Haghighipour & Winter (2016) examined the appearance and intensity of the secular resonances of Jupiter and Saturn for different distribution of protoplanetary bodies and different orbital eccentricities of Jupiter and Saturn. These authors demonstrated that while the intensities of the secular resonances stay unaffected by the surface density of the protoplanetary disk and the orbital eccentricities of Jupiter and Saturn, the mass, orbital assembly and water contents of the final terrestrial planets are strongly influenced by the two secular resonances.

The discovery of circumstellar giant planets in binary stars with separations smaller than 40 AU motivated us to extend the above studies to these systems as well. We, therefore, simulated terrestrial planet formation in binary systems with two circumstellar giant planets and used the general theory recently developed by Haghighipour (2023) to determine the locations of their secular resonances. Our goal was to determine how these secular resonances affect the formation of terrestrial planets around the planet-hosting star, and how the results would compare with those in single-star systems. We studied the dynamical evolution of these resonances and investigate their effects on the formation of the final system.

1.1. *The Model and Initial Set up*

We considered a four-body system consisting of the planet-hosting star (primary), two giant planets, and the secondary star. The semimajor axis of the binary was taken from a range of 20 AU to 40 AU, and its eccentricity was varied between 0 and 0.5. The planet-hosting star was considered to be solar-mass and the mass of the secondary star was taken to be between 0.4 and 1.3 times the mass of the Sun. The inner and outer planets were taken to be Jupiter- and Saturn-mass with semimajor axes of 1.6 AU and 2.94 AU, respectively. These semimajor axes were chosen to ensure that for all choices of the orbital parameters of the binary and the mass of the secondary star, the two giant planets would always stay within the stable zone around the primary. They also resemble the near 5:2 communicability between Jupiter and Saturn in our solar system. All other orbital parameters of the two planets were taken to be equal to those of Jupiter and Saturn. We also distributed 120 moon- to Mars-mass planetary embryos and 2000 planetesimals from 0.5 AU to 4.5 AU. The surface density of the disk was taken to follow an $r^{-1.5}$ radial profile. The choice of the radial extent of the disk was made to ensure the truncation effect of the secondary star.

1.2. *Numerical Simulations and Results*

We integrated our systems for 100 Myr using a version of the N-body integration package MERCURY that has been designed to integrate circumstellar bodies in binary star systems (Chambers et al. 2002). The appearance of both mean-motion and secular resonances can be seen in all simulations. Figure 1 shows two examples of the results. The binary separation in both systems is 40 AU. The secondary star in the left panel is a 0.4 solar-mass M star and in the right panel, it is a 1.3 solar-mass F star. The two top panels show the initial locations of the secular resonances of the inner and outer planets (g_1 and g_2, respectively) based on the general theory of Haghighipour (2023). The two bottom panels show the evolution of the protoplanetary disk for the first 0.5 Myr of integrations. They also show the appearance of the mean-motion resonances of the inner planet (dashed lines) and the secular resonance of the outer planet (g_2). As shown here, the secular resonance of the outer planet appears very close to its theoretical value. A comparison with the simulations of disk evolution in Haghighipour & Winter (2016) demonstrates that the secular resonance of the outer planet has been severely suppressed (see also figure 2). We note that in the above two cases, the location of the secular resonance of the inner planet falls interior to the inner edge of the disk. As a result, this resonance does not appear in the simulations. However, we do expect this resonance to have been suppressed too.

Figure 2 shows another important characteristic of secular resonances, namely, their inward migration. As demonstrated by Ward (1981), in a system with giant planets, when a small object such as a planetesimal or planetary embryo is embedded in a massive disk, the response of the disk to the perturbation of the giant planets acts as another

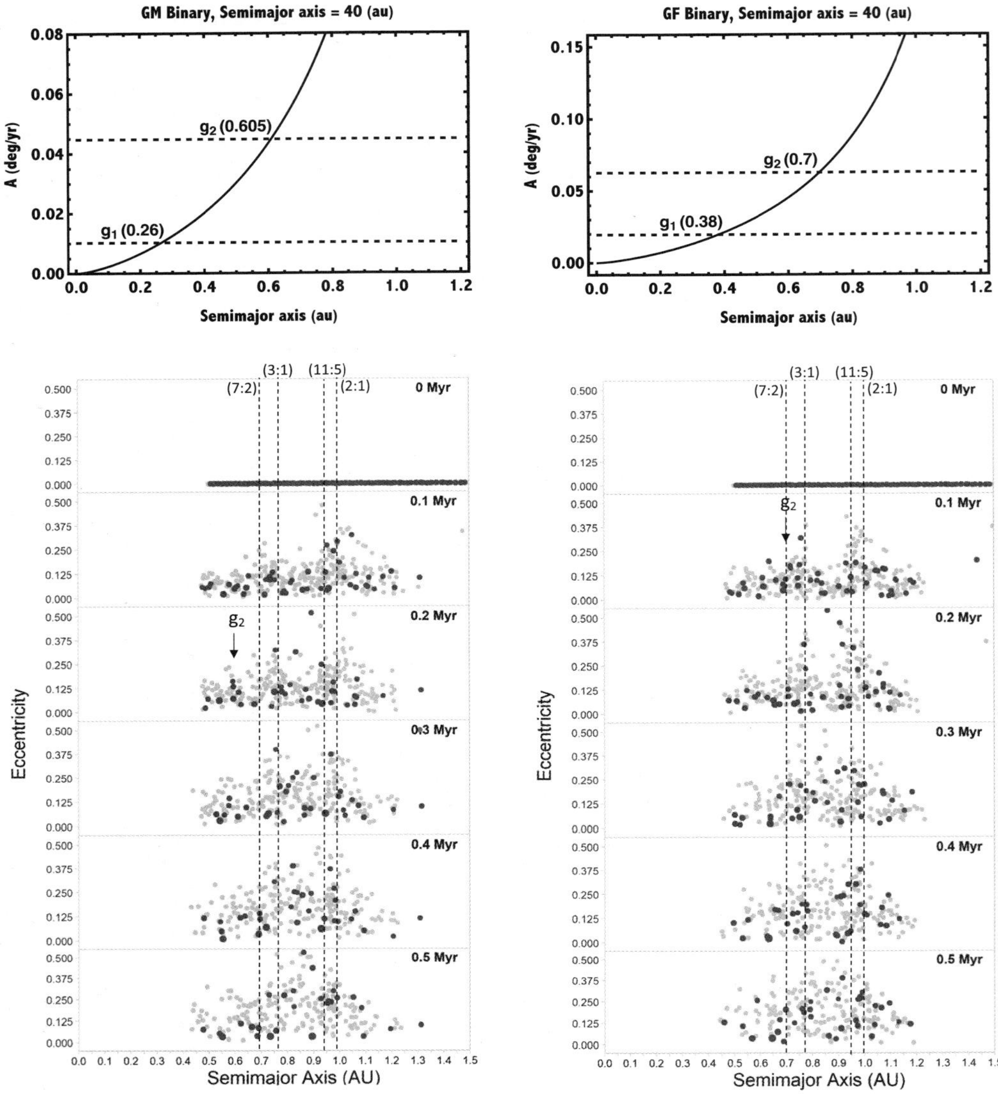

Figure 1. Location of the secular resonance of the inner planet (g_1) and the outer planet (g_2) in a GM (left) and a GF binary with a separation of 40 AU. The top panels show the theoretical locations of these resonances obtained from the general theory of planetary secular resonances in multiple star system (Haghighipour 2023). The solid line shows the variation of the rate of the precession of the longitude of perihelion (A) of a small body with semimajor axes. The dashed lines correspond the the same quantity for the inner and outer planets shown by indeces 1 and 2, respectively. The bottom panels show the appearance of secular resonances during the evolution of the protoplanetary disk interior to the inner giant planet. The planet-hosting star (primary) is at the origin. The secondary star and the two giant planets are not shown. The red circles represent planetary embryos and the gray dots are the background planetesimals. The dashed lines in the bottom panels correspond to the locations of mean-motion resonances with the inner planet.

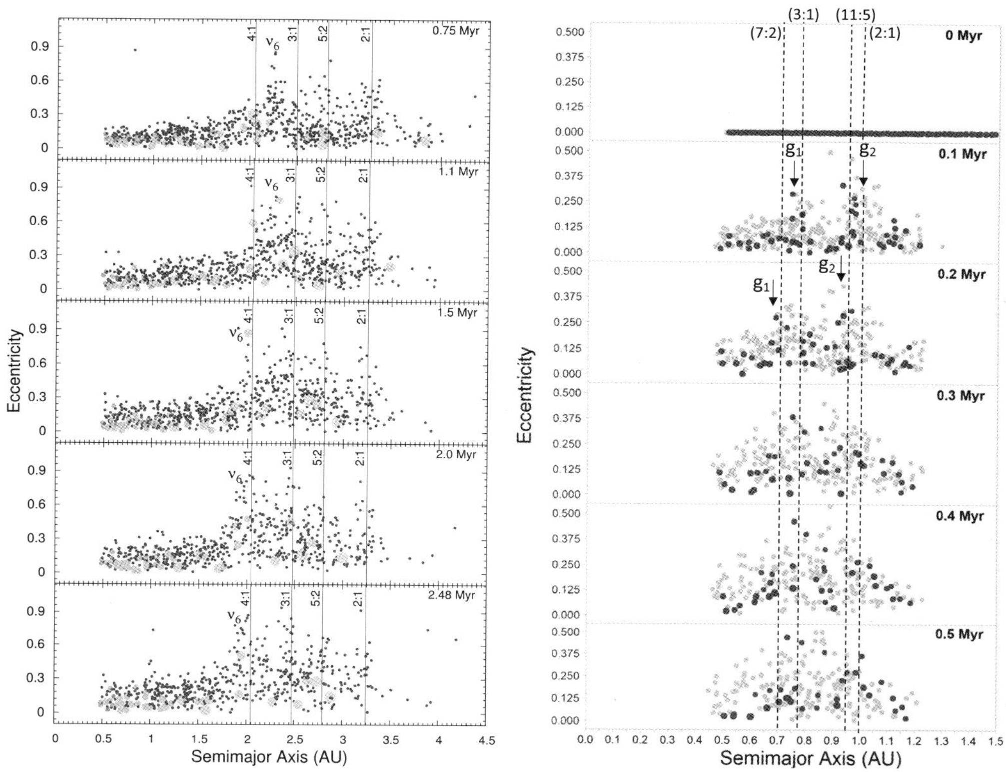

Figure 2. The inward displacement of secular resonances. The panel on the left (from figure 9 of Haghighipour & Winter 2016) shows the inward displacement of the secular resonance of Saturn (ν_6) in our solar system. The circles in cyan and light green represent planetary embryos and purple dots are the background planetesimals. The panel on the right shows the inward motion of the secular resonances of the inner planet (g_1) and outer planet (g_2) in a 20 AU GG binary in our simulations. Color coding in this panel is similar to that in figure 1.

perturbation that affects the precession of the orbit of the small body. This author showed that as the disk loses mass due to the effect of secular resonances, its response weakens which causes the locations of these resonances to vary in time. Levison & Agnor (2003) noticed that in their simulations of the formation of terrestrial planets in our solar system, when the disk is considerably massive, the locations of the secular resonances of Jupiter and Saturn move outward. Haghighipour & Winter (2016) showed that as the perturbation of these planets causes the disk to lose mass in its outer regions, the outward migration of secular resonances stop. These authors also showed that the growing embryos in the disk's inner regions create their own perturbing effects which causes smaller embryos in their vicinity to be captured in temporary secular resonances with these growing bodies. As the latter raises the eccentricity of smaller objects to higher values causing them to be scattered out of their orbits, the transfer of angular momentum to the bodies in the inner regions where the disk maintains more mass, causes the secular resonances to naturally move toward these regions. The left panel in figure 2 shows this for the secular resonance of Saturn in our solar system (shown by ν_6, Haghighipour & Winter 2016). As shown here, Saturn's secular resonance moves inward sweeping the region between 2.4 AU and 1.9 AU. The right panel of figure 2 shows the inward migration of the secular resonance of the outer planet for a 20 AU binary with two solar-mass stars. As shown here, within 100,000 years, g_2 moves slightly inward. Because in our

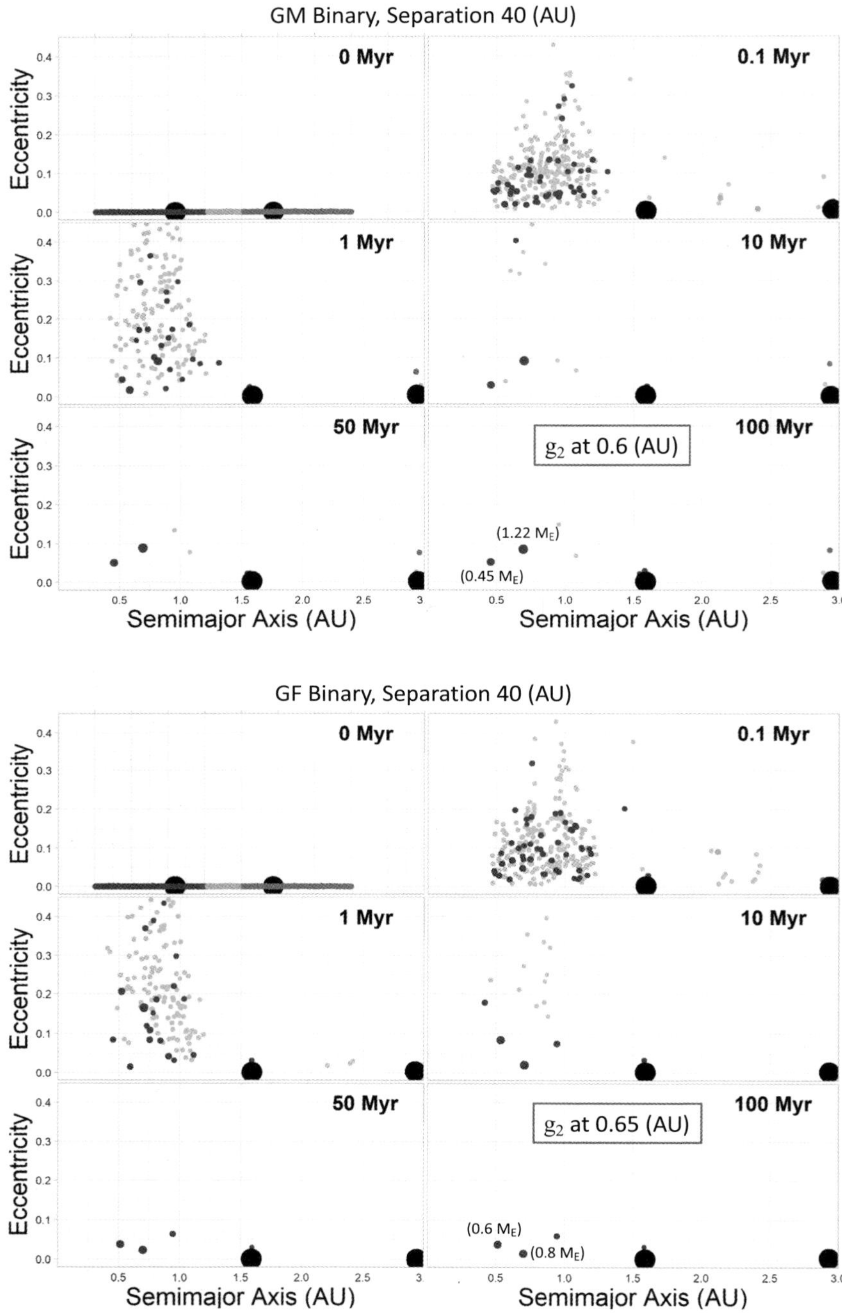

Figure 3. Snapshots of the formation of terrestrial planets interior to the inner planet in a GM (left) and a GF binary (right) with a separation of 40 AU. To show both planets at the start of the simulations, we have used smaller scales in the two $t = 0$ panels. In these two pnels, each scale on the semimajor axis corresponds to 0.25 AU. The planetary embryos are distributed between 0.5 AU and 4 AU with the color coding representing their water-mass ratio. Red corresponds to dry embryos, light green corresponds to embryos with 1% water content, and blue corresponds to embryos with 5% water-mass ratio. The last panel at 100 Myr shows the masses of the final terrestrial planets and the final location of the secular resonance of the outer giant body (g_2). As shown here, the final planets carry no water and unlike in our solar system, terrestrial planets can form interior and exterior to the location of g_2.

binary systems, secular resonances are weaker than those in our solar system, the inward migration of g_2 does not result in as sever of a mass-loss as that caused by the secular resonance of Saturn. In fact, in our binaries, most of the mass-loss is due to the mean-motion resonance of the inner planet. For that reason, the rate and amount of the inward migration of the secular resonance of the outer planet is substantially small.

The fact that the secular resonance of the outer planet is not strong enough to scatter many objects out of the disk has an immediate implication for the formation of small planets interior to the inner giant planets. Unlike in our solar system where the secular resonance of Saturn confines the formation of terrestrial planet to inside 2.1 AU, in our binary systems, small planets can form both inside and outside the final location of the outer planet's secular resonance. Figure 3 shows the latter for the two systems of figure 1. In the top panel, corresponding to a 40 AU GM binary, the last location of the outer planet's secular resonance is at 0.6 AU. As show here, at the end of the simulations, a planet with a mass of 1.22 Earth-masses is formed at 0.7 AU. The bottom panel shows the situation for a 40 AU GF binary. In this system, the final location of the outer planet's secular resonance is at 0.65 AU while two planets with masses of 0.6 and 0.8 Earth-masses have formed at 0.51 AU and 0.7 AU, respectively.

2. Summary and Concluding Remarks

We have studied the occurrence of secular resonances and their effects on the formation of terrestrial planets around one star of a moderately closed binary. We considered a four-body system consisting of the primary star, two giant planets and the secondary star, and simulated the formation of terrestrial planets interior to the inner planet. Results demonstrated the appearance of secular resonances in very close proximity to their theoretical locations, and also showed that

- Unlike in single star systems where the secular resonances of giant planets are strong and play an important role in shaping the dynamical architecture of system, in binary stars, these resonances are severely suppressed. The perturbation of the secondary star in these systems weakens the strength and effect of secular resonances. For instance, our simulations demonstrated that the secular resonance of the outer planet was hardly able to increase the eccentricities of planetesimals and planetary embryos to values larger than 0.5, whereas in our solar system, the secular resonance of Saturn has had significant effect on depleting the inner part of the asteroid belt;
- Similar to the resonances in the solar system, secular resonances take longer to appear than mean-motion resonances. As a result, in binary star systems, mean-motion resonances of giant planets play a more effective role in removing material from interior regions;
- The above two results imply that unlike in our solar system where the secular resonance of Saturn confined the formation of terrestrial planets to the region interior to 2.1 AU, the weak nature of these resonances in binary star systems does not have significant effects on the formation of planets interior to their locations allowing terrestrial-class planets to form inside and outside the location of secular resonances.

3. Acknowledgment

NH acknowledges support from NASA XRP through the grants 80NSSC21K1050 and 80NSSC23K0270, and NSF grant AST-2109285. The authors would like to thank the Information Technology section at the Institute for Astronomy, University of Hawaii-Manoa for maintaining computer clusters on which the simulations of this study were carried out.

References

Chambers, J. E., Quintana, E. V., Duncan, M. J. & Lissauer, J. J. 2002, AJ, 123, 2884
Haghighipour, N. & Winter, O. C. 2016, Cel. Mech. Dyn. Ast., 124, 235
Haghighipour, N. 2023, submitted
Levison, H. F. & Agnor, c. 2003, AJ, 125, 2692
Milani, A. & Knežević, Z. 1992, Icarus, 98, 211
Milani, A. & Knežević, Z. 1994, Icarus, 107, 219
Ward, W. R. 1981, Icarus, 47, 234

Complex Planetary Systems II: Latest Methods for an Interdisciplinary Approach
Proceedings IAU Symposium No. 382, 2024
A. Lemaitre & A.-S. Libert, eds.
doi:10.1017/S1743921323004957

Positive Weathering Feedback Compensates Carbonates at Shallow Ocean Depths

Hakim Kaustubh

KU Leuven, Institute of Astronomy, Celestijnenlaan 200D, 3001 Leuven, Belgium
Royal Observatory of Belgium, Ringlaan 3, 1180 Brussels, Belgium
email: kaustubh.hakim@kuleuven.be

Abstract. Continental silicate weathering and seafloor carbonate precipitation are key steps in the carbonate-silicate cycle to draw down CO_2. Contrary to the classic understanding of negative feedback, silicate weathering can exhibit positive feedback at high temperatures. Taking into account this positive feedback, the compensation depth (CCD) in exoplanet oceans becomes shallower, implying a potential instability in the carbonate-silicate cycle at high temperatures.

Keywords. Carbon Cycle, Carbonate Compensation Depth, Positive Weathering Feedback

1. Introduction

The carbonate-silicate cycle is a geochemical cycle that is thought to have enabled temperate climates during the history of Earth by recycling CO_2 through planetary carbon reservoirs (Walker et al. 1981; Berner et al. 1983; Krissansen-Totton et al. 2018). The regulation of atmospheric CO_2 ensures that the greenhouse warming effect of CO_2 is sufficient to maintain a clement surface temperature T and a liquid water reservoir on the surface, which are vital for long-term habitability Catling & Zahnle (2020). During the Archean, when the sun was only 70% as bright as today (Sagan & Mullen 1972), a warmer climate was ensured by a higher CO_2 partial pressure P_{CO_2} and a lower intensity of silicate weathering (Kasting et al. 1993). Walker et al. (1981) discovered that the weathering of continental silicate rocks provides negative feedback to create a thermostat-like effect. If outgassing increases, P_{CO_2} rises and consequently weathering intensifies to increase the CO_2 drawdown. Alternatively, if outgassing decreases, weathering diminishes until P_{CO_2} is high enough to attain a new balance between weathering and outgassing. Once the silicate weathering products are transported to oceans by rivers, the precipitation of carbonates occurs on the seafloor for eventual subduction of carbonates into the mantle (Sleep & Zahnle 2001). The carbonate compensation depth (CCD) is a critical ocean depth below which carbonates are unstable and dissolve into the ocean (Broecker & Peng 1987; Zeebe & Westbroek 2003). If the ocean depth exceeds the CCD, the carbonates cannot be removed from the ocean-atmosphere system and the carbonate-silicate cycle halts.

2. Positive weathering feedback

With the application of a multi-regime silicate weathering model ($W = f(P_{CO_2}, T)$) of (Maher & Chamberlain 2014), Hakim et al. (2021) find that weathering exhibits negative feedback at low temperatures ($W \propto T$), but positive feedback at high temperatures ($W \propto 1/T$, Fig. 1a). There is a transition from the classic kinetic limit (Walker et al. 1981) to the thermodynamic limit. The transition temperature is sensitive to rock composition

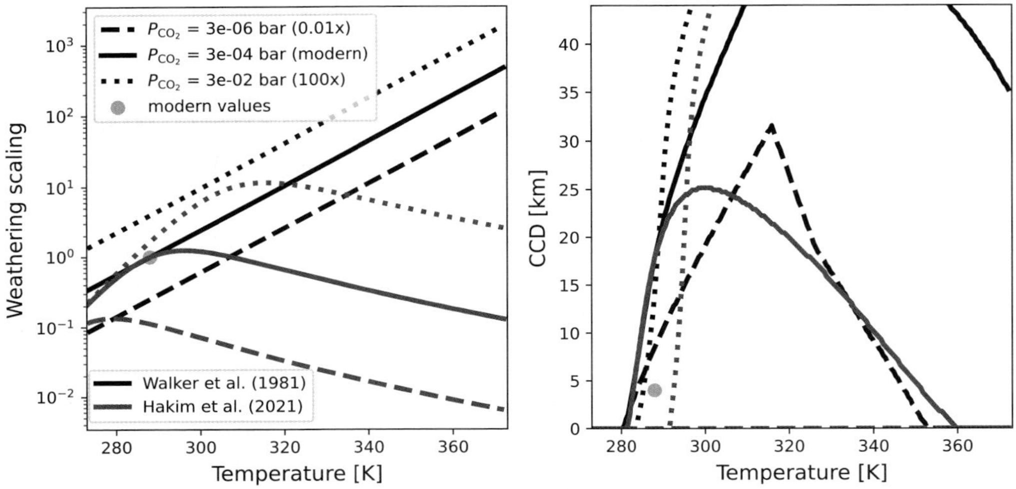

Figure 1. (a) Continental weathering scaling relations of Walker et al. (1981) and Hakim et al. (2021) for three P_{CO_2} values. (b) Carbonate compensation depth (CCD) in exoplanet oceans for the two weathering models is based on the OCRA ocean chemistry model (Hakim et al. 2023).

(e.g., kinetic and thermodynamic properties of granite, basalt or peridotite) and soil transport properties (e.g., flowpath length, soil age) (Hakim et al. 2021). The transition temperature also increases with P_{CO_2} (Fig. 1a calculations for 3 constant values of P_{CO_2}). It is important to note that P_{CO_2} should be coupled to T via a climate model (Krissansen-Totton et al. 2018), but it is kept constant here to isolate the effect of temperature.

3. Shallow Carbonate Compensation Depths

Fig. 1b shows that the CCD first increases and then decreases with T (ocean temperature = surface temperature). For the modern model (black solid line, Fig. 1b), the CCD sharply increases from about 0 km at 288 K to >45 km at 310 K. Beyond 360 K, the CCD decreases due to silicates becoming more stable than carbonates (Hakim et al. 2023). The CCD becomes deeper at 100× modern P_{CO_2} and shallower at 0.01× modern P_{CO_2}. When the multi-regime weathering model from Hakim et al. (2021) is used (blue lines, Fig. 1b), the CCD becomes shallower at high T than that in classic kinetic weathering cases. This is a result of the transition from the negative weathering feedback at low T to the positive weathering feedback at high T (Fig. 1a). This effect is strong enough to ensure that no carbonates are stable in oceans when P_{CO_2} is 0.01× of the modern value. Shallower CCDs at higher T imply a potential instability in the carbonate-silicate cycle. Future studies need to evaluate the effect of the coupled climate and multi-regime weathering models on the CCD.

References

Berner, R. A., Lasaga, A. C., & Garrels, R. M. 1983, American Journal of Science, 283, 641. doi:10.2475/ajs.283.7.641

Broecker, W. S. & Peng, T.-H. 1987, Global Biogeochemical Cycles, 1, 15. doi:10.1029/GB001i001p00015

Catling, D. C. & Zahnle, K. J. 2020, Science Advances, 6, eaax1420. doi:10.1126/sciadv.aax1420

Hakim, K., Bower, D. J., Tian, M., et al. 2021, Planetary Science Journal, 2, 49. doi:10.3847/PSJ/abe1b8

Hakim, K., Tian, M., Bower, D. J., et al. 2023, Astrophysical Journal Letters, 942, L20. doi:10.3847/2041-8213/aca90c

Kasting, J. F., Whitmire, D. P., & Reynolds, R. T. 1993, Icarus, 101, 108. doi:10.1006/icar.1993.1010

Krissansen-Totton, J., Arney, G. N., & Catling, D. C. 2018, Proceedings of the National Academy of Science, 115, 4105. doi:10.1073/pnas.1721296115

Maher, K. & Chamberlain, C. P. 2014, Science, 343, 1502. doi:10.1126/science.1250770

Sagan, C. & Mullen, G. 1972, Science, 177, 52. doi:10.1126/science.177.4043.52

Sleep, N. H. & Zahnle, K. 2001, Journal of Geophysical Research, 106, 1373. doi:10.1029/2000JE001247

Walker, J. C. G., Hays, P. B., & Kasting, J. F. 1981, Journal of Geophysical Research, 86, 9776. doi:10.1029/JC086iC10p09776

Zeebe, R. E. & Westbroek, P. 2003, Geochemistry, Geophysics, Geosystems, 4, 1104. doi:10.1029/2003GC000538

Complex Planetary Systems II: Latest Methods for an Interdisciplinary Approach
Proceedings IAU Symposium No. 382, 2024
A. Lemaitre & A.-S. Libert, eds.
doi:10.1017/S1743921323004842

Hazardous Asteroids and Space Situational Awareness – Do Look UP !

Daniel Hestroffer

Observatoire de Paris, France
email: daniel.hestroffer@obspm.fr

Abstract. We present in the following the introductory talk on "Hazardous asteroids and the Hera mission", made during the round table on Space Awareness. It reminds the context of our awareness for near-Earth objects, the characterisation of risk, current international surveillance programmes, and mitigation measures in particular with space missions, and last, a *rendez-vous* with Apophis to note in your agenda for 2029.

Keywords. Asteroids, Space Situational Awareness, DART, Apophis

1. Some examples of asteroid impacts

Near-Earth Objects are asteroids and comets orbiting in the vicinity of the Earth, defined more precisely by a perihelion distance of less than 1.3 astronomical unit. NEOs, as for asteroids, are of general scientific interest and typically present no threat to Earth. Nevertheless we have numerous records on astronomical/geological time scales of asteroids impacts, with craters present on the surface of the Earth and other planetary bodies in the Solar System. At the beginning of the 20th century, the Tunguska event didnt provoke a crater, but a vast zone of Siberian land was devastated by a supposed blast from the hypervelocity entry of a celestial object. The international asteroid day, now held every June 30th, is an anniversary of this event.

Closer in time, additional events are also hitting the head-lines of our present time news and media. One example is the impact of comet SL9 fragments on the gas-giant Jupiter in July 1994, about 30 years ago†. Infrasound and optical monitoring of bolides entry show that many meter-sized objects are entering our atmosphere, sometimes hitting the ground and ending as meteorites exposed in our museums of natural history. Or sometimes hitting a car trunk, like in 1992 Ms. Michelle Knapp's Chevy in the USA‡.

Last, I will finish this short and very limited introductory inventory by the event of February 2013 in Chelyabinsk, Russian Federation. A little less than 20 meter large asteroid unpredictably entered the atmosphere at a speed of approximately 19 km/s; Its catastrophic disruption at 30km altitude generated a bright light and a blast on ground early morning over the city of Chelyabinsk. Some constructions were damaged from the high pressure blast, many building windows were broken, and about twelve hundreds people were injured mostly from glass breaking in their face, burns, and falls, with nevertheless no direct fatalities. The overall cost of the damages was estimated to more than thirty million euros.

† Let us pretend that it is recent event, since I was writing my PhD manuscript at that time !
‡ Presently in France, in this beginning of July 2023, asteroid/meteorite falls are not the largest threat to cars

2. Is there a risk?

Among the Near-Earth Objects, a class named Potentially Hazardous Asteroids (PHAs) can get closer to Earth, with an inter-orbit distance (MOID) of less than 0.2 Lunar distance. Those NEO asteroids show many similarities with space debris situational awareness presented just before by Dr. Aaron Rosengren: In tracking, detecting, monitoring, orbit computation, orbit & uncertainty propagation, risk assessment, and avoidance manoeuvres... I will not repeat them here, or point the differences and necessary adaptations. Our common dictionaries define risk as "a situation involving exposure to danger" or "the possibility that something bad or dangerous will happen". A more physical standardised view of risk (although not unique!) combines the likelihood and severity of a hazardous event.

Are we at risk of an asteroid impact? Yes, indeed, as it was in the recent past. Thus, asteroid impact risk is the product of the frequency of an impact event with its cost or our vulnerability to such impact. Tiny pebbles hitting the Earth on millennium scales is zero risk, while big killer-asteroids hitting the Earth every decade is a very-very high risk!... Of course these are unrealistic hypothetical illustrative cases. With asteroids originating from a dynamically collision population, the size distribution of NEOs is such that we are actually in-between those extreme situations. Some 50-80 tons of extraterrestrial material are falling on Earth every day, as small mostly micro-meteorites and dust, still totally harmless.

3. Know your enemy, near Earth Asteroids surveys

So, astronomers started to look for the largest NEOs, larger than approximately 1 km, to assess their impact probability, before looking for those of most probable risk. Lets remind that this population of Near-Earth asteroids—with the first one (433) Eros discovered in 1898—was still mostly unknown at the end of the '90s. One key element here is the detection of PHAs, and we now have curves of detection rate from dedicated surveys showing that almost all asteroids larger than approximately 1 km have been catalogued. Moreover, their orbits present no impact risk in the coming Century hence no major catastrophic mass extinction event is foreseen in the near future. On the other side of the (size-) detection spectra, we have presently a record of seven imminent impactors. These are generally small (1-5 meter sized) harmless† asteroids, discovered in space less than 24 hours before their predicted entry in the atmosphere. The last in date—asteroid 2023 CX1 in February this year discovered in space with a T1m Schmidt telescope in Hungary—did a luminous bolide seen and recorded by many observers that were alerted of the entry. The asteroid ended its course in France Normandy, making many colleagues happy with several meteorites collected now at the museum of natural history MNHN in Paris, France. The rate of such imminent impactors detection should increase in the future with current and new dedicated surveys (from ground and space).

The situation is still not favourable for intermediate size objects, showing risks at regional/continental level, the next goal is to detect more than 90% of objects larger than approximately 140 m (or say absolute magnitude $H < 20$). In fact, only 50% of NEOs capable of doing regional/continental damages have been discovered so far. Surveys—mostly driven by the USA—presently detect about 3000 new NEOs per year, with a total cumulative number of 36,000 NEOs end August 2023. Keeping in mind that the 10,000th NEO was discovered mid-June 2013, we see the progress made in just ten years. This rate of detection and catalogue completion will even considerably increase, thanks

† Or a priori harmless : in contrast to the micro-meteorites, some decimeter large meteorite might hit a car trunk, construction roof, animal or part of a human body, as recorded in the past.

to the venue of the ground-based Vera Rubin Telescope LSST survey in 2025, and future NEO-surveyor space mission: if we are looking for them... we'll get them !

4. Mitigation of a threatening NEA

Many activities are hence devoted to detections, surveys, and catalogue completions. These are completed by orbit and uncertainty computations, with risks assessment (see for instance CNEOS at NASA/JPL or NEOCC at ESA) and international coordination through UNOOSA. Now, would we face a real threat of a future impacting event much larger than the small imminent impactor, one would need to put in place mitigation measures. For 1/ a very dangerous impact, i.e. with major effects on ground (from blast, or impact crater, or tsunami), and 2/ with enough warning time, a mitigation in space is the most likely option. The aim of such mitigation mission is to slightly modify the asteroids orbit. A so-called kinetic impactor will transfer a small amount of linear momentum, preferably along the motion. With this principle, the asteroids orbital period is modified, so that the future collision can be avoided, even if the trajectories still intersect. Several studies have analysed the mitigation scenario; at the European level we can mention NEOshield 1&2, Stardust 1&R, and NEOmap.

5. The DART mission

In addition to these research works and risk assessment, one needs to ensure that all steps of a risk management chain can be mastered. In particular the space mission for the mitigation process. The role of the Asteroid Impact and Deflection Assessment (AIDA) collaboration is to probe the kinetic impactor mitigation technique – starting with the Double Asteroids Redirection Test (DART) mission. It is a test on real scale, to measure the linear momentum imparted to an asteroid through the change in the orbital period of the Didymos/Dimorphos binary systems relative orbit. It mimics all aspect of a real mitigation mission that would change the heliocentric orbital period, while the sensitivity of momentum transfer is much higher on the relative orbit. A 600 kg probe was launched en-route toward the moon Dimorphos which it impacted on November 2022. The mission was a great success hence validating the kinetic impact as planetary defence option. Results of the collision were impressive with autonomous targeting of a 160 m large target at 6 km/s speed, an ejecta plume observed by cubesat LiciaCUBE, a large beta factor momentum enhancement, and a significant change of the orbital period, ... "This marks humanitys first time purposely changing the motion of a celestial object and the first full-scale demonstration of asteroid deflection technology" says NASA's PR. More insight will be given on the craterisation and the secondarys mass thanks to the follow-on mission Hera that will visit the asteroid end 2026, so stay tuned.

6. Rendez-vous with Apophis 2029

Last, let me invite you to take a rendez-vous with the near-Earth asteroid (99942) Apophis. This PHA will have a close encounter with the Earth, at a distance below the altitude of our geostationary satellites 36,000 km. It is not the first time a NEO gets very close to the Earth, but Apophis is approximately 400 m large in size, large enough on that occasion and distance to be visible by naked eye (during night in Europe and part of Northern hemisphere). The rendez-vous date is on 2029 April, 13, a Friday !..

7. A final note

I have made a reference in my presentations title to the recent box-office movie "Dont look UP" (Adam McKay dir., 2022) which scenario is explicitly based on such an impact threat, and another reference to an older issue of the Italian magazine Domenica del

Corriere (21 August 1966). The cover page of this magazine titled "La fine del Mondo non ci sara" shows a comet miraculously missing its impact over Rome. Moreover, and of special interest to us, there is an article of a young high-school student that just got a national award: this teenager is Andrea Milani (1948–2018); one of Heras cubesat is named in his honour.

8. Bibliography

- NASA/JPL CNEOS `https://cneos.jpl.nasa.gov`
- ESA NEOCC `https://neo.ssa.esa.int/`
- DART `https://dart.jhuapl.edu/`
- AIDA `https://www.esa.int/Space_Safety/Hera/`
- Asteroid_Impact_Deflection_Assessment_AIDA_collaboration
- NASA PD `https://www.nasa.gov/planetarydefense`
- UNOOSA `https://www.unoosa.org/oosa/en/ourwork/topics/neos/index.html`

Complex Planetary Systems II: Latest Methods for an Interdisciplinary Approach
Proceedings IAU Symposium No. 382, 2024
A. Lemaitre & A.-S. Libert, eds.
doi:10.1017/S1743921323003885

Secular Resonance Maps

Zoran Knežević

Serbian Academy of Sciences and Arts, Belgrade, Serbia
email: zoran@aob.rs

Abstract. A complete list of combinations of the rates of asteroid perihelia and nodes and the corresponding fundamental frequencies of planets, giving rise to secular resonances and involving up to 4 frequencies, is known from the previous work, while for the resonances with 6 frequencies a systematically derived comprehensive list is given here for the first time. There are 28 divisors in the theory of degree up to 4, not all of which can give rise to resonances, while at degree 6 there are (at least) 33 such possibly resonant frequency combinations.

Mapping the secular resonances by plotting the resonant lines in the phase space of proper elements or of secular frequencies, possibly also against the background of known asteroids, enables to straightforwardly identify resonances causing large long periodic variations of asteroid orbital elements, resonances that interact with known families, those that bound the dynamically distinct regions, deplete or disturb asteroids in these regions, etc.

Keywords. secular resonances, proper elements, asteroid families

1. Introduction

Secular resonances correspond to singularities in the phase space of orbital elements where integer combinations of frequencies of the longitude of perihelion and the longitude of node of perturbed (asteroids) and perturbing (planets) bodies, appearing in divisors of the perturbing terms in the equations of motion, tend to zero. Permitted combinations are determined by D'Alembert rules, but not all of them give rise to secular resonances located in the region of interest.

Roughly, secular resonances can be divided into two broad categories: the linear ones, which are the strongest and most important in terms of the effects they produce in the motion of asteroids, and the nonlinear ones, which affect the motion to a lesser extent, depending on the degree of these resonances themselves. Linear resonances appear in the equations of motion as divisors of the terms linear in eccentricity and/or (sine of) inclination (that is, of degre 2 in the perturbing Hamiltonian) and they include the difference of just two frequencies - one of the asteroid and another of a perturbing planet, while all others are nonlinear resonances, corresponding to divisors of terms with amplitudes containing higher degrees of $(e, \sin I)$ and involving combinations of 4, 6, 8, and so on, frequencies. Limiting our dynamical model to include only the most massive perturbing planets - Jupiter and Saturn - the linear secular resonances are just three: $g - g_5, g - g_6, s - s_6$, while examples of degree 4 and 6 nonlinear ones are, e.g.: $g + s - g_6 - s_6, g + 2s - g_5 - 2s_6$; here g, s denote the secular frequencies of the longitude of perihelion and the longitude of node of the asteroid's orbit, while $g_5 = 4.26$ arcsec/y, $g_6 = 28.25$ arcsec/y, $s_6 = -26.34$ arcsec/y are the corresponding frequencies for Jupiter and Saturn.

Knowledge of the exact positions of secular resonances in the phase space of proper orbital elements or frequencies encompassing the asteroid belt, is of paramount importance for a number of reasons; these include the possibility to reliably assess the interaction with asteroids inside or near the resonant region, the reshaping of asteroid families interacting with resonances, the transport of objects inside or outside the asteroid belt, and so on. Mapping the secular resonances is the most effective tool to pin down and visualize their positions, and to appreciate their importance in terms of the asteroid dynamics.

The first secular resonance maps in the region of asteroid belt were presented by Williams and Faulkner (1981), who plotted secular resonance surfaces projected to $(a, \sin I)$ plane. They showed the lines corresponding to a number of different values of eccentricity, representing the centers of libration only for the three above mentioned linear resonances with Jupiter and Saturn (which they labeled ν_5, ν_6, ν_{16}, respectively, and for one nonlinear resonance located in the very high inclination region. Subsequently, Milani and Knežević (1990), and later Milani and Knežević (1992) and Milani and Knežević (1994), presented improved version of these maps, with resonant positions marked by the contour lines approximately indicating also the width of the resonant strip. They upgraded the mapping by showing the positions of a number of most significant nonlinear secular resonances against the background of known asteroids in the proper elements space, illustrating the interaction of asteroid families with these resonances, studying dynamical behavior of asteroids affected by resonances, etc. Knežević et al. (1991) further extended the representation of positions of linear secular resonances to the interval of semimajor axes from 2 to 50 au, followed by Michel and Froeschle (1997) who mapped the positions of linear secular resonances in the region of semimajor axes less than 2 au. The most recent entries into the literature on secular resonance positions pertain to an attempt to precisely locate the $g - g_5$ secular resonance Knežević (2020), and to a survey of positions of all the resonances in the asteroid belt up to degree 4, as well as of an incomplete list of degree 6 ones and of several resonances of degree 8 Knežević (2022).

To be able to map the secular resonances, one needs to know which are the combinations of frequencies, allowed by the D'Alembert rules, giving rise to resonances. For this purpose in the following we consider the lists of resonant combinations and assess their completeness.

2. Secular resonance census

The most obvious way to assess the completeness of a list of resonances is on a degree by degree basis. We proceed from the linear, degree 2 resonances, involving combinations of two frequencies only, and continue afterwards with resonances of degrees 4, 6, and so on, involving the corresponding number of frequencies, while checking for all of them whether they can achieve zero value in the region of interest, that is, in the asteroid belt.

As already explained above, if we limit ourselves to consider a dynamical model with only Jupiter and Saturn as perturbing planets, we see that the degree two resonances are just three, while in the full planetary system model, with 8 perturbing planets, there are 15 such combinations.

The situation is not so trivial already at the next level, in the degree 4 case. In Table 1 we give a list of 28 divisors appearing in the analytical theory of Milani and Knežević (1990), not all of which, however, can give rise to a resonance - obviously, this is the case with divisor $g_5 - g_6$, which is simply a constant, but also divisors Nos. $22 - 27$ cannot be zero, as frequency s has always a negative value throughout the asteroid belt. Bearing in mind that this list does not include the "fake" divisors, involving s_5, which are removed from the theory by the change of coordinates to express inclinations with respect to the invariable plane, one may state that this list of resonances is complete.

Table 1. Secular resonances of degree 2 and 4

No	Divisor	No	Divisor
1	$g - g_5$	15	$2g - 2s_6$
2	$g - g_6$	16	$g + g_5 - s - s_6$
3	$s - s_6$	17	$g + g_6 - s - s_6$
4	$g - g_5 + s - s_6$	18	$g + g_5 - 2s_6$
5	$g - g_6 + s - s_6$	19	$g + g_6 - 2s_6$
6	$2g - 2s$	20	$s - s_6 + g_5 - g_6$
7	$g - 2g_5 + g_6$	21	$s - s_6 - g_5 + g_6$
8	$g + g_5 - 2g_6$	22	$2s - 2g_5$
9	$2g - g_5 - g_6$	23	$2s - 2g_6$
10	$g - g_5 - s + s_6$	24	$2s - g_5 - g_6$
11	$g - g_6 - s + s_6$	25	$s + s_6 - 2g_5$
12	$2g - s - s_6$	26	$s + s_6 - 2g_6$
13	$g + g_5 - 2s$	27	$s + s_6 - g_5 - g_6$
14	$g + g_6 - 2s$	28	$g_5 - g_6$

While for resonances up to degree 4 we could use the results available from the previous work, for the extension to degree 6 resonances there is no such possibility. Thus, a dedicated effort is needed to get the full list of these resonances suitable for their mapping.

Before we venture to address the problem at hand, let us clarify whether this is a worthwhile effort. Let us, namely, consider whether the degree 6 secular resonances are at all important in the asteroid dynamics: are their effects large enough to be easily recognized and do they offer plausible explanations for some peculiar features of the motion of asteroids and of their distribution in the phase space. For simplicity, we shall recall here only a couple of examples of secular resonances affecting the shapes of asteroid families in the space of proper elements, because these are easy to recognize and understand.

One of the first ever examples of the observable effect of a secular resonance, and to that matter of a degree 6 resonance, on the shape of an asteroid family is due to Bottke et al. (2001): they showed that the so-called "Prometheus surge", consisting of members of the Koronis asteroid family exhibiting systematically larger proper eccentricities than the rest of the family, is a consequence of the interaction of family members migrating outwards from the Sun due to Yarkovsky effect with the secular resonance $g + 2g_5 - 3g_6$.

Another example pertains to Hansa family Milani et al. (2019) affected by the degree 6 secular resonance $2g - g_5 - g_6 + s - s_6$. The effect of this resonance explains a large scatter of proper eccentricity e in the part of the family with semimajor axis $a < a_0$, where a_0 is the proper a of the parent body. The members on this side of the family, also originally ejected in the same direction as suggested by the positive V-shape of the family, are moving preferentially towards the Sun due to the Yarkovsky effect, so that many members currently outside of the secular resonance zone must have passed through it in the past. The eccentricities of the family members being low and inclinations high, the scattering in this case affects much more the proper e than the proper $\sin I$ because of the D'Alembert rule, thus the changes in e are much larger than those in $\sin I$.

There are other examples of this kind, but even just these two are enough to conclude that secular resonances of degree 6 are important and thus worth an effort of identifying and mapping them..

2.1. *Identification of secular resonances of degree 6*

The identification of degree 6 secular resonances is not a straightforward thing to do: the analytical theory based on the expansion of perturbing function beyond degree 4 terms in eccentricity and sine of inclination, but with inclinations expressed with respect to a common coordinate plane, is not available Ellis and Murray (2000), and a computer

Table 2. Secular resonances of degree 4 as combinations of linear and Kozai ones. Columns gs and ν give the same divisor written in two ways

No	gs	ν	No	gs	ν
1	$g - g_5$	ν_5	15	$2g - 2s_6$	$2\nu_{16} + 2K$
2	$g - g_6$	ν_6	16	$g + g_5 - s - s_6$	$2K - \nu_5 + \nu_{16}$
3	$s - s_6$	ν_{16}	17	$g + g_6 - s - s_6$	$2K - \nu_6 + \nu_{16}$
4	$g - g_5 + s - s_6$	$\nu_5 + \nu_{16}$	18	$g + g_5 - 2s_6$	$2K - \nu_5 + 2\nu_{16}$
5	$g - g_6 + s - s_6$	$\nu_6 + \nu_{16}$	19	$g + g_6 - 2s_6$	$2K - \nu_5 + 2\nu_{16}$
6	$2g - 2s$	Kozai	20	$s - s_6 + g_5 - g_6$	$\nu_6 - \nu_5 + \nu_{16}$
7	$g - 2g_5 + g_6$	$2\nu_5 - \nu_6$	21	$s - s_6 - g_5 + g_6$	$\nu_5 - \nu_6 + \nu_{16}$
8	$g + g_5 - 2g_6$	$2\nu_6 - \nu_5$	22	$2s - 2g_5$	$2\nu_5 - 2K$
9	$2g - g_5 - g_6$	$\nu_5 + \nu_6$	23	$2s - 2g_6$	$2\nu_6 - 2K$
10	$g - g_5 - s + s_6$	$\nu_5 - \nu_{16}$	24	$2s - g_5 - g_6$	$\nu_5 + \nu_6 - 2K$
11	$g - g_6 - s + s_6$	$\nu_6 - \nu_{16}$	25	$s + s_6 - 2g_5$	$2\nu_5 - \nu_{16} - 2K$
12	$2g - s - s_6$	$\nu_{16} + 2K$	26	$s + s_6 - 2g_6$	$2\nu_6 - \nu_{16} - 2K$
13	$g + g_5 - 2s$	$2K - \nu_5$	27	$s + s_6 - g_5 - g_6$	$\nu_5 + \nu_6 - \nu_{16} - 2K$
14	$g + g_6 - 2s$	$2K - \nu_6$	28	$g_5 - g_6$	$\nu_6 - \nu_5$

algebra system like TRIP Gastineau and Laskar (2011) is both quite demanding to use and available only in a limited version, in any case not yet applied for our purpose. There is, however, a simple and efficient "empirical" way to find most, if not all, secular resonances of degree 6 (and higher, if neeeded) and produce their list; whether such a list is also complete remains to be verified by one of the above methods.

Carruba and Michtchenko (2009) proposed a new notation for nonlinear secular resonances that employs the ν_k labels introduced by Williams and Faulkner (1981); the new notation was based on the idea that nonlinear resonances can be expressed as simple combinations of linear ones (c.f. $g + g5 - 2g_6 = 2\nu_6 - \nu_5$; $2g - s - g_5 - g_6 + s_6 = \nu_5 + \nu_6 - \nu_{16}$). Although this conjecture soon proved not to be entirely correct since only 10 resonances of degree 4 could be constructed by means of the three linear ones and 14 of them could not (e.g. resonances like $g - 2s + g6$ or $s + s6 - 2g5$, appearing in Table 1, cannot be obtained as a combination of linear ones), the idea lingered around, to eventually become useful providing an appropriate correction is added. It turned out, namely, that all divisors in Table 1 can be recovered if also Kozai's resonance $2K = 2g - 2s$ is included in the arithmetic. This is demonstrated in Table 2, where all the resonances of degree 4 listed in Table 1 are expressed as combinations of either the linear ones only, or as combinations that involve Kozai's resonance as well.

It is now straightforward to assume that resonances of higer degree, in particular these of degree 6, can be identified in the same way. The procedure may be considered an extrapolation, but note that D'Alembert rules are at work here, making the higher degree terms logically connected with the lower degree ones. In Table 1 we thus list all 33 resonances of degree 6 identified as combinations of linear resonances only (1–20), and as combinations with Kozai's resonance included too (21–33).

3. Mapping of resonances

Once we have a list of secular resonances at our disposal, it is easy to map them and visualize their positions and interactions with asteroid families and background objects in a suitable phase space, be it a proper elements space or, perhaps more appropriate in this context, a secular frequencies space. In the former case, one uses frequencies computed on a regular grid by means of the polynomial fitting procedure developed by Knežević and Milani (2019), and represents them by means of the contour lines in the plane of two proper elements (e.g. $a, \sin I$), while keeping the third one (e) fixed Knežević (2022). In the latter case, resonance loci are simply plotted as straight lines in the corresponding (g, s) plane. Adding in the same figures the positions of asteroids selected to

Table 3. Secular resonances of degree 6 as combinations of linear and Kozai ones. Columns gs and ν give the same divisor written in two ways

No	gs	ν	No	gs	ν
1	$3g - 2g_5 - g_6$	$2\nu_5 + \nu_6$	21	$3g - 2s - g_5$	$\nu_5 + 2K$
2	$3g - g_5 - 2g_6$	$\nu_5 + 2\nu_6$	22	$3g - 2s - g_6$	$\nu_6 + 2K$
3	$2g + s - g5 - g6 - s6$	$\nu_5 + \nu_6 + \nu_{16}$	23	$3s - 2g - s_6$	$\nu_{16} - 2K$
4	$2g - s - g_5 - g_6 + s_6$	$\nu_5 + \nu_6 - \nu_{16}$	24	$3s - g - g_5 - s_6$	$\nu_5 + \nu_{16} - 2K$
5	$2g + s - 2g_5 - s_6$	$2\nu_5 + \nu_{16}$	25	$3g - s - g5 - s6$	$\nu_5 + \nu_{16} + 2K$
6	$2g - s - 2g_5 + s_6$	$2\nu_5 - \nu_{16}$	26	$3s - g - g_6 - s_6$	$\nu_6 + \nu_{16} - 2K$
7	$2g + s - 2g_6 - s_6$	$2\nu_6 + \nu_{16}$	27	$3g - s - g_6 - s_6$	$\nu_6 + \nu_{16} + 2K$
8	$2g - s - 2g_6 + s_6$	$2\nu_6 - \nu_{16}$	28	$2g - 2s - g_5 + g_6$	$\nu_5 - \nu_6 + 2K$
9	$g + 2s - g_5 - 2s_6$	$\nu_5 + 2\nu_{16}$	29	$2g - 2s + g_5 - g_6$	$\nu_6 - \nu_5 + 2K$
10	$g - 2s - g_5 + 2s_6$	$\nu_5 - 2\nu_{16}$	30	$2s - g - 2g_5 + g_6$	$2\nu_5 - \nu_6 - 2K$
11	$g + 2s - g_6 - 2s_6$	$\nu_6 + 2\nu_{16}$	31	$2s - g + g_5 - 2g_6$	$2\nu_6 - \nu_5 - 2K$
12	$g - 2s - g_6 + 2s_6$	$\nu_6 - 2\nu_{16}$	32	$g + 2s - g_5 - 2g_6$	$\nu_5 + 2\nu_6 - 2K$
13	$2g - 3g_5 + g_6$	$3\nu_5 - \nu_6$	33	$g + 2s - 2g_5 - g_6$	$2\nu_5 + \nu_6 - 2K$
14	$2g + g_5 - 3g_6$	$3\nu_6 - \nu_5$			
15	$g + s - 2g_5 + g_6 - s_6$	$2\nu_5 - \nu_6 + \nu_{16}$			
16	$g + s + g_5 - 2g_6 - s_6$	$2\nu_6 - \nu_5 + \nu_{16}$			
17	$g - s - 2g_5 + g_6 + s_6$	$2\nu_5 - \nu_6 - \nu_{16}$			
18	$g - s + g_5 - 2g_6 + s_6$	$2\nu_6 - \nu_5 - \nu_{16}$			
19	$2s - g_5 + g_6 - 2s_6$	$\nu_5 - \nu_6 + 2\nu_{16}$			
20	$2s + g_5 - g_6 - 2s_6$	$\nu_6 - \nu_5 + 2\nu_{16}$			

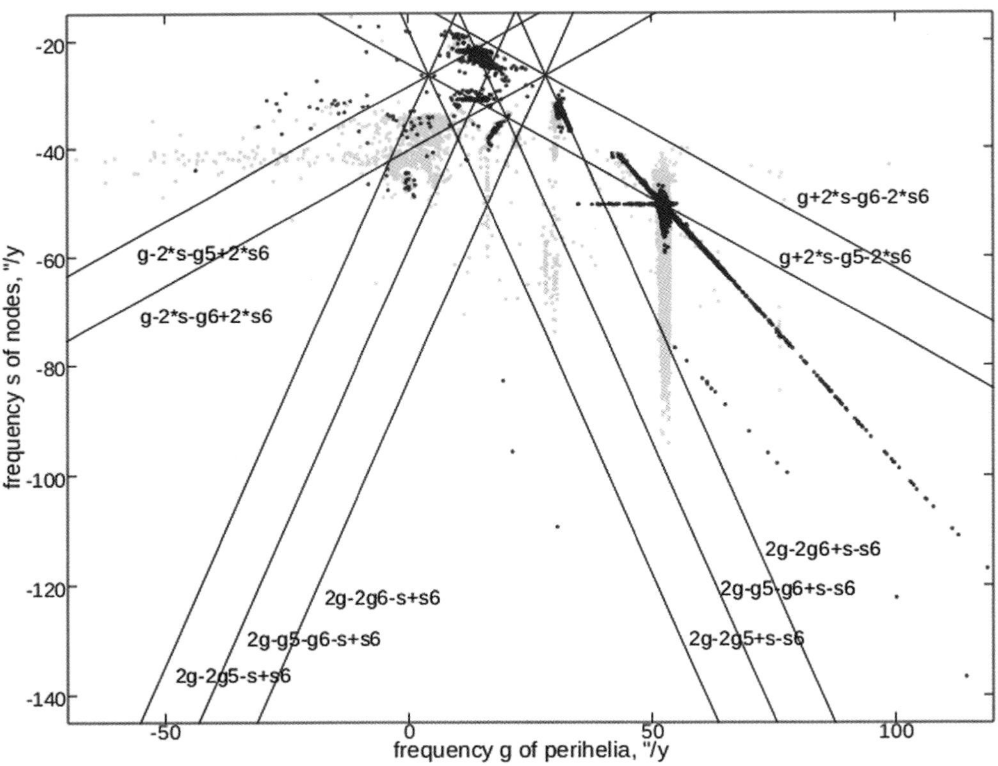

Figure 1. Sample of secular resonances of degree 6 plotted in the plane of frequencies (g, s). Overlapped are asteroids with deteriorated proper elements: $\sigma e > 0.005$ (green dots) and $\sigma \sin I > 0.003$) (blue dots).

demonstrate interaction of the resonance with a particular asteroid population, (families, asteroids with deteriorated proper elements, or similar), one also enables analysis of the resonant dynamics and effects these interactions cause.

In Figure 1 a sample of secular resonances of degree 6 is shown in the plane of frequencies (g, s). Also shown are asteroids with deteriorated proper elements - these with $\sigma e > 0.005$ (green dots) and with $\sigma \sin I > 0.003$ (blue dots). Although the strips with lower accuracy data align mostly with resonances of different degrees not shown in the figure, in at least one case $(2g - 2g_6 + s - s_6$, at $g \simeq 30$ arcsec/y, $s \simeq -30$ arcsec/y) the alignment of blue dots with the resonance is obvious.

4. Conclusions

In conclusion one can state that the complete list of resonances of degree 6 (and possibly higher) is indeed necessary, since many of them have important effects on the dynamics of asteroids, and that their exact positions must be known to pursue the research. Whether the simple scheme presented in this work solves the issue of completeness of their census remains to be verified in the future with some more sophisticated approaches.

References

Bottke, W. F., Vokrouhlický, D., Broz, M., Nesvorný, D., Morbidelli, A. 2001, *S*cience, 294, 1693

Carruba, V., Michtchenko, T.A. 2009, *A&A*, 493, 267

Ellis, K., Murray, C.D. 2000, *I*carus, 147, 129

Gastineau, M., Laskar, J. 2011 *ACM Commun. Comput. Algebra*, 44, 194

Knežević, Z. 2020, *M*NRAS, 497, 4921

Knežević, Z. 2022, *S*erbian Astron.J., 204, 1

Knežević, Z., Milani, A. 2019, *Cel. Mech. Dyn. Astron.*, 131:27

Knežević, Z., Milani, A., Farinella, P., Froeschle, Cl., Froeschle, Ch. 1991, *I*carus, 93, 316

Michel, P., Froeschle, Ch. 1997, *I*carus, 128, 230

Milani, A., Knežević, Z. 1990, *Cel. Mech. Dyn. Astron.*, 49, 347

Milani, A., Knežević, Z. 1992, *Icarus*, 98, 211

Milani, A., Knežević, Z. 1994, *Icarus*, 107, 219

Milani, A., Knežević, Z., Spoto, F., Paolicchi, P. 2019, *A&A*, 622, A47

Williams, J.G., Faulkner, J. 1981, *Icarus*, 46, 390

Complex Planetary Systems II: Latest Methods for an Interdisciplinary Approach
Proceedings IAU Symposium No. 382, 2024
A. Lemaitre & A.-S. Libert, eds.
doi:10.1017/S1743921323003630

On the retrograde planar co-orbital asteroid motion with Jupiter

George Voyatzis and Thomas Kotoulas

Aristotle University of Thessaloniki, Department of Physics, Post Code: 54124, Thessaloniki, Greece
email: tkoto@physics.auth.gr

Abstract. We study the motion of an asteroid being in retrograde 1/1 resonance with Jupiter (co-orbital motion). We consider the planar case (i=180°) and Jupiter is on a circular or elliptic orbit ($e' = 0.048$). In the cirular model we compute families of symmetric periodic orbits and their stability type. In the elliptic model we have isolated periodic orbits which affect the orbital modes of motion as it is shown by the FLI dynamical maps.

Keywords. asteroids, restricted 3-body problem, retrograde co-orbital motion

1. Model and results

We use the Elliptic Restricted Three Body Problem (ERTBP) with primaries Sun and Jupiter on the rotating $Ox-$axis and of masses $1-\mu$ and $\mu = 0.001$, respectively. We consider planar retrograde motion of an asteroid with osculating orbital elements: the semimajor axis a, the eccentricity e, the longitude of pericenter ω and the mean anomaly M (prime quantities refer to Jupiter). For the planar 1/1 retrogade motion the resonant angle is defined as $\phi = \lambda - \lambda' - 2\omega$, where λ is the mean longitude.

The planar circular problem has been studied by Huang (2018) and Morais (2019) who distinguished three types of motion called *mode 1, 2* and *3*. We recomputed the families of periodic orbits of these modes and found the periodic orbits of period multiple of 2π, which are generating orbits for families of periodic orbits in the elliptic model with the eccentricity e' as parameter. There are two initial configurations: Jupiter at perihelion ($\omega' = 0°$) or at aphelion ($\omega' = 180°$). The characteristic curves of the families are shown in Fig. 1. At $e' = 0.048$ we have isolated periodic orbits ($1p$, $1a$, etc). Kotoulas (2020) computed such families for several retrograde resonances in the main asteroid belt. In Fig. 2 we present dynamical FLI maps for $e' = 0.048$. Compared with the case $e' = 0$, the maps reveal relatively slight differences. The region of stable orbits of mode 1 (Morais (2019)) is quite large. The white region indicates strongly unstable orbits which are not apparent for $e' = 0$. The stable regions of mode 3 and mode 2 are centered around stable periodic orbits, which exist at small and very large eccentricities, respectively.

2. Conclusions

We explored the phase space structure of the planar elliptic RTBP ($e' = 0.048$) in the 1/-1 resonance by computing families of periodic orbits and dynamical maps. Compared with the results of the circular model, the eccentricity of Jupiter affects mainly the modes 2 and 3 of motion. The mode 1 is the most stable motion but a zone of strong instability appears at high eccentricities.

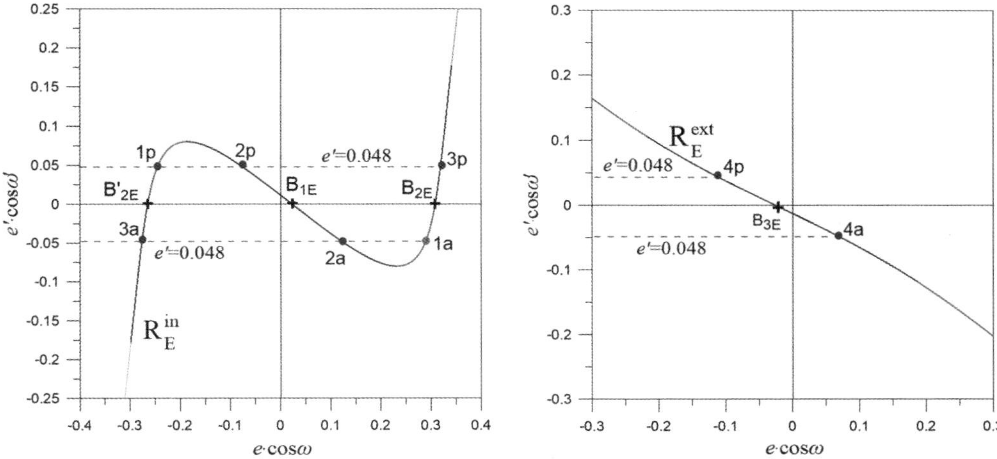

Figure 1. Families of 1/1 retrograde periodic orbits in the elliptic model. Numbered dots indicate the isolated orbits for the Jupiter's eccentricity $e' = 0.048$.

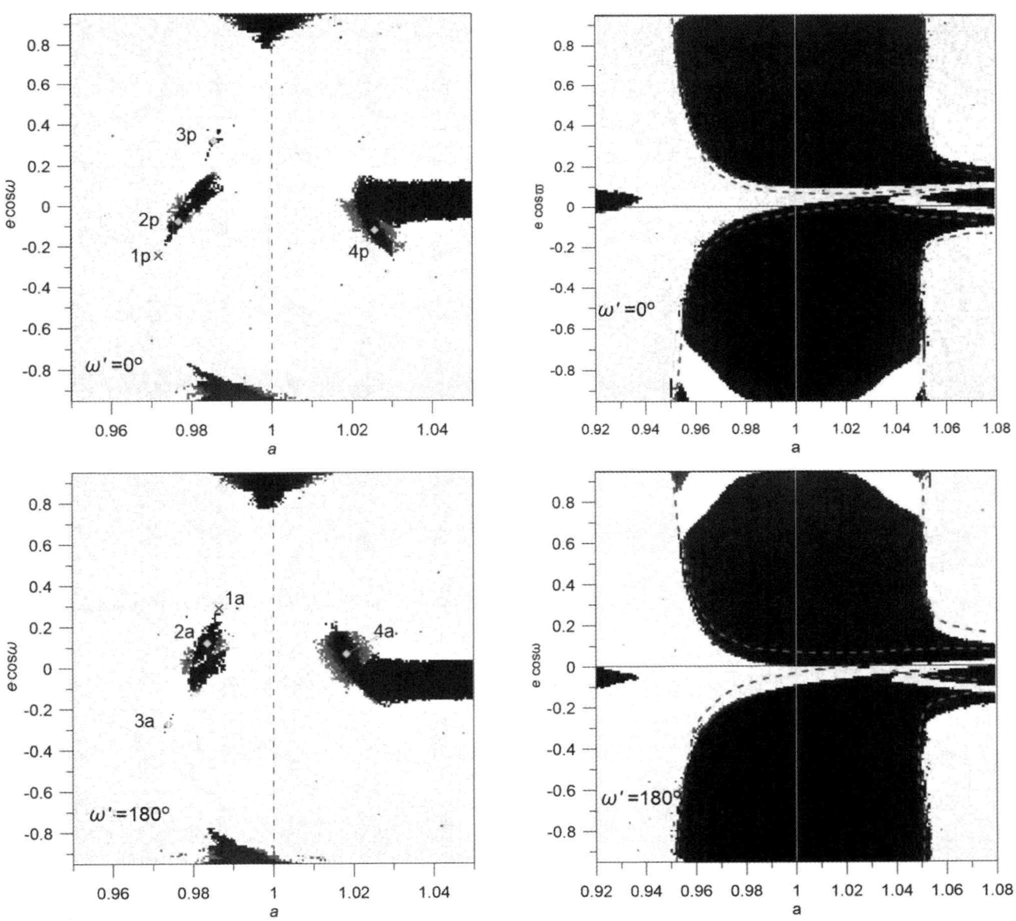

Figure 2. FLI maps for the planar elliptic problem ($e' = 0.048$), $\omega' = 0°$ (top) and $\omega' = 180°$ (bottom). Left column refers to modes 2 and 3 and the right one to mode 1. Dark colors indicate low FLI values, i.e. stable motion.

Supplementary material

To view supplementary material for this article, please visit http://doi.org/10.1017/S1743921323003630

References

Huang, Y., Li, M., Li, J., Gong, S. *et al.* 2018, *AJ*, 155, 262
Kotoulas, T.A., Voyatzis, G., 2020, *Planet. Space Sci.*, 182, 1
Morais, M.H.M., Namouni, F. 2019, *MNRAS*, 490, 3799

Complex Planetary Systems II: Latest Methods for an Interdisciplinary Approach
Proceedings IAU Symposium No. 382, 2024
A. Lemaitre & A.-S. Libert, eds.
doi:10.1017/S1743921323004891

Sun-planets tidal interactions: an extended catalog

Sergey M. Kudryavtsev[1], Rodolfo G. Cionco[2] and Willie W.-H. Soon[3,4]

[1]M.V. Lomonosov Moscow State University, Sternberg Astronomical Institute, 13, Universitetsky Pr., Moscow, Russia
email: ksm@sai.msu.ru

[2]CONICET and Grupo de Estudios Ambientales UTN, Colón 332, San Nicolás, Buenos Aires, Argentina

[3]Center for Environmental Research and Earth Sciences (CERES), Salem, MA 01970, USA

[4]Institute of Earth Physics and Space Science (ELKH EPSS), H-9400, Sopron, Hungary

Abstract. An extended catalog of long-periodic terms of the Sun tide-generating potential (STGP ver.2) is released. It contains 2.5 times of such terms than in the first version of the STGP catalog (Cionco et al. 2023). The analysis of the new data is done in order to re-examine the existence of tidal forces of $\approx$ 11.0-yr and $\approx$ 22.0-yr period able to excite the observed solar-activity cycles. In several recent studies, such tidal forcings are claimed to be a result of certain combinations of Venus, Earth and Jupiter mean motions with the multipliers (6, -10, 4) and (3, -5, 2), respectively. So, in this contribution, we specifically look for the periodic terms related to these combinations. As a result of these additional investigation we do not confirm any noticeable tidal forcing factors of $\approx$ 11.0-yr nor $\approx$ 22.0-yr period in the extended STGP spectrum either.

Keywords. Solar Cycle; Models; Oscillations; Ephemerides

1. Introduction

There is a long-running hypothesis that the planetary tides may have a modulating effect on the solar activity cycles. In particular, several studies assume that certain quasi-alignments between Venus, Earth and Jupiter ("V-E-J configurations") provide a periodicity of $\approx$ 11.0 yr and/or $\approx$ 22.0 yr, and that the operation of solar dynamo can be synchronized with these configurations. Specifically, these periods appear as a result of combinations of planetary mean motions $(6n_2 - 10n_3 + 4n_5)$ and $(3n_2 - 5n_3 + 2n_5)$, where n_2, n_3 and n_5 are the mean motions of Venus, Earth and Jupiter, respectively. The relevant bibliography can be found in Cionco et al. (2023). Nevertheless, the evidence behind this proposed tidal forcing is still debatable. In Cionco et al. (2023), we first made a development of the STGP represented by accurate harmonic series, but no "V-E-J configurations" that lead to any noticeable tidal force was found. However, in some succeeding researches (e.g. Stefani *et al.* 2023) such "V-E-J configurations" are still being considered as an essential source of tidal effects on the Sun. Therefore, in the present study we made a deeper investigation of that important issue. Specifically, we decreased the searched threshold level for the long-periodic terms of the STGP development from 10^{-7} m^{2}s^{-2} to 10^{-8} m^{2}s^{-2} in order to cover even more tidal terms. (By "long-periodic" terms we conventionally understand the terms which do not include the rate of the Sun's axial rotation, even when the period of some such terms is less than the solar rotation period.) The obtained extended series were re-analyzed for the existence of tidal forces of $\approx$ 11.0-yr and $\approx$ 22.0-yr period. Finally, we explicitly calculated the amplitudes of the tidal terms at the debatable combinations of frequencies $(6n_2 - 10n_3 + 4n_5)$ and $(3n_2 - 5n_3 + 2n_5)$.

2. Outlines of method

The Sun tide-generating potential is developed in terms of accurate harmonic series. The new STGP series are built over 13,000 BC–17,000 AD time interval, they clearly identify and separate the effects of various planetary configurations on the STGP. The latest long-term planetary ephemeris DE-441 (Park *et al.* 2021) is adopted as a source. The series represent the STGP value, $V(t)$, at an arbitrary point $M(r, \phi, \lambda)$ on the Sun's surface as follows:

$$V(t) = \sum_{n=1}^{n_{max}} \left(\frac{r}{R_\odot}\right)^n \sum_{m=0}^{n} \bar{P}_{nm}(\sin\phi) \sum_{i=1}^{i_{max}(n,m)} \left[C_{nm_i}(t)\cos A_{nm_i}(t) + S_{nm_i}(t)\sin A_{nm_i}(t)\right],$$

where r are the heliocentric distance and ϕ, λ are, respectively, the heliographic latitude and longitude of point M; $R_\odot$ is the Sun mean equatorial radius; $\bar{P}_{nm}$ are associated Legendre functions of degree n and order m; n_{max} is the maximum degree of the series, it was found to be equal to 4 for the selected level of accuracy of the new STGP development.

The amplitudes C_{nm_i}, S_{nm_i} of the series terms are obtained as time polynomials of the fifth degree. (Note, in the first STGP development (Cionco et al. 2023) the amplitudes were accurate to the third degree polynomials of time.) The arguments A_{nm_i} have a general form

$$A_{nm_i}(t) = m\,[W(t) + \lambda] + \sum_{j=1}^{8} k_{ij} l_j(t),$$

where $W(t)$ is the Sun rotational angle (Archinal *et al.* 2018), k_{ij} are some integer multipliers and $l_j(t)$ are mean mean longitudes of eight major planets represented by time polynomial of up to the sixth degree (Simon *et al.* 1994). For the long-periodic terms m = 0.

A more detailed description of the development method and the results of its application to various planetary tasks can be found in Kudryavtsev (2004, 2007, 2016, 2017), Kudryavtsev & Kudryavtseva (2009), Cionco et al. (2021, 2023).

3. Results

We have obtained an extended catalog of 326 terms that presents the long-periodic part of the STGP at the advanced level of accuracy (`http://sai.msu.ru/neb/ksm/tgp_sun/STGPv2-LongPeriods.zip`). In the previous version of the STGP reported in Cionco et al. (2023), the number of such terms was 128.

In the new catalog, we re-focused on the STGP terms related to various V-E-J configurations and especially on the terms with periods of $\approx$ 11.0 yr and $\approx$ 22.0 yr. We did not find, once again, any term of such periods. The V-E-J configurations did not produce any discernible terms in the extended STGP spectrum either.

We calculated the amplitudes of periodic terms related to the specific combinations of Venus, Earth and Jupiter mean motions with the multipliers (3, −5, 2) and (6, −10, 4). The amplitudes of those terms were found to be, respectively, only 1.5×10^{-9} m^2s^{-2} and 2.3×10^{-10} m^2s^{-2}.

4. Conclusion

An $\approx$ 11.0-yr and/or $\approx$ 22.0-yr tidal period with a direct physical relevance for the 11-year-like solar-activity cycle is confirmed to be highly improbable.

Supplementary material

To view supplementary material for this article, please visit http://doi.org/10.1017/S1743921323004891

References

Archinal, B. A., Acton, C. H., A'Hearn, M. F., *et al.* 2018, *Celest. Mech. Dyn. Astron.*, 130, 22
Cionco, R. G., Kudryavtsev, S. M., Soon, W. W.-H. 2021, *Earth and Space Sci.*, 8, e2021EA001805
Cionco, R. G., Kudryavtsev, S. M., Soon, W. W.-H. 2023, *Sol. Phys.*, 298:70
Kudryavtsev, S. M. 2004, *J. Geodesy*, 77, 829
Kudryavtsev, S. M. 2007, *A&A*, 471, 1069
Kudryavtsev, S. M. 2016, *MNRAS*, 456, 4015
Kudryavtsev, S. M. 2017, *MNRAS*, 466, 2675
Kudryavtsev, S. M., Kudryavtseva, N. S. 2009, *Celest. Mech. Dyn. Astron.*, 105, 353
Park, R. S., Folkner, W. M., Williams, J. G., Boggs, D. H. 2021, *AJ*, 161:105
Simon, J.-L., Bretagnon, P., Chapront, J., *et al.* 1994, *A&A*, 282, 663
Stefani, F., Horstmann, G.M., Klevs, M., *et al.* 2023, arXiv:2309.00666v2

Complex Planetary Systems II: Latest Methods for an Interdisciplinary Approach
Proceedings IAU Symposium No. 382, 2024
A. Lemaitre & A.-S. Libert, eds.
doi:10.1017/S1743921323003988

New semi-analytical calculation of lunar, solar and planetary perturbations in motion of Earth satellites

Sergey M. Kudryavtsev

M.V. Lomonosov Moscow State University, Sternberg Astronomical Institute,
13, Universitetsky Pr., Moscow, Russia
email: ksm@sai.msu.ru

Abstract. We suggest an advanced algorithm for semi-analytical calculation of orbital perturbations of Earth artificial satellites caused by the gravity attraction of the "3rd-bodies" (the Moon, the Sun, major planets). A new accurate analytical series for the relevant perturbation function is developed. It is obtained through a careful spectral analysis of the long-term DE406 planetary/lunar ephemerides and valid over 2000 years, 1000-3000. The series is used in the author's semi-analytical model of satellite motion. The results of the motion prediction of several Earth satellites obtained by means of the semi-analytical model and a numerical integration method are compared.

Keywords. Satellite Motion; Semi-Analytical Model; Perturbations; Sun; Moon; Major Planets

1. Introduction

When predicting the motion of planetary (both natural and artificial) satellites by an analytical/semi-analytical integration method the first task is to represent all perturbation functions by exact analytical expressions or accurate approximating series. In the present study we deal with the perturbation function caused by the gravity attraction of the "3rd-bodies" (the Moon, the Sun, major planets). In many practical analytical models of satellite motion (e.g., Ivanov *et al.* 1988; Kolyuka *et al.* 1991) the relevant perturbation function is derived from some available analytical theories of the "3rd-bodies' motion. However, presently the accuracy of such analytical theories does not match the precision of the current numerical ephemerides of the Moon and major planets. Thus, the development of new analytical series that accurately approximate the perturbation function caused by the "3rd-bodies" is an actual task.

Here we present a new series for the "3rd-bodies" perturbation function for Earth satellites and provide some examples of its use in the author's semi-analytical model of satellite motion.

2. The perturbation function caused by the "3rd-bodies"

Following (e.g., Giacagalia 1974; Emelyanov 1980) the perturbation function caused by the gravity attraction of the "3rd-bodies", R, can be expanded as follows

$$R = \sum_{l=2}^{\infty} \sum_{m=0}^{l} \sum_{p=0}^{l} \sum_{q=-\infty}^{\infty} \left(\frac{a}{R_s}\right)^l \bar{F}_{lmp}(i) X_{l-2p+q}^{l,l-2p}(e) \left[\bar{A}_{lm} \cos\psi_{lmpq} + \bar{B}_{lm} \sin\psi_{lmpq}\right],$$

where a, e, i, Ω, π, λ are Keplerian elements of the satellite orbit, $\bar{F}_{lmp}(i)$ is an inclination function, $X_{l-2p+q}^{l,l-2p}(e)$ is a Hansen coefficient (a function of the eccentricity of a satellite orbit), R_s is a scaling factor that is assumed to be 43,000 km in our study in order to have the relation $a{R_s}^{-1} < 1$ true for the majority of Earth satellites,

$$\psi_{lmpq} = (l - 2p + q)\lambda - q\pi + (m + 2p - l)\Omega \,,$$

$$\bar{A}_{lm} = \begin{cases} \bar{C}_{lm} \,, & \text{if } l - m \text{ is even} \\ -\bar{S}_{lm} \,, & \text{if } l - m \text{ is odd} \,, \end{cases} \qquad \bar{B}_{lm} = \begin{cases} \bar{S}_{lm} \,, & \text{if } l - m \text{ is even} \\ \bar{C}_{lm} \,, & \text{if } l - m \text{ is odd} \,, \end{cases}$$

$$\bar{C}_{lm}(t) = \frac{1}{2l+1} \sum_j \frac{\mu_j}{R_s} \left(\frac{R_s}{r_j(t)} \right)^{l+1} \bar{P}_{lm}\left(\sin \delta_j(t)\right) \cos m\alpha_j(t) \,,$$

$$\bar{S}_{lm}(t) = \frac{1}{2l+1} \sum_j \frac{\mu_j}{R_s} \left(\frac{R_s}{r_j(t)} \right)^{l+1} \bar{P}_{lm}\left(\sin \delta_j(t)\right) \sin m\alpha_j(t) \,,$$

and μ_j, r_j, α_j, δ_j are the gravitational parameter, geocentric distance, right ascension and declination of the jth perturbing body at epoch t, resp.; $\bar{P}_{lm}$ is an associated Legendre function.

3. Development of $\bar{C}_{lm}$, $\bar{S}_{lm}$ to approximating series

Step 1: Numerical values for the $\bar{C}_{lm}(t)$, $\bar{S}_{lm}(t)$ coefficients were tabulated with a one day's step over the 2000 years' time interval centered at epoch J2000. The Moon, the Sun, Mercury, Venus, Mars, Jupiter and Saturn were taken as the attracting bodies. The long-term numerical ephemerides DE406 (Standish 1998) were used as the source of the planetary/lunar positions.

Step 2: The tabulated values for the coefficients have been developed to approximating series by using a modified spectral analysis method (Kudryavtsev 2004, 2007). A feature of this method is that amplitudes and frequencies of the series' terms are obtained in the form of time polynomials. It allows us to perform the spectral analysis over a long interval of time, up to several thousand years. In this way the terms of close frequencies are better separated and the series have a higher accuracy.

In the present study the form of series for every coefficient is

$$\bar{C}_{lm}(t) \left[\text{or } \bar{S}_{lm}(t)\right] \approx \sum_k \left\{ \left[A_{k0}^c + A_{k1}^c t + A_{k2}^c t^2\right] \cos \omega_k(t) + \left[A_{k0}^s + A_{k1}^s t + A_{k2}^s t^2\right] \sin \omega_k(t) \right\} ,$$

where $\omega_k(t) = \nu_{k1} t + \nu_{k2} t^2 + \cdots + \nu_{k4} t^4$ are some linear combinations of integer multipliers of Delaunay arguments and expressions for the time polynomial part of the planetary/lunar mean orbital longitudes (Simon *et al.* 1994); A_{k0}^c, A_{k1}^c, ..., A_{k2}^s and ν_{k1}, ..., ν_{k4} are some constants derived in the course of the development. The minimum amplitude of terms in the series for $\bar{C}_{lm}(t)$, $\bar{S}_{lm}(t)$ was chosen to be equivalent to $10^{-6}\,\mathrm{m^2 s^{-2}}$ over the entire time interval (or $\sim 10^{-8}$ of the maximum values of the coefficients). The maximum degree l of terms with an amplitude above that threshold is eight. Table 1 gives the number of terms in the development of every $\bar{C}_{lm}(t)$, $\bar{S}_{lm}(t)$ coefficient. The total number of terms in the final series is 38,585.

Table 1. The number of terms in the development of $\bar{C}_{lm}(t)$ / $\bar{S}_{lm}(t)$ coefficients.

l \ m	0	1	2	3	4	5	6	7	8
2	2268 / -	2358 / 2644	2789 / 2652						
3	1160 /-	1212 / 1200	1443 / 1370	1307 / 1354					
4	694 /-	726 / 678	855 / 777	841 / 881	768 / 749				
5	401 /-	427 / 435	489 / 450	494 / 525	510 / 492	397 / 402			
6	195 /-	206 / 215	214 / 225	236 / 253	279 / 267	262 / 269	217 / 215		
7	70 /-	73 / 70	69 / 85	95 / 101	104 / 98	104 / 105	104 / 103	80 / 80	
8	19 /-	20 / 20	24 / 26	21 / 25	34 / 31	31 / 32	28 / 28	28 / 28	24 / 24

Table 2. Comparison of satellite orbital positions calculated by semi-analytical and numerical methods. Motion model: gravity perturbations from the Moon, the Sun and major planets.

Satellite	Semi-major axis, km	The number of terms[1]	Time interval[2]	RMS deviation, m
ETALON-1	25,500	22,216	1 year	8.8
LAGEOS-1	12,300	7,544	1 year	1.5
STARLETTE	7,330	3,040	1 month	0.2

Notes: [1] The number of the "3rd-bodies" terms in the semi-analytical model; [2] Time interval where the comparison is done.

4. Semi-analytical calculation of the "3rd-bodies" effect on satellite motion

To calculate the "3rd-bodies" effect on satellite positions, an algorithm of semi-analytical integration of Lagrange motion equations (Kudryavtsev 1995, 1997, 2002) was employed. It uses the above presented development of the "3rd-bodies" perturbation function. However, the algorithm requests the right-hand parts of the motion equations to be represented by a sum of pure trigonometric functions with numerical values for amplitude, frequency and phase. Thus, the derived series for the "3rd-bodies" perturbation function were further approximated by some trigonometric series. To do it we used the fact of the relative smallness of the obtained values for high-order amplitudes $A_{k1}^{c,s}$, $A_{k2}^{c,s}$ and frequencies ν_{k2}, ν_{k3}, ν_{k4}. So that one can write

$$A_{k1}^{c,s} t \approx \sin A_{k1}^{c,s} t, \qquad A_{k2}^{c,s} t^2 \approx 2\left(1 - \cos\sqrt{A_{k2}^{c,s}}\, t\right),$$

$$\cos\omega_k(t) \approx \cos\nu_{k1}t - \nu_{k2}t^2 \sin\nu_{k1}t, \qquad \sin\omega_k(t) \approx \sin\nu_{k1}t + \nu_{k2}t^2\cos\nu_{k1}t, \qquad \text{etc.}$$

Then the subsequent multiplications of two and more trigonometric functions produces a new trigonometric series to be further used in the right-hand sides of Lagrange motion equations. Our analyses proves that for all obtained values of $A_{k1}^{c,s}$, $A_{k2}^{c,s}$ and ν_{k2}, ν_{k3}, ν_{k4} the approximating trigonometric series keep the accuracy of the original development ($10^{-6}\,\mathrm{m^2 s^{-2}}$) over a time interval of up to 50 years back and forward from any preset initial epoch of the approximation.

The accuracy of the series and algorithm was estimated. First, positions of a low-attitude STARLETTE, mid-attitude LAGEOS-1 and high-attitude ETALON-1 satellite by the 14th-order Everhart numerical integration method were calculated. An exact model of the lunar, solar and planetary perturbations based on the DE405 ephemeris (Standish 1998) was used. Then the calculated satellites positions were assumed as fictitious observations and processed by the semi-analytical integration method with the use of the derived series. The algorithm takes into account the "3rd-bodies" perturbations of up to the 2nd order inclusive. The results of the comparison of the satellites positions calculated by both methods are given in Table 2.

5. Conclusions

• We developed a new semi-analytical series that represents the "3rd-bodies" perturbation function acting on Earth satellites. It is valid over 1000–3000 and has an accuracy compatible with that provided by the current numerical ephemerides of the Moon and planets.

• The series is included in our semi-analytical model of satellite motion. A comparison of satellites coordinates, obtained by using the semi-analytical model and a numerical integration method, show close results. Presently the semi-analytical model calculates the perturbations from the "3rd-bodies" of up to the 2nd order. We plan to further improve the model quality by developing and taking into account the relevant perturbations of higher orders.

Supplementary material

To view supplementary material for this article, please visit http://doi.org/10.1017/S1743921323003988

References

Emelyanov, N. V. 1980, *Trudy GAISH*, 49, 122 [in Russian]
Giacagalia, G. E. O. 1974, *Celest. Mech.*, 9, 239
Ivanov, N. M., Kolyuka, Yu. F., Kudryavtsev, S. M., Tikhonov, V. F. 1988, *Sov. Astron. Lett.*, 14, 405
Kolyuka, Yu. F., Kudryavtsev, S. M., Tarasov, V. P., *et al.* 1991, *Planet. Space Sci.*, 39, 349
Kudryavtsev, S. M. 1995, *Celest. Mech. Dyn. Astron.*, 61, 207
Kudryavtsev, S. M. 1997, *Celest. Mech. Dyn. Astron.*, 67, 131
Kudryavtsev, S. M. 2002, *Celest. Mech. Dyn. Astron.*, 82, 301
Kudryavtsev, S. M. 2004, *J. Geodesy*, 77, 829
Kudryavtsev, S. M. 2007, *A&A*, 471, 1069
Simon, J.-L., Bretagnon, P., Chapront, J., *et al.* 1994, *A&A*, 282, 663
Standish, E. M. 1998, *JPL Interoffice Memorandum*, 312.F–98–048

Complex Planetary Systems II: Latest Methods for an Interdisciplinary Approach
Proceedings IAU Symposium No. 382, 2024
A. Lemaitre & A.-S. Libert, eds.
doi:10.1017/S1743921323004210

Can the stellar dynamical tide destabilize the resonant chains of planets formed in the disk?

Leon Ka Wang Kwok[1], **Emeline Bolmont**[1], **Alexandre Revol**[1], **Stéphane Mathis**[2], **Aurélie Astoul**[3], **Corinne Charbonnel**[1] **and Sean Raymond**[4]

[1]Observatoire de Genève, Université de Genève, 51 Chemin des Maillettes, 1290 Sauverny, Switzerland
email: leon.kawang.kwok@gmail.com

[2]Université Paris-Saclay, Université Paris Cité, CEA, CNRS, AIM, 91191, Gif-sur-Yvette, France

[3]Department of Applied Mathematics, School of Mathematics, University of Leeds, Leeds LS2 9JT, UK

[4]Laboratoire d'Astrophysique de Bordeaux, CNRS and Université de Bordeaux, Allée Geoffroy St. Hilaire, 33165 Pessac, France

Abstract. Evolution models of planetary systems find that resonant chains of planets often arise from the formation within protoplanetary disks. However, the occurrence of observed resonant chains, such as the notable TRAPPIST-1 system, is relatively low. This suggests that the majority of these chains become destabilized after the dissipation of the protoplanetary disk. Stellar tides, especially the wavelike dynamical tide, could be proposed as potential contributors to the destabilization of resonant chains. The dissipation of the dynamical tide, because of the frequency-dependant tidal excitation of stellar oscillation eigenmodes, potentially leads to a boost in migration for the close-in planets and disrupts the fragile stability of resonant chains. Thus, we investigate the influence of the stellar dynamical tide on multi-planet systems with taking their dissipation into account in the N-body code Posidonius. Notably, this research represents the first exploration of the impact of frequency-dependent dynamical tides on multi-planet systems.

Keywords. star-planet interactions, planetary dynamics, mean motion resonances (MMRs), resonant chains, tidal interactions, stellar tides, dynamical tides

1. Introduction

From the theories of planetary formation, the emergence of resonant chains of exoplanets is a common outcome, such as in the Trappist-1 system (Gillon *et al.* 2017). (Izidoro *et al.* 2017) showed that the cumulative distribution of the resonant chains resulting from formation mechanisms is higher than the statistical observations. While other mechanisms have been invoked (e.g. planet-planet instability, (Pichierri & Morbidelli 2020)), we focus here on the impact the tidal response of the host star to the presence of a close companion (Bolmont & Mathis 2016) could have on breaking the chains.

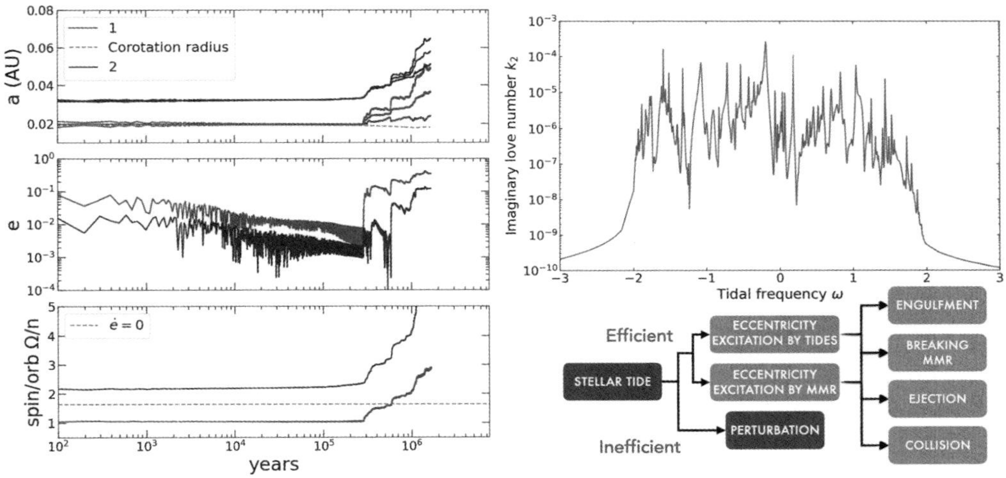

Figure 1. Left: Two hot Jupiters in 2:1 MMR experiencing an outward migration, with the inner planet initially located at $a = 0.02$ AU. Starting from the top panel: the evolution of semi-major axis a (including periastron and apastron), eccentricity e and spin-orbit ratio Ω/n. **Right**: The tidal response spectrum, illustrating the connection between tidal dissipation (quantified by the love number k_2) and tidal frequency, and a brief summary on the influence of the stellar dynamical tide on pairs of planets in first-order resonances.

In the convective envelope of low-mass stars hosting planets, the most efficient dissipation of tidal flows to allow the transfer of angular momentum from the star to planetary orbits is that of tidally-excited inertial modes (Ogilvie & Lin 2007; Bolmont & Mathis 2016). Their restoring force is the Coriolis acceleration. They are dissipated because of the turbulent friction applied by stellar convection. The resulting torque strongly depends on the tidal frequency depending on the strength of the coupling between the tidal force and stellar inertial modes.

As the classical equilibrium tide models (e.g. (Hut 1981)) often used in celestial mechanics are not designed to deal with complex frequency dependencies, we developed for this work a model which uses the Kaula formalism (Kaula 1964), which allows us to take into account any frequency-dependent tidal torque. We implemented it in the N-body code Posidonius (Blanco-Cuaresma & Bolmont 2017) in order to investigate this question.

2. Results

We performed simulations with different planetary masses (5 Earth masses $M_\oplus$ & 1 Jupiter mass M_J), and different initial stellar spin rates Ω, semi-major axis a and states of mean motion resonance (MMR). Figure 1 shows the result of a simulation with a pair of 1 M_J planets initially in 2:1 MMR, demonstrating the breaking of MMR from the dynamical tide. For the lower left panel, the grey dashed line indicates whether the stellar tide can excite the planetary eccentricities (above) or damp them (below).

Due to the resonant characteristics of the dynamical tide, the inner planet experiences several boosts of migration, as the frequency of the tidal interaction coincides with a resonant frequency of the star as we are close to a resonance of the tidal torque (different peaks of the tidal spectrum, see Fig. 1). Due to the migration boosts of the inner planet and the stellar tide, the outer one experiences boosts both in migration and eccentricity resulting mainly from the self-stabilization of the MMR (apart from the torque of the stellar tide). The MMR exhibits an intriguing property: by exciting the planet's eccentricities as the ratio of the orbital period decreases, the MMR configuration is reinforced. Note that the simulation is still ongoing but escape of MMR is expected due to the orbits crossing.

The general influence of dynamical tides on pairs of planets in first-order resonances is given in the lower right panel of Fig. 1. To conclude, if the dissipation of stellar dynamical tide is strong (massive or close-in planets), this interaction would be able to break the MMRs, potentially resulting in engulfment, ejections, and collisions; otherwise, its influence on the system is minimal. An upcoming article will present cases featuring more planets and ultra-short-period planets, with a more intricate analysis of the resonance states (Deck 2013).

References

Blanco-Cuaresma, S., & Bolmont, E. 2017, in EWASS 4, Star-planet interactions (EWASS-SS4-2017)

Bolmont, E., & Mathis, S. 2016, Celest Mech Dyn Astr, 126, 275

Deck, K. M. et al. ApJ 774, 129

Gillon, Michael, et al. Nature, vol. 542, no. 7642, Feb. 2017, pp. 456–60

Hut, P. Astronomy and Astrophysics, vol. 99, June 1981, pp. 126–40

Izidoro, André, et al. Monthly Notices of the Royal Astronomical Society, vol. 470, no. 2

Kaula, William M. Reviews of Geophysics and Space Physics, vol. 2, Jan. 1964, pp. 661–85

Ogilvie, G. I., and D. N. C. Lin. The Astrophysical Journal, vol. 661, June 2007, pp. 1180–91

Pichierri G., & Morbidelli A., 2020, MNRAS, 494, 4950

Complex Planetary Systems II: Latest Methods for an Interdisciplinary Approach
Proceedings IAU Symposium No. 382, 2024
A. Lemaitre & A.-S. Libert, eds.
doi:10.1017/S1743921323004003

Medium Earth Orbit Secular Resonances and Navigation Satellites

Edoardo Legnaro

University of Padova. Department of Mathematics "Tullio Levi Civita"
Aristotle University of Thessaloniki. Department of Physics
email: edoardo.legnaro@gmail.com

Abstract. Inclination-only dependent lunisolar resonances shape the dynamics of MEO (Medium Earth Orbit) objects over secular time scales (i.e. several decades). Their main effect is to increase an object's eccentricity, possibly up to a value where the orbit's perigee meets the Earth's atmosphere and friction will determine the object's re-entry. Thus, understanding this mechanism allows the design of low-cost end-of-life disposal strategies which exploit the resonant dynamics. In this proceeding, we will summarize our results in developing an analytic theory for lunisolar resonances and the characterization of diffusion properties along them. On this topic, the techniques proposed are of interest in most problems of secular resonances encountered in Celestial Mechanics.

Keywords. Diffusion, Lunisolar Resonances, Navigation Satellites

1. Introduction

The MEO (Medium Earth Orbit) region spans from an altitude of 2000 km up to the Geosynchronous orbit, 35786 km, and is the home of various artificial satellites. Among those, we have the GNSS (Global Navigation Satellites System) constellations: the European GALILEO ($a \sim 29600$ km, $i \sim 55°$), the American GPS ($a \sim 26560$ km, $i \sim 55°$) the Chinese BEIDOU ($a \sim 27906$ km, $i \sim 55°$) and the Russian GLONASS ($a \sim 25510$ km, $i \sim 64.8°$). As many studies have highlighted, the MEO region is characterized by the presence of a network of secular lunisolar resonances which govern the dynamics (see Rosengren et al. (2015), Daquin et al. (2016), Daquin et al. (2022), Legnaro and Efthymiopoulos (2022) and references therein). A *lunisolar secular resonance* occurs whenever

$$\exists\, \vec{k} = (k_1, k_2, k_3) \in \mathbb{Z}^3 \backslash \{0\} \qquad \text{s. t.} \qquad k_1\dot{\omega} + k_2\dot{\Omega} + k_3\dot{\Omega}_L = 0, \tag{1}$$

where ω and Ω are respectively the argument of perigee and node of the satellite, while Ω_L is the node of the Moon. If $k_3 = 0$, the resulting resonances are independent of a and e and thus are located at a critical value of the inclination $i_\star$ found by solving

$$k_1\left(5\cos^2 i - 1\right) - 2k_2 \cos i = 0. \tag{2}$$

So, these resonances are called *inclination-only dependent lunisolar resonances* (or ILSRs). The main effect of such resonances is to increase an object's eccentricity, potentially up to a value so high that the orbit's perigee reaches the atmosphere and then friction determines the re-entry (see Figure 1 and Rossi (2008), Alessi et al. (2014), Gkolias et al. (2016), Alessi et al. (2016), Armellin and San-Juan (2018) and Skoulidou et al. (2019)).

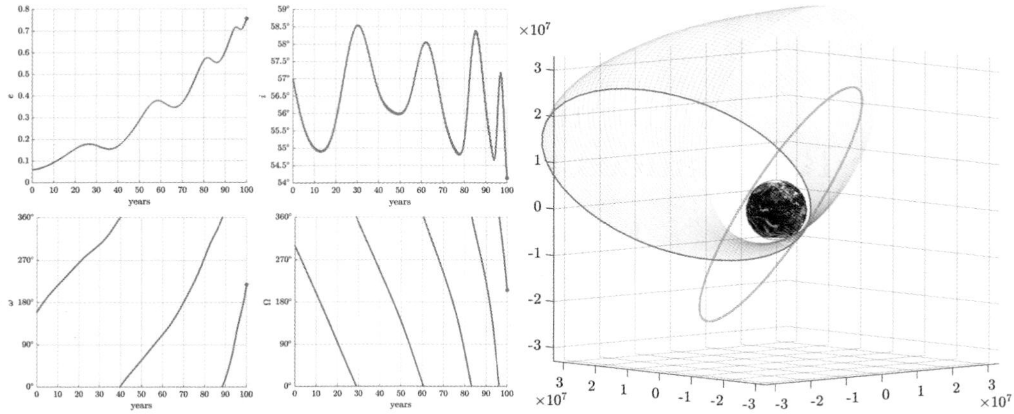

Figure 1. Numerical integration of a satellite at Galileo altitude with $A/m = 0.01$, $a = 29600$ km, $e = 0.06$, $i = 57°$, $E = 10°$, $\omega = 160°$, $\Omega = 300°$. The orbit is propagated using a cartesian integrator in the EME2000 reference frame. Left: evolution of the Keplerian elements e, i, ω, Ω. Right: 3D view of the satellite's orbital plane in cartesian Earth-centered coordinates with axis' unit in km at time $t = 100$ years, with initial circular inclined orbit portrayed in orange and trace showing the orbital position over the previous year, with solid blue line indicating the final orbit. From an almost circular orbit, in 100 years the satellite's eccentricity grows to a value where the satellite's perigee reaches the Earth's atmosphere, so that drag starts the satellite's decay. The minimum lifetime τ of a satellite in the resonant domain of the $2g + h$ resonance approximately follows a logarithmic law, and an object with these initial conditions would have $\tau = 168$y for $e_0 = 0.01$, $\tau = 262$y for $e_0 = 0.001$, $\tau = 369$y for $e_0 = 0.0001$. More details on this can be found in Legnaro (2023).

Nevertheless, as highlighted by these numerical studies, the MEO region and graveyards of navigation satellites are unstable, meaning that a slight difference in initial conditions may lead to dramatically different evolutions of the eccentricity.

This is the reason why having an analytic theory that can provide insights into finding which initial conditions will lead to the re-entry of an object can allow the design of low-cost end-of-life strategies for navigation satellites and space debris.

This proceeding is structured as follows. In Section 2, we will outline the steps needed to study a given lunisolar resonance. In Section 3, the Melnikov and Landau-Teller theories will be presented as a tool to obtain semi-analytical estimates of the speed of chaotic diffusion for an ensemble of points moving in the domain of a lunisolar resonance. In addition, such an approach can be useful in most problems encountered in Celestial Mechanics where a system is driven by the separatrix-like stochastic layers of a secular resonance, such as the ν_6 secular resonance between asteroids and Saturn or secular resonances in the main belt. Finally, Section 4 will summarize the results presented.

2. Analytic Theory for Secular Lunisolar Resonances

The details of our results on the development of an analytic theory for secular lunisolar resonances in the MEO region can be found in Daquin et al. (2022) and Legnaro et al. (2023).

The steps needed to study a given lunisolar resonance are the following. We first derive a set of resonant variables (J_R, u_R, J_F, u_F) adapted to the particular resonance at hand, which involves a 'fast' and a 'resonant' component. This allows to derive a Resonant Hamiltonian $H_R(X, Y, J_F, u_F, A, \Omega_L)$, having Poincaré variables X, Y and a dummy action A conjugated to the node of the Moon Ω_L. Finally, by finding a suitable analytic approximation of the fast component we arrive at an integrable Hamiltonian H which

• allows us to understand and predict the structures seen in Fast Lyapunov Indicator (FLI) cartography while changing the initial values of Ω_L or of the satellite's perigee and node ω, Ω,

• provides insights into the eccentricity growth mechanism, allowing to understand the different behaviours of the maximum eccentricity reachable along a lunisolar resonance as a function of the initial phases and eccentricity,

• allows to reliably predict the domain of lunisolar resonances.

In particular, if we define the *center manifold* as the normally hyperbolic invariant manifold where the motion of circular orbits takes place, it is possible to see that this manifold has a large measure of invariant tori, and the initial choice of Ω shifts the resonance along the tori limiting its domain. The initial choice of Ω_L changes the shape of the center manifold. We define *fast drift plane* (FDP) as the locus of points where the fast action J_F is constant. The (i, e) plane is foliated by FDP, and each FDP is associated with an integrable phase portrait which shows the resonance's separatrix. The combination of the initial values of ω and Ω defines a scanning direction, and its intersection with the integrable separatrix gives an eccentricity value for that specific fast drift plane. In addition, for every FDP, the extension of the figure-8 separatrix dictates the theoretical maximum eccentricity value e_{max} that an object would reach moving along the resonance. This can explain numerical results that find different behaviours of the maximum eccentricity as a function of the initial inclination and node, like Alessi et al. (2016).

3. Diffusion along lunisolar resonances

What can be done to characterize the diffusion properties of an ensemble of points along a lunisolar resonance? In Legnaro et al. (2023), we show how it is possible to find estimates on the diffusion speed adopting Melnikov or Landau-Teller approaches. Also, the presented framework can be applied to most secular resonances encountered in Celestial Mechanics.

The integrable model derived in Section 2 for lunisolar resonances belongs to the Second Fundamental Model of Resonance (SFM, see Henrard and Lemaitre (1983)). To illustrate the concepts of the theory, consider the following 'archetype model' for lunisolar resonances

$$H_a = -S\cos 2\sigma + S^2\lambda\left[1 + \varepsilon\cos(2\sigma - \phi) + \varepsilon\cos(2\sigma - \phi_2)\right] + A\Omega + A_2\Omega_2, \qquad (3)$$

where $(A, A_2) \in \mathbb{R}^2$, $S \geq 0$, $\sigma \in \mathbb{S}^1$, and $\varepsilon, \Omega, \Omega_2 \geq 0$. Consider Poincaré variables $X = \sqrt{2S}\sin\sigma$, $Y = \sqrt{2S}\cos\sigma$. In systems like this, which involve a resonant degree of freedom coupled to oscillators, the evolution of the adiabatic actions is not uniform in time, but rather proceeds with a sequence of discrete 'jumps' (see Figure 2). The adiabatic action A captures the evolution of the energy of the system over time, thus we wish to model the diffusion on this variable.

Consider N trajectories $A_i(t)$. If the evolution of the i-th trajectory proceeds as a sequence of consecutive discrete jumps $\Delta A_i^{(j)}$ with mean period $< T >$, then there are $n = [t/ < T >]$ jumps taking place per each trajectory. Setting the initial condition $A_i(0) = 0$ for all $i = 1, \ldots, N$, we have

$$A_i(t = n < T >) = \sum_{j=1}^{n} \Delta A_i^{(j)}. \qquad (4)$$

If the jumps are not correlated, i.e., the individual jumps $\Delta A_i^{(j)}$ are randomly chosen from a distribution $p(\Delta A)$ with mean $\mu = 0$ and dispersion $\sigma_{\Delta A}$, then by the central limit theorem follows $A_i \sim \mathcal{N}\left(0, n\sigma^2_{\Delta A}\right)$ for n large. Since $\mu = 0$, we have $\sigma^2_{\Delta A} =< \Delta A^2 >$.

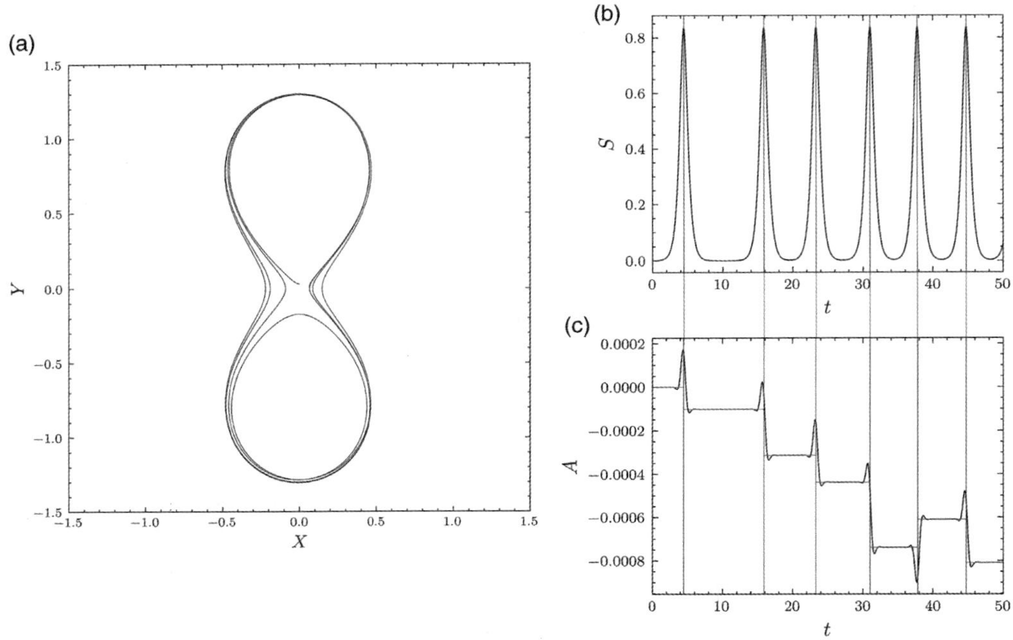

Figure 2. Evolution of an orbit with initial conditions $(X_0, Y_0) = (0, 0.3)$ and $(A, A_2, \phi, \phi_2) = (0, 0, 0, 0)$ under the flow of Hamiltonian (3) with parameters $\varepsilon = 0.01, \lambda = 1.2, \Omega = 5, \Omega_2 = \sqrt{5}$. The jumps in the adiabatic action A (c) correspond to homoclinic loops in the X, Y variables (a), and the jumps occur at the top of the loop, as can be seen from the evolution of $S(t)$ (b) which essentially measures the distance from the center.

Hence, the linear diffusion coefficient for the ensemble satisfies the relation:

$$D = \frac{n\sigma^2_{\Delta A}}{n < T >} = \frac{< \Delta A^2 >}{< T >} \tag{5}$$

In this way, the task of estimating the diffusion speed reduces to finding an estimate of $< \Delta A^2 >$ and $< T >$. For the first quantity, it is possible to adopt a *Melnikov Approach.* Consider $\dot{A} = -\frac{\partial H}{\partial \phi} = -\varepsilon\lambda S^2 \sin(2\sigma - \phi)$. We approximate $S(t)$, $\sigma(t)$, as well as $\phi(t)$, $\phi_2(t)$, with the analytic solution $\hat{S}(t)$, $\hat{\sigma}(t)$, $\hat{\phi}(t)$, $\hat{\phi}_2(t)$ of the separatrix of H_a for $\varepsilon = 0$. This yields

$$A(t) \approx A(t_0) - \int_{t_0}^{t} \varepsilon\lambda\hat{S}^2(u) \sin\left(2\hat{\sigma}(u) - (\phi(t_0) + \Omega u)\right) du. \tag{6}$$

The approximation is based on the idea that the evolution of the resonant variables (X, Y) (or (S, σ)) for initial conditions close to the separatrix of the integrable model is quite similar to the full one. Finally, the jump $\Delta A(\phi_0) = A(t_1) - A(t_0)$ between consecutive plateaus of $A(t)$ is approximated by the Melnikov integral

$$\Delta A \approx \Delta A_M = -\varepsilon\lambda \int_{-\infty}^{\infty} \hat{S}^2_\infty(u) \sin\left(2\hat{\sigma}_\infty(u) - (\phi_0 + \Omega u)\right) du. \tag{7}$$

In the particular example at hand, since $S_\infty(t)$, $\sigma_\infty(t)$ are odd and even functions of t respectively, we readily find:

$$\Delta A_M(\phi_0) = \varepsilon\lambda \sin\phi_0 I_1 \tag{8}$$

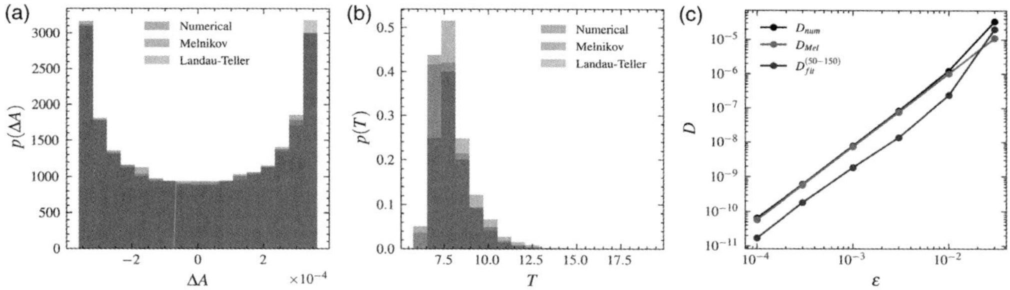

Figure 3. (a) Numerical (grey) histogram of the probability distribution $P(\Delta A)$ of the individual jumps in the value of the adiabatic action $A(t)$, collected by the data of the complete ensemble of 10000 simulated trajectories for $\varepsilon = 0.001, \lambda = 1.2, \Omega = 5, \Omega_2 = \sqrt{5}$ with initial X, Y data in the square $[-0.05, 0.05]^2$, on top of the corresponding histogram by the Melnikov approximation (blue, Eq.((8))). A third estimate of the same histogram (red) is obtained independently by the stochastic Landau-Teller model. (b) Comparison of the three histograms obtained as in (a) for the PDF of the periods $p(T)$. (c) The estimates on the diffusion coefficients with the assumptions of normal diffusion and uncorrelated phases $D = <\Delta A^2> / <T>$ for the numerical data (black) and by the Melnikov approach (red), as a function of ε. The blue curve shows the numerical diffusion coefficient as a function of ε obtained by linearly fitting the dispersion.

where

$$I_1 = \int_{-\infty}^{\infty} \hat{S}_{\infty}^2(u) \cos\left(2\hat{\sigma}_{\infty}(u) - \Omega u\right) du,$$

which in turn allows to find a theoretical estimate for the probability density function (PDF) of the jumps, which implies $< \Delta A_M^2 > = \varepsilon^2 I_1^2/2$. Alternatively, a similar estimate of the PDF of the jumps can be found adopting what we call a *stochastic Landau-Teller model*, which consists of using the Fourier representation of finite-time homoclinic pulses along with a stochastic process based on random values of the integrable energy obtained through a suitably defined distribution.

For the mean period, it is possible to estimate the initial width of the chaotic layer by computing the *first subharmonic Melnikov integral*, which essentially computes the displacement between the stable and unstable manifolds of the hyperbolic point. Finally, it is possible to approximate $< T >$ by considering a uniform distribution of points inside the estimated chaotic layer and then computing for each point the analytic period.

It is now possible to estimate the diffusion coefficient with $D_{Mel} = \Delta A_M^2 / < T >$, which approximates very well the numerically computed $D_{num} = <\Delta A^2> / <T>$ (see Figure 3 (a-b)). Still, in these systems, the presence of temporary stability islands and partial barriers generates long-term correlations of the 'phases' due to 'stickiness' effects (see Contopoulos and Harsoula (2010)). This can be addressed by introducing a *reduction factor* affecting all estimates on the speed of diffusion of the adiabatic actions given by the model, as also done in Chirikov (1979) (compare with Figure 3 (c)).

The same framework can be applied in the system of navigation satellites for the semi-analytical characterization of diffusion properties. We will focus on the $2g + h$ resonance since it is the most interesting for applications. We are interested in the diffusion along FDPs since they are linked to e_{max}, as discussed in Section 2. By averaging over the angles u_F and Ω_L of the resonant Hamiltonian and keeping just the leading terms, we get the integrable Hamiltonian which belongs to the SFM. Using the Melnikov approach and estimating the average period by computing the first subharmonic Melnikov integral, we obtain semi-analytically a diffusion coefficient

$$D(J_F)_M := \left\langle \Delta J_F^2 \right\rangle / \langle T \rangle \tag{9}$$

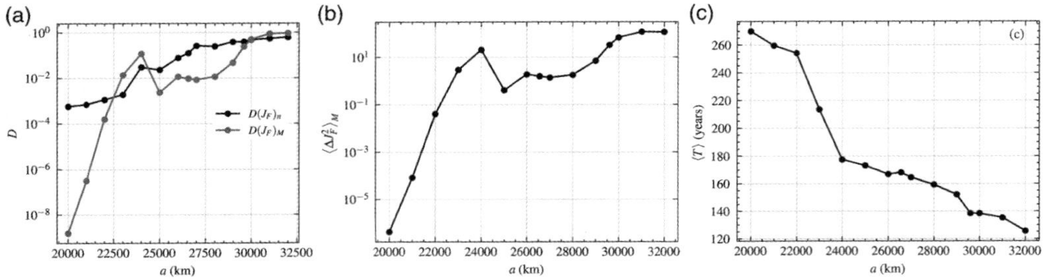

Figure 4. (a) Comparison between the semi-analytical diffusion coefficient $D(J_F)_M$ (red) and the numerically computed one $D(J_F)_n$ (black). (b) The value of $\langle \Delta J_F^2 \rangle$ is derived using the Melnikov approach. (c) The value of $\langle T \rangle$ estimated computing the first subharmonic Melnikov integral.

which can be compared with the numerically computed one $D(J_F)_n$ found by fitting the $\sigma^2(J_F)$ curve (see Figure 4).

The semi-analytical prediction works best for higher altitudes. This is to be expected, since lower altitudes would require additional normalization steps, while at higher altitudes no additional perturbation step would make the remainder smaller and so the original Hamiltonian is already the optional one.

4. Conclusions

In this proceding, we presented our results in the development of an analytic theory for secular lunisolar resonances. This allows to understand the dynamics of navigation satellites and so can be applied to the design of low-cost end-of-life disposal strategies. In addition, we provided a framework to semi-analytically characterize the diffusion in the MEO region. Such a framework can be applied to study the diffusion in most systems with secular resonances encountered in Celestial Mechanics.

References

Alessi, E. M., Deleflie, F., Rosengren, A. J., Rossi, A., Valsecchi, G. B., Daquin, J., & Merz, K. 2016, A numerical investigation on the eccentricity growth of GNSS disposal orbits. *Celestial Mechanics and Dynamical Astronomy*, 125(1), 71–90.

Alessi, E. M., Rossi, A., Valsecchi, G., Anselmo, L., Pardini, C., Colombo, C., Lewis, H., Daquin, J., Deleflie, F., Vasile, M., *et al.* 2014, Effectiveness of GNSS disposal strategies. *Acta Astronautica*, 99, 292–302.

Armellin, R. & San-Juan, J. F. 2018, Optimal Earth's reentry disposal of the galileo constellation. *Advances in Space Research*, 61(4), 1097–1120.

Chirikov, B. V. 1979, A universal instability of many-dimensional oscillator systems. *Physics reports*, 52(5), 263–379.

Contopoulos, G. & Harsoula, M. 2010, Stickiness effects in conservative systems. *International Journal of Bifurcation and Chaos*, 20(07), 2005–2043.

Daquin, J., Legnaro, E., Gkolias, I., & Efthymiopoulos, C. 2022, A deep dive into the $2g + h$ resonance: separatrices, manifolds and phase space structure of navigation satellites. *Celestial Mechanics and Dynamical Astronomy*, 134(1), 1–31.

Daquin, J., Rosengren, A. J., Alessi, E. M., Deleflie, F., Valsecchi, G. B., & Rossi, A. 2016, The dynamical structure of the MEO region: long-term stability, chaos, and transport. *Celestial Mechanics and Dynamical Astronomy*, 124(4), 335–366.

Gkolias, I., Daquin, J., Gachet, F., & Rosengren, A. J. 2016, From order to chaos in Earth satellite orbits. *The Astronomical Journal*, 152(5), 119.

Henrard, J. & Lemaitre, A. 1983, A second fundamental model for resonance. *Celestial mechanics*, 30(2), 197–218.

Legnaro, E. & Efthymiopoulos, C. 2022, A detailed dynamical model for inclination-only dependent lunisolar resonances. Effect on the "eccentricity growth" mechanism. *Advances in Space Research*,, S0273117722006871.

Legnaro, E., Efthymiopoulos, C., & Harsoula, M. 2023,. Semi-analytical estimates for the chaotic diffusion in the second fundamental model of resonance. Application to Earth's navigation satellites.

Legnaro, E. A. 2023,. *Orbital Dynamics and Diffusion at the resonances in the near-Earth space environment.* PhD thesis, Aristotle University of Thessaloniki.

Rosengren, A. J., Alessi, E. M., Rossi, A., & Valsecchi, G. B. 2015, Chaos in navigation satellite orbits caused by the perturbed motion of the Moon. *Monthly Notices of the Royal Astronomical Society*, 449(4), 3522–3526.

Rossi, A. 2008, Resonant dynamics of medium earth orbits: space debris issues. *Celestial Mechanics and Dynamical Astronomy*, 100(4), 267–286.

Skoulidou, D. K., Rosengren, A. J., Tsiganis, K., & Voyatzis, G. 2019, Medium Earth Orbit dynamical survey and its use in passive debris removal. *Advances in Space Research*, 63(11), 3646–3674.

Complex Planetary Systems II: Latest Methods for an Interdisciplinary Approach
Proceedings IAU Symposium No. 382, 2024
A. Lemaitre & A.-S. Libert, eds.
doi:10.1017/S1743921323004763

Investigation of the dynamical evolution of planetary systems with isotropically varying masses

M. Zh. Minglibayev[1,2], **A. N. Prokopenya**[3] **and A. B. Kosherbayeva**[1]

[1]Al-Farabi Kazakh National University, Almaty, Kazakhstan

[2]Fesenkov Astrophysical Institute, Almaty, Kazakhstan
email: Minglibayev@gmail.com

[3]Warsaw University of Life Sciences, Warsaw, Poland

Abstract. In this work, the secular evolution of exoplanetary systems is investigated, when the variability of the masses of celestial bodies is the leading factor of dynamical evolution. The masses of the parent star and the planets change due to the particles leaving the bodies and falling on them. At the same time, bodies masses are assumed to change isotropically at different rates. The law of mass change is considered to be known and given function of time. The relative motions of the planets are investigated by the methods of the canonical perturbation theory in the absence of resonances. It is assumed that the orbits of the planets do not intersect. Evolutionary equations in analogues of Poincaré variables $(\Lambda_i, \lambda_i, \xi_i, \eta_i, p_i, q_i)$ are obtained and used to study the *K2-3* exoplanetary system. All analytical and numerical calculations are performed with the aid of the Wolfram Mathematica.

Keywords. variable mass, canonical variables, perturbation theory, exoplanetary system, Poincaré variables.

1. Introduction

To date, more than 4,000 exoplanetary systems are known (Exoplanet Exploration 2023). The variability of the masses of real celestial bodies significantly affects the possible trajectories of their motion and determines the rich variety of evolutionary orbits. In our work, within the framework of the classical problem of $n+1$ bodies with variable masses, the relative motion of n planets in quasi-elliptical non-intersecting orbits is investigated. It is assumed that the celestial bodies are spherically symmetrical and attract each other according to Newton's law of gravitation.

2. Problem statement

In the considered case of the n planets problem, it is convenient to write the differential equations in the following form (see Minglibayev (2012)):

$$\ddot{\vec{r}}_i + f\frac{(m_0+m_i)}{r_i^3}\vec{r}_i - \frac{\ddot{\gamma}_i}{\gamma_i}\vec{r}_i = grad_{\vec{r}_i} f\sum_{j=1}^{n} m_j\left(\frac{1}{r_{ij}} - \frac{\vec{r}_i\cdot\vec{r}_j}{r_j^3}\right) - \frac{\ddot{\gamma}_i}{\gamma_i}\vec{r}_i, \quad i\neq j, \tag{1}$$

here $\vec{r}_i$ is the radius-vector of the considered planet relative to the center of mass of the parent star, f is the gravitational constant, m_0, m_i are the masses of the parent star

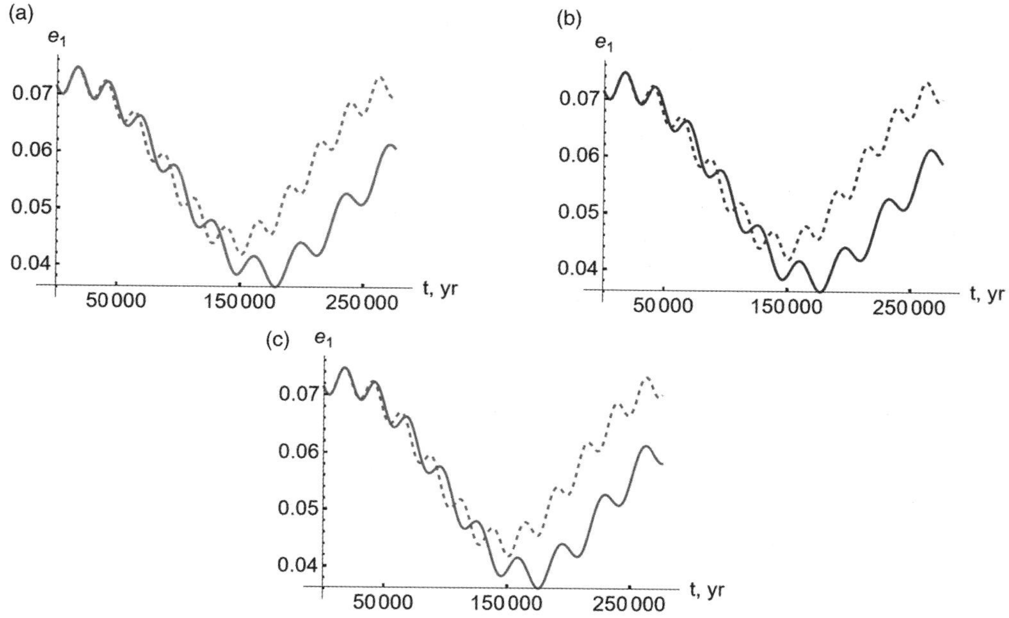

Figure 1. Comparison of the obtained results for the eccentricities in the constant mass case (dashed line) and in the cases of variable mass (continuous line). The Eddington-Jeans laws are considered when the degree of the mass law for the parent star are a) $n_0 = 5/2$; b) $n_0 = 3$ and c) $n_0 = 7/2$.

and planet, respectively, $\vec{r}_{ij} = \vec{r}_j - \vec{r}_i$, $\gamma_i = \frac{m_0(t_0)+m_i(t_0)}{m_0(t)+m_i(t)} = \gamma_i(t)$, t_0 is the initial instant of time.

3. Secular perturbations

In the system of differential equations (2.1), if the right-hand side is zero, we get aperiodic motions along the quasi-conic sections (as in see Minglibayev (2012) for more information), which are used as unperturbed motion. The corresponding canonical theory of perturbations in analogues of Poincaré variables has been developed in Prokopenya et al. (2022) and in Kosherbayeva et al. (2023).

4. Dynamical evolution of the system *K2-3*

The *K2-3* exoplanetary system is considered as an example (Almenara et al. 2015). Numerical solutions of the evolutionary equations written in dimensionless variables in Kosherbayeva et al. (2023) are obtained, and the dependence of the analogues of the osculating elements on time is found. Let's assume that the mass of the parent star changes over time according to the Eddington-Jeans law $\dot{m}_0 = -\alpha_0 \ m_0^{n_0}$, where $0.4 < n_0 < 4$ and parameter α_0 is chosen from the condition $\dot{m}_0 = -10^{-5} m_0/yr$ at the initial instance of time. Masses of the planets are assumed to increase linearly with time with the rates $\dot{m}_1 = \frac{m_1}{40000} yr^{-1}$, $\dot{m}_2 = \frac{m_2}{70000} yr^{-1}$ and $\dot{m}_3 = \frac{m_3}{80000} yr^{-1}$. Figure 1 shows the results obtained for the analogue of the eccentricity of the first planet during the interval of 250 000 yr.

5. Conclusion

The resulting system of evolutionary equations includes $4n$ linear non-autonomous differential equations, the coefficients of which are various complicated functions of time.

In this work, the evolutionary equations for $n=3$ are explicitly obtained for the system *K2-3*. Various tracks of osculating elements' evolution are studied by numerical methods.

6. Acknowledgements

This research was funded by the Committee of Science of the Ministry of Science and Higher Education of the Republic of Kazakhstan (Grant No. AP14869472).

References

Exoplanet Exploration, url: https://exoplanets.nasa.gov/, Last update: October 16, 2023.

Minglibayev, M. Zh., 2012, LAP LAMBERT Academic Publishing, 224 (in Russ.)

Prokopenya, A. N., Minglibayev, M. Zh., Kosherbaeva, A. B., 2022, Programming and Computer Software, 48(2),107–115, DOI: 10.1134/S0361768822020098

Kosherbayeva, A. B., Minglibayev, M. Zh. 2023, Proceedings of the International Astronomical Union, 17, S370: Winds of Stars and Exoplanets, 283–284, DOI:10.1017/S1743921322003611

Almenara, J. M., et al. 2015, A&A, 581, id.L7, 6, DOI:10.1051/0004-6361/201525918

Complex Planetary Systems II: Latest Methods for an Interdisciplinary Approach
Proceedings IAU Symposium No. 382, 2024
A. Lemaitre & A.-S. Libert, eds.
doi:10.1017/S1743921323003915

On the long-term orbital evolution of a satellite revolving around an oblate body

Gabriela-Ana Nadabaică

Faculty of Mathematics, Al. I. Cuza University of Iaşi
Blvd. Carol I, no. 11, 700506 Iaşi, Romania
email: nadabaicagabriela@yahoo.com

Abstract. In the context of a perturbed two body problem, in which the Keplerian motion of the small object (the satellite) is perturbed by the oblateness of the central body (the asteroid) and the attraction of a third body (the Sun), we discuss the long-term evolution of the orbital elements of a satellite orbiting an oblate body, with a particular focus on the behavior of the inclination and the longitude of the ascending node. We derive analytically the position of the Laplace plane as a function of several parameters and use this solution to analyse the long-term evolution of distant circular orbits. The analytical study is complemented by numerical tests, performed in the context of both Cartesian and Hamiltonian frameworks. The results give a description of the orbital dynamical environment of asteroids and reveal the parameters that play a key role in the long-term stability of distant circular orbits.

Keywords. Laplace plane, obliquity, Hamiltonian approach, Cartesian approach

1. Introduction

Recent and future missions have the purpose of orbiting and landing on the surface of small bodies, like asteroids, comets and planetary satellites. Some of them are sample return missions (e. g. OSIRIS-REx mission). Such missions rely on a good knowledge of the dynamics around the explored object since a detailed study of the motion on various time scales can provide optimal trajectories for the spacecraft, give stable orbits or even describe the space environment. Indeed, a cartographic study of the dynamics around minor bodies might assess the possibility of existence of accompanying swarms of particles. On the other hand, post mission stability analysis can provide valuable information on the body system, opening windows into the past of the Solar system and creating opportunities for future human explorations.

In this work we are using both the Cartesian equations of motion and the Hamiltonian formalism to investigate the long–term evolution of the orbits of a particle orbiting an asteroid within the context of a model taking into account the attraction of the central body (the asteroid), including the perturbations induced by the J_2, J_3 and J_4 harmonic terms, and the attraction of the Sun. In particular, we determine the location of the Laplace plane, also called the invariable plane (see Allan & Cook (1964)), whose normal vector is located in the plane determined by the normal vector of the equatorial plane of the asteroid and the normal vector of the Sun's orbital plane, and use this solution to analyse the long-term evolution of distant initially circular orbits.

We perform a parametric study to identify those parameters whose variation leads to large amplitude librational motions of the inclination and the longitude of the ascending node of initially circular orbits. In this sense, we show that the obliquity of the minor body is a key parameter that plays a major role in the long-term evolution of the inclination and the longitude of the ascending node. We illustrate this aspect by analysing two cases, namely the dynamics around Vesta (the obliquity of Vesta has a moderate value $\varepsilon = 15.66^o$, according to Bills & Nimmo (2011)) and, respectively, around Eros (the obliquity of Eros has the extreme value of $\varepsilon = 89^o$, see Souchay et al. (2003)).

We also reveal the role played by the secular resonances and the J_3 harmonics terms in the long-term evolution of the eccentricity of initially circular orbits and based on this analysis we discuss the conditions in which the asteroids might have accompanying satellites or swarms of particles.

2. Models and methods

We consider an infinitesimal particle (natural satellite or spacecraft) orbiting an oblate minor body (asteroid or comet) which is uniformly rotating around a fixed axis, aligned with the body axis (the principal axis with the highest moment of inertia), and revolves around the Sun on an elliptical orbit.

In *Cartesian coordinates*, the motion of the particle, referred to a centred body system, is described by:

$$\ddot{\mathbf{r}} = R_3(-\theta)\ \nabla V(\mathbf{r}) - Gm_S\left(\frac{\mathbf{r}-\mathbf{r}_S}{|\mathbf{r}-\mathbf{r}_S|^3} + \frac{\mathbf{r}_S}{|\mathbf{r}_S|^3}\right), \tag{2.1}$$

where $\mathbf{r}$ is the position vector of the particle, $\mathbf{r}_S$ is the position vector of the Sun, V is the gravitational potential of the oblate body (see Kaula (1961)), m_S is the Sun's mass, G is the gravitational constant, ∇ is the gradient operator, θ is an angle describing the rotation of the minor body and $R_3(\theta)$ is the rotation matrix of angle θ around the third axis.

Using the action–angle *Delaunay variables* $(L, G, H, M, \omega, \Omega)$, which are related to the orbital elements $(a, e, i, M, \omega, \Omega)$ by $L = \sqrt{\mu a}$, $G = L\sqrt{1-e^2}$, $H = G\cos i$, where $\mu = Gm$, m being mass of the minor oblate body, the dynamics is described by *the Hamiltonian*

$$\mathcal{H} = -\frac{\mu^2}{2L^2} - \mathcal{R}_A - \mathcal{R}_S\ , \tag{2.2}$$

where $\mathcal{R}_A$ (see Kaula (1961)) and $\mathcal{R}_S$ (see Kaula (1962)) are the disturbing functions due to the minor body and respectively the Sun.

Kaula (1961) gives a Fourier expansion of $\mathcal{R}_A$ in terms of the orbital elements and rotation angle θ:

$$\mathcal{R}_A = -\frac{\mu_A}{a}\sum_{n=2}^{\infty}\sum_{m=0}^{n}\left(\frac{R_A}{a}\right)^n J_{nm}\sum_{p=0}^{n} F_{nmp}(i)\sum_{q=-\infty}^{\infty} G_{npq}(e)\, cs_{nm}\Big(\Psi_{nmpq}(M,\omega,\Omega,\theta)\Big)\,, \tag{2.3}$$

where R_A is the reference radius of the minor body, J_{nm} are the harmonic coefficients, F_{nmp}, G_{npq} are the inclination and eccentricity functions, cs_{nm} is the cosine (sine) function if $n-m$ is even (odd) and

$$\Psi_{nmpq}(M,\omega,\Omega,\theta) = (n-2p)\omega + (n-2p+q)M + m(\Omega-\theta) - m\lambda_{nm}.$$

Kaula (1962) derived the expansion of the third body disturbing function, as a function of the orbital elements of both the perturbed and perturbing bodies, in the form:

$$\mathcal{R}_S = \sum \mathcal{A}^S_{k_1k_2k_3k_4k_5k_6}(a/a^*_S, e, i, e^*_S, i^*_S)\cos(k_1M + k_2M^*_S + k_3\omega + k_4\omega^*_S + k_5\Omega + k_6\Omega^*_S),$$

respectively, with k_j integers, where the elements of the satellite and the third-body are referred to the celestial equator. Since the motion of the minor body is assumed to be elliptic, it follows that, for the Sun, the angles ω^*_S and Ω^*_S are constant and M^*_S varies linearly in time.

Focusing on the study of distant circular orbits (see also Allan & Cook (1964)), we consider a double averaged model that includes the effects of J_2 and the attraction of the Sun. Expressed in *the Milankovitch elements* $\mathbf{e} = e\mathbf{P}$ and $\mathbf{h} = (1-e)^2\mathbf{R}$, where $\mathbf{P}$ is a unit vector in the orbital plane and directed toward pericentre, while $\mathbf{R}$ is the unit vector along the positive normal to the orbital plane, the equations of motion has the form:

$$\dot{\mathbf{h}} = \mathbf{h} \times \frac{\partial V^*}{\partial \mathbf{h}} + \mathbf{e} \times \frac{\partial V^*}{\partial \mathbf{e}}, \qquad \dot{\mathbf{e}} = \mathbf{e} \times \frac{\partial V^*}{\partial \mathbf{h}} + \mathbf{h} \times \frac{\partial V^*}{\partial \mathbf{e}}, \tag{2.4}$$

where V^* is the potential averaged over the mean anomalies.

3. Results

The second equation of the system (2.4) is identically satisfied with $e = 0$. So, assuming that the orbit is circular, the first equation of (2.4) becomes:

$$\dot{\mathbf{R}} = -\omega_0(\mathbf{R}\cdot\mathbf{R}_0)(\mathbf{R}_0 \times \mathbf{R}) - \omega_1(\mathbf{R}\cdot\mathbf{R}_1)(\mathbf{R}_1 \times \mathbf{R}), \tag{3.1}$$

where $\mathbf{R_0}$ is the unit vector along the normal to the equatorial plane of the asteroid, $\mathbf{R_1}$ is the unit vector along the normal to the orbital plane of the Sun, ω_0 is a constant depending on J_2 and ω_1 a constant depending on the mass of the Sun.

The identification of the equilibrium points of the dynamical problem (3.1) is effectively a problem of eigenvalues and eigenvectors (see Allan & Cook (1964)):

$$(\omega_0\mathbf{R}_0\mathbf{R}_0 + \omega_1\mathbf{R}_1\mathbf{R}_1)\mathbf{V} = \lambda\mathbf{V},$$

where $\mathbf{AB}$ is the dyadic product of the vectors $\mathbf{A}$ and $\mathbf{B}$. The eigenvalues λ are obtained by finding the roots of the following equation

$$det(\omega_0\mathbf{R}_0\mathbf{R}_0 + \omega_1\mathbf{R}_1\mathbf{R}_1 - \lambda\underline{I}) = 0,$$

where $\underline{I}$ is the idemtensor. It is easy to see that one eigenvalue is null, let us denote it by λ_1, $\lambda_1 = 0$, while the other two, namely λ_2 and λ_3 might be expressed in terms of the constants ω_0 and ω_1. By computing the corresponding eigenvectors, we determine, in fact, the equilibrium positions for the considered dynamical system. The inclination and the longitude of the ascending node of the equilibrium points are deduced from the Cartesian coordinates of the unit vectors associated to the eigenvectors.

In view of the above remarks, it follows that the equation (3.1) admits 3 equilibrium points in the domain $i \in [0^0, 90^0]$, $\Omega \in [0^0, 180^0]$, two stable and one unstable. The position of the equilibria can be determined by drawing the level sets as in Figure 1, where two cases have been depicted, namely the dynamics around Vesta (top panels) and around Eros (bottom panels).

In the (i, Ω) plane, there are possible two types of motions, namely librational and rotational, and since the vectorial equation (3.1) can be integrated (it can be reduced to a set of equations similar to the Euler equations for the motion of a rigid body) the position of the equilibria as well as the amplitude and the period of librations are provided by analytical formulas.

For instance, the position of the stable equlibrium point located at $\Omega = 0^o$, which, in facts, gives the position of the Laplace plane (the angle between the Laplace plane and the equatorial plane of the minor body), is shown in Figure 2 as a function of the distance (expressed in equivalent radii) from the central body. The amplitude of the librational island depends on many parameters, such as the orbital elements of the Sun, the value of J_2, in fact - the mass of the minor body and its obliquity. We identified the obliquity as a key parameter in the sense that the larger the value of the obliquity is, the larger the amplitude of the libration should be (see Figure 1).

The results are validated by numerical studies obtained by integrating the Cartesian equations of motion (2.1) and the canonical equations associated to the Hamiltonian

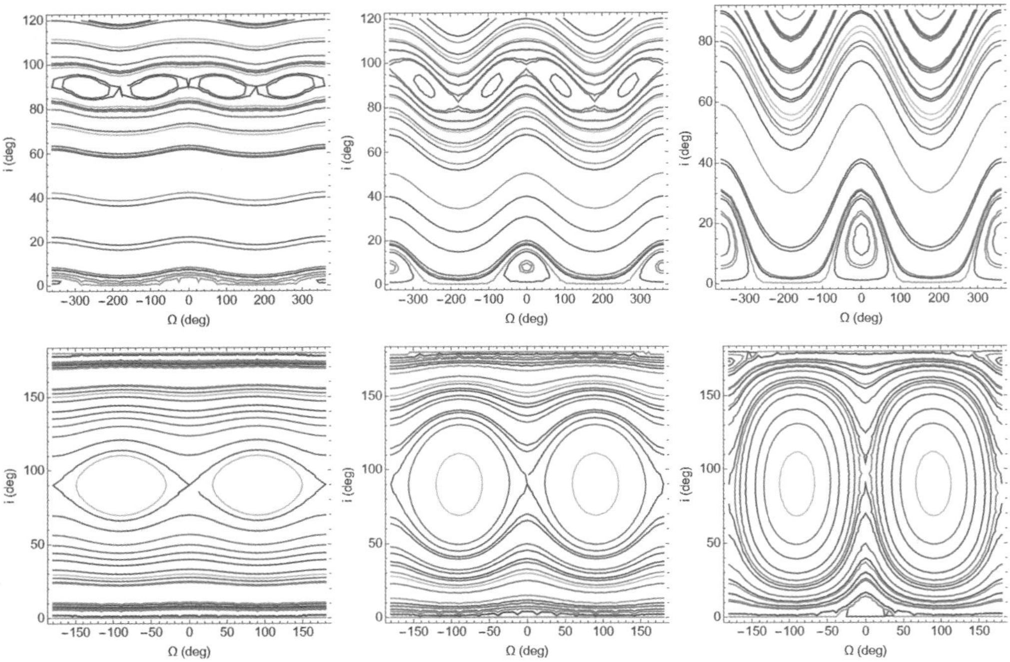

Figure 1. The level sets for the case of Vesta: $a = 6000$ km (top left), $a = 9000$ km (top middle), $a = 15000$ km (top right) and for the case of Eros: $a = 200$ km (bottom left), $a = 250$ km (bottom middle), $a = 280$ km (bottom right).

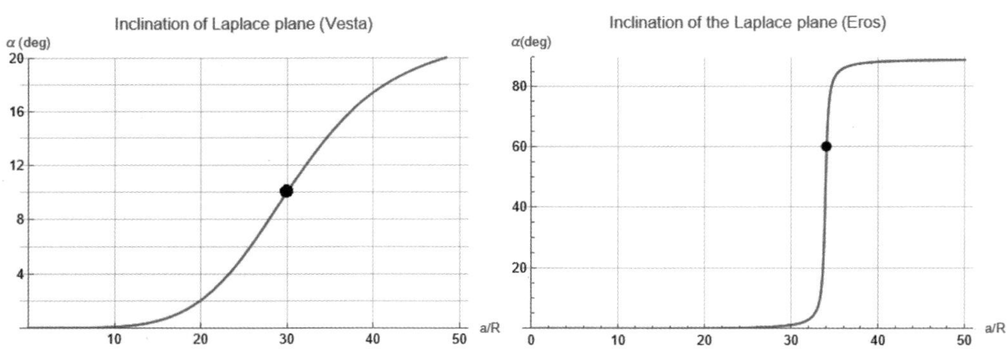

Figure 2. The location of the Laplace plane in the cases of Vesta (left) and Eros (right).

(2.2) and by considering a more complete model that includes all zonal harmonic terms up to degree $n = 4$.

Although the inclination and the longitude of the ascending node have the behaviour predicted by the above analytical arguments, the eccentricity does not remain constant. As effect of J_3, which induces a slight variation of the eccentricity (even if the initial eccentricity is zero), and of the secular solar resonances, which might induce very large variations of eccentricity for non–circular orbits, the long-term behaviour of the orbits is complex. Indeed, if an initially circular orbit suffers a variation in inclination large enough to cross a secular solar resonance, then, as effect of J_3, the eccentricity becomes larger than zero making possible for the orbit to be captured in the libration region of the secular resonances, which induces large excursions in eccentricity.

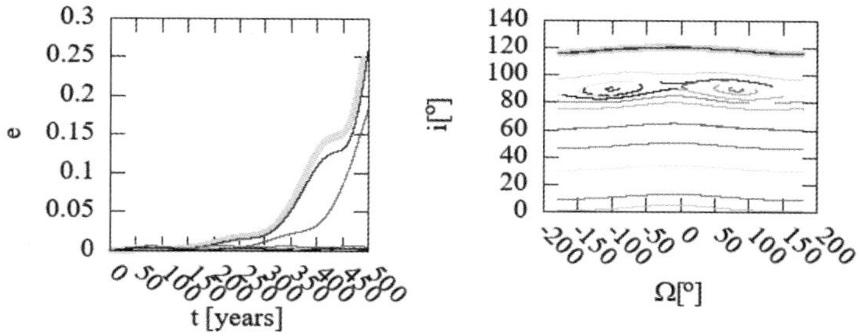

Figure 3. Evolution of the orbital elements for the case of particles orbiting Vesta, located at the distance 6000 km, with the initial eccentricity $e = 10^{-4}$. The green and yellow thick lines represent orbits propagated using the Cartesian approach and the thin lines are obtained using the Hamiltonian formalism. In both frameworks, the effects of the J_2, J_3 and J_4 harmonic terms is taken into account.

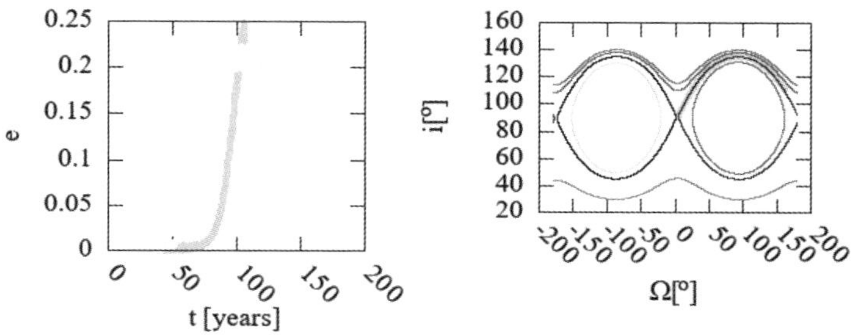

Figure 4. Evolution of the orbital elements for the case of particles orbiting Eros, located at the distance 250 km, with the initial eccentricity $e = 10^{-6}$. The green and yellow thick lines represent orbits propagated using the Cartesian approach (considering all 3 harmonic terms) and the thin lines are obtained by using the Hamiltonian formalism (considering only the J_2 term).

For relatively small inclinations (the secular resonances are absent), the circular orbits around Vesta are stable (remain almost circular) even at appreciable distances (see Figure 3). At $i = 63.4^o$ and $i = 63.4^o$ and $i = 116.4^o$ are located the critical inclination resonances. The curves in green, blue and red (Figure 3) cross one of these resonances, and as described above the eccentricity increases considerably.

On the other hand, the initially circular orbits around Eros do not share the same properties, due to the large obliquity of Eros, about 89°. At distances, let say larger than 250 km, the inclinations of the orbits perform large excursions (see Figure 1, bottom panels) and as a consequence, they cross the regions where secular regions are located. As effect of the J_3 harmonic term the eccentricity becomes larger than zero, and the secular resonances induces a large increase in eccentricity (see Figure 4).

4. Implications

In this work we have done an analysis of the equilibrium points (in the (Ω, i) plane) and we have obtained some preliminary results: the position of the equilibrium points and the amplitude of the libration islands depend on a variety of parameters (the semi-major axis of the asteroid, its eccentricity, its obliquity and so on). The simulations show that a key parameter is the obliquity of the minor body. The results are important for the determination of the orbital dynamical environment of asteroids and, in particular, in the analysis of the long-term stability of circular orbits. A preliminary conclusion is the following: Vesta might possess satellites even at appreciable distances, provided the inclination is less than 45^0. However, in the case of Eros, for altitudes larger than 250 km, the large Eros' obliquity of 89° (see Souchay et al. (2003)) induces large excursions in inclinations that cross multiple secular resonances. Eros cannot have satellites on circular orbits at distances larger than 250 km.

Acknowledgements. This work was co-funded by the European Social Fund, through Operational Programme Human Capital 2014-2020, project number POCU/993/6/13/153322, project title "Educational and training support for PhD students and young researchers in preparation for insertion into the labor market". Acknowledgment is given to infrastructure support from the Operational Program Competitiveness 2014-2020, Axis 1, under POC/448/1/1 Research infrastructure projects for public R&D institutions/Sections F 2018, through the Research Center with Integrated Techniques for Atmospheric Aerosol Investigation in Romania (RECENT AIR) project, under grant agreement MySMIS no.127324.

References

R.R. Allan, & G.E. Cook 1964, *The long period motion of the pane of a distant circular orbit*, Proc. R. Soc. Lond. A 280, 97–109.

B.G. Bills, & F. Nimmo 2011, *Forced obliquities and moments of inertia of Ceres and Vesta*, Icarus, 213, 2, 496–509.

W.M. Kaula 1961, *Analysis of Gravitational and Geometric Aspects of Geodetic Utilization of Satellites*, Geophysical Journal International, 5, 2, 104–133.

W.M. Kaula 1962, *Development of the lunar and solar disturbing functions for a close satellite*, Astronomical Journal, 67, 300–303.

J. Souchay, H. Kinoshita, H. Nakai, & S. Roux 2003 *A precise modeling of Eros 433 rotation*, ICARUS 166, 285–296.

Complex Planetary Systems II: Latest Methods for an Interdisciplinary Approach
Proceedings IAU Symposium No. 382, 2024
A. Lemaitre & A.-S. Libert, eds.
doi:10.1017/S1743921323003629

Food for thought

Yaël Nazé

FNRS/University of Liège, Belgium
email: ynaze@uliege.be

Abstract. The space domain has undergone several changes in the last decade (aka "new space era"), and will continue to do so in coming years. This led to several challenges which should be recognized and tackled by all actors of this domain, including scientists.

Keywords. General

Over the two thirds of a century since the launch of the first satellite, the space domain has evolved drastically in many ways. In this contribution, we propose some "food for thought" regarding several current issues, often requiring urgent consideration for action.

Communication A simple search on the word "space", on internet or social media, usually yields answers linked to scientific discoveries in astronomy or space missions (with or without human beings). Such activities are of course important, but one should recognize that they represent a really tiny fraction of what is done in the space domain. The majority of space activities is rather linked to commercial or military uses. They are much less publicized, so that the public, politicians, or medias have a biased vision of space. Key issues (see below) can then be difficult to comprehend for them: communication regarding space therefore should evolve to better reflect the actual situation and its challenges.

Sustainability In a decade, the number of launches has more than doubled - there has already been about a hundred of them in the first half of 2023. In parallel, the satellites have become smaller. There is thus an exponential increase in the number of satellites. This has of course several consequences, especially regarding sustainability. First, having more objects in space implies an unavoidable increase in debris numbers and more generally an increase of the overall "orbital traffic". Passive satellites and uncontrolled debris cannot escape collisions, further increasing the number of debris. In contrast, thanks to performant ground-based radars, collision alerts can be issued and avoidance maneuvers are then scheduled for working satellites. However, the number of alerts is increasing and will continue to do so in coming years (e.g. Starlink satellites had to swerve $\sim$25 000 times between December 2022 and May 2023 - this currently represents a doubling of the number every six months). With every maneuver requiring energy, one may wonder when this situation will become untenable (one maneuver every month? every week? every day?) hence low orbit unusable, to the detriment of all. A second question is related to atmospheric pollution. Indeed, launches and satellite/debris re-entries are not without impact on the atmosphere (notably regarding the ozone layer and climate change). With more detailed simulations performed in this field, it is time to discuss what level of pollution would be deemed acceptable. Finally, there are also consequences for scientific activities. Indeed, satellites are a source of (optical) light and radio pollution. Some companies have darkened their satellites, but that only prevents them from being visible to the naked eye: professional telescopes can still pick them up. Satellite strikes then renders images unusable, although the exact fraction of lost data depends on the nature of the telescope program (an all sky survey like that of the Vera Rubin telescope being more impacted than pointed observations with e.g. ESO telescopes). In a similar way, satellite radio emissions interfere with radiotelescope observations. Reversing the direction of observation, Earth observation satellites also suffer from satellites at lower orbits crossing their line-of-sight. For example, some humidity diagnostics could be affected by radio interference from satellites, impairing weather predictions. One may here wonder which limit for such interferences to scientific work could be accepted.

Regulation At the international level, things are still mostly ruled by the Outer Space Treaty from 1967, which has been signed by many countries, and in particular all space-faring nations. However, it is much less stringent than e.g. the subsequent Moon Treaty (ratified by only few countries) and it now appears quite outdated since it mostly considers space activities run by states and not by private companies. In the last decades, several countries (United States, Luxemburg, United Arab Emirates...) have adopted national laws which are clearly supportive of private space exploitation. However, many issues can only be solved by international agreements. Space debris constitute a good example: not only the consequences of collisions should be settled (who's responsible?) but also guidelines should be elaborated for debris removal (can one country evacuate the "debris" associated to another nation?). There is also the question of space tourists - there is an international agreement for state astronauts, but it does not account for the recent development of space tourism. Moreover, space weaponization is increasing, and becoming more mixed with private interests: e.g. some national "space forces" guidelines explicitly mention the defense of private interests of companies linked to that country, while some private services are now used by armies with consequences on the battlefields (e.g. the military use of Starlink in Ukraine, which was limited after a personal decision of E. Musk). Finally, space exploitation also causes problems. For example, water extraction in polar craters of the Moon attracts the attention from various countries with different interests (e.g. China and United States), laying the ground for future conflicts. More philosophically, one could wonder about the limits of exploitation: is it ethically acceptable to destroy a small asteroid to exploit it? or a big one? or to change the face of the Moon forever and for all terrestrial beings to see? or to fully use non-rewable ressources (such as the lunar water), leaving nothing to exploit or study for the next generation?

As scientists and citizens, many questions related to the current and future space activities arise. Let's not be afraid of them and let's discuss them, to ensure everyone will enjoy space in the future.

Complex Planetary Systems II: Latest Methods for an Interdisciplinary Approach
Proceedings IAU Symposium No. 382, 2024
A. Lemaitre & A.-S. Libert, eds.
doi:10.1017/S174392132300385X

On the computation of resonant tidal dissipation in the liquid layers of planets and stars

Jérémy Rekier[1,2] **and Santiago A. Triana**[1]

[1]Royal Observatory of Belgium, 3 avenue circulaire, 1180 Brussels, Belgium
email: jeremy.rekier@observatory.be

[2]Dpt. of Earth & Planetary Science, University of California, Berkeley, California 94720, USA

Abstract. Planets and stars have liquid layers that can support internal gravity waves and inertial waves respectively restored by the buoyancy and Coriolis forces. Both types of waves are excited by tides, leading to resonantly amplified dissipation. We review the theoretical formalism to compute these resonances and present some challenges and methods to overcome them.

Keywords. liquid layers - buoyancy - Coriolis - tides - resonances

---

1. Introduction

The long-term orbital evolution of stars, planets and moons around a central object is directly affected by energy dissipation in both the orbiting object and the central body. This mechanism is mediated by tidal gravitational forces as they deform periodically the rigid layers of the given body, generating heat in the process. Fluid layers, if present, respond differently to periodic tidal forces since they can support inertial waves, i.e. waves whose main restoring force is the Coriolis force, as well as internal gravity waves, i.e. whose main restoring force is buoyancy. Whenever the tidal forcing frequency is close to one of the normal mode frequencies associated with inertial or internal gravity modes of the fluid layer, resonant amplification might occur. This amplified fluid motion is usually accompanied by comparatively large energy dissipation rates, ultimately impacting the long-term evolution of the orbit. Thus it is crucial to determine the resonant mode frequencies in these fluid layers as well as their damping rates. Together with observations, this knowledge also allows us to better characterize the internal structure of the fluid layers in the same way as done in asteroseismology, helioseismology or planetary ring seismology. Computing the normal modes of a rotating fluid body can be challenging. The use of a spherical geometry, together with the anelastic or Boussinesq approximation to eliminate acoustic waves (which due to their much higher frequency are unlikely to be involved in tidally forced systems), leads a numerically tractable problem. In the following we present in some detail the formalism needed to obtain a set of differential equations suitable for numerical studies.

2. Mathematical formalism

2.1. *Momentum equation and hydrostatic equilibrium*

The momentum equation governing the motion in the rotating fluid layers reads:

$$\rho\left(\partial_t \mathbf{v} + (\mathbf{v}\cdot\nabla)\mathbf{v} + 2\boldsymbol{\Omega}\times\mathbf{v} + \boldsymbol{\Omega}\times(\boldsymbol{\Omega}\times\mathbf{r})\right) = -\nabla P - \rho\nabla\Phi + \mathbf{f}, \tag{2.1}$$

where $\mathbf{r}$ and $\mathbf{v}$ are the position vector and flow velocity measured in the frame rotating at constant angular velocity $\boldsymbol{\Omega}$ (henceforth *planetary frame*), ρ is the fluid's mass density, P is the isotropic pressure, and $\mathbf{f}$ represents the sum of additional forces (viscosity, Lorentz, etc.). The gravitational potential Φ must satisfy Poisson equation:

$$\nabla^2 \Phi = -4\pi G\rho. \tag{2.2}$$

This may be written as the sum of an equilibrium and a time-varying contributions (see eq. (2.26) below). The equilibrium potential satisfies the hydrostatic condition obtained by setting $\mathbf{v} = \mathbf{0}$ (and assuming $\mathbf{f} = \mathbf{0}$) into eq. (2.1):

$$\nabla P_0 = \rho_0 \mathbf{g}_0, \tag{2.3}$$

$$\text{with,} \quad \mathbf{g}_0 = \nabla \left(\Phi_0 + \frac{1}{2} |\boldsymbol{\Omega} \times \mathbf{r}| \right), \tag{2.4}$$

where $\mathbf{g}_0$ is the equilibrium gravity acceleration in which we include the centrifugal potential. Taking the curl of eq. (2.3) we see that the equipotentials of ρ_0 and P_0 coincide (baroclinic fluid) and are everywhere perpendicular to $\mathbf{g}_0$.

2.2. *First order perturbations*

In the limit of small velocities, the advection term can be omitted from eq. (2.1). In general, the fluid motion will cause variations in the pressure and density fields:

$$P = P_0 + \delta P, \qquad \rho = \rho_0 + \delta\rho, \tag{2.5}$$

where δP and $\delta\rho$ denote the *Eulerian* increments of pressure and density, respectively. Those are related to their *Lagrangian* counterparts, ΔP and $\delta\rho$, via (e.g. Dahlen and Tromp 1998):

$$\Delta P = \delta P + \mathbf{x} \cdot \nabla P_0, \qquad \Delta\rho = \delta\rho + \mathbf{x} \cdot \nabla \rho_0, \tag{2.6}$$

where $\mathbf{x}$ is the displacement vector related to the velocity via $\mathbf{v} = \partial_t \mathbf{x}$ (different to the position vector $\mathbf{r}$). To first order in the perturbations and in the displacement, the momentum eq. (2.1) reduces to:

$$\partial_t \mathbf{v} + 2\boldsymbol{\Omega} \times \mathbf{v} = -\frac{1}{\rho_0} \nabla \delta P + \frac{\delta\rho}{\rho_0} \mathbf{g}_0 - \nabla \delta\Phi + \frac{1}{\rho_0} \mathbf{f}. \tag{2.7}$$

The increment of gravitational potential is the sum of internal and external perturbations and satisfies:

$$\nabla^2 \delta\Phi = 4\pi G \delta\rho. \tag{2.8}$$

The Eulerian increment of density, $\delta\rho$, must also satisfy the continuity equation expressing mass conservation (e.g. Dahlen and Tromp 1998):

$$\delta\rho = -\nabla \cdot (\rho_0 \mathbf{x}). \tag{2.9}$$

In principle eq. (2.9) can be used into eqs. (2.7) and (2.8) to solve for $\mathbf{x}$ and $\delta\Phi$. This method however proves challenging and additional simplifications are desirable.

2.3. *Anelastic approximation*

The anelastic approximation assumes that the Eulerian increments of density and pressure are negligible in eq. (2.6):

$$\Delta P \approx \mathbf{x} \cdot \nabla P_0, \qquad \Delta\rho \approx \mathbf{x} \cdot \nabla \rho_0. \tag{2.10}$$

Moreover, from eqs. (2.6) and (2.9), we find $\Delta\rho/\rho_0 = -(\nabla \cdot \mathbf{x})$, and the continuity equation reduces to:

$$\nabla \cdot (\rho_0 \mathbf{x}) = 0. \tag{2.11}$$

This equation effectively filters out the rapid oscillations in density and pressure typically associated with acoustic waves of frequencies much higher than typical tidal frequencies (Dintrans and Rieutord 2001). After some rearrangements, we can rewrite the momentum equation as:

$$\partial_t^2 \mathbf{x} + 2\mathbf{\Omega} \times \partial_t \mathbf{x} = -\nabla \left(\frac{\delta P}{\rho_0} + \delta\Phi \right) - N^2 \frac{\mathbf{g}_0}{|\mathbf{g}_0|^2} (\mathbf{g}_0 \cdot \mathbf{x}) + \frac{1}{\rho_0} \mathbf{f}, \tag{2.12}$$

where the second term on the right-hand side is the buoyancy force per unit mass, where N is the *Brunt-Väisälä frequency* defined as:

$$N^2 = -\mathbf{g}_0 \cdot \left(\frac{\mathbf{g}_0}{c^2} - \frac{\nabla \rho_0}{\rho_0} \right), \tag{2.13}$$

and where we have introduced the *isentropic* speed of sound (squared): $c^2 \equiv \Delta P/\Delta\rho$. Dintrans and Rieutord (2001) observed that the pressure and gravity increments can be eliminated by taking the curl of eq. (2.12). The resulting equation coupled with the constraint eq. (2.11) form a complete system in the displacement $\mathbf{x}$. Once solved, the gravity increment can be recovered from Poisson eq. (2.8) which reduces to:

$$\nabla^2 \delta\Phi = 4\pi G \rho_0 \frac{N^2}{|\mathbf{g}_0|^2} (\mathbf{g}_0 \cdot \mathbf{x}). \tag{2.14}$$

2.4. *Incompressible fluid limit*

When the fluid is taken as incompressible, we have $\Delta\rho = 0$, corresponding to the limit of $c \to \infty$. From eqs. (2.10) and (2.11) we then find:

$$\nabla \cdot \mathbf{x} = 0, \tag{2.15}$$

while the Brunt-Väisälä frequency reduces to:

$$N^2 = \mathbf{g}_0 \cdot \frac{\nabla \rho_0}{\rho_0}. \tag{2.16}$$

Notice that eq. (2.15) combined with eq. (2.9), and compared to eq. (2.10) gives:

$$\delta\rho = \mathbf{x} \cdot \nabla \rho_0 \approx 0. \tag{2.17}$$

Imposing eq. (2.17) exactly would make the buoyancy force vanish entirely. This force is proportional to $\delta\rho/\rho_0$ which, though small, is multiplied by $\mathbf{g}_0$ which might be large. This is the physical reason behind the common prescription known as the *Boussinesq approximation* according to which one should neglect density gradient and retain the Eulerian increment of density in the momentum equation only where it is multiplied by gravity acceleration. By the same token, Poisson eq. (2.14) reduces to a Laplace equation in the perturbation:

$$\nabla^2 \delta\Phi \approx 0. \tag{2.18}$$

In this limit, the gravity potential increment inside the fluid therefore reduces to a harmonic function of the coordinates, related to the external potential by boundary conditions (see eqs. (2.21 and (2.22) below).

2.5. *Viscous dissipation*

Dissipation in the fluid layers can have many origins (viscous, ohmic, etc.). For simplicity, we focus on the effect of viscosity, introduced into eq. (2.12) as an additional force: $\mathbf{f} = \nabla \cdot \boldsymbol{\tau}$, where the *deviatoric stress tensor*, $\boldsymbol{\tau}$ writes:

$$\boldsymbol{\tau} = \zeta(\nabla \cdot \mathbf{v})\mathbb{1} + \mu \left[\nabla\mathbf{v} + \nabla\mathbf{v}^{\mathsf{T}} - \frac{2}{3}(\nabla \cdot \mathbf{v})\mathbb{1}\right], \tag{2.19}$$

with ζ and μ respectively denoting the *bulk* and *dynamic* viscosities ($\mathbb{1}$ is the identity). ζ conveniently disappears from the curl of the momentum equation, leaving only μ as a parameter. The power dissipated by viscous forces can then be shown equal to:

$$\mathcal{D}_{\text{visc.}} = 2\nu \int_{\mathcal{V}} \widehat{\nabla\mathbf{v}} : \widehat{\nabla\mathbf{v}}, \tag{2.20}$$

where $\widehat{\nabla\mathbf{v}} = \frac{1}{2}(\nabla\mathbf{v} + \nabla\mathbf{v}^{\mathsf{T}})$, and : denotes the double tensor contraction. $\nu = \mu/\rho_0$ is the kinematic viscosity. The value of this parameter is typically not well known. Its molecular value can safely be assumed as very small–of the order of that of water $10^{-6}\text{m}^2/\text{s}$–in planetary and stellar interior where it is generally treated as a constant.

2.6. *Boundary conditions*

The gravity potential must satisfy the following constraints at the equilibrium surface (Dahlen and Tromp 1998):

$$[\delta\Phi]_-^+ = 0 \tag{2.21}$$

$$[\hat{\mathbf{n}} \cdot (\nabla\delta\Phi + 4\pi G\rho_0\mathbf{x})]_-^+ = 0, \tag{2.22}$$

where $\hat{\mathbf{n}}$ is the unit normal vector and the notation $[\cdot]_-^+$ denotes the difference in the values of the enclosed quantity across the boundary. The remaining conditions on the displacement field depend on the nature of the boundary. A free boundary will move so as to minimise stresses at its surface. To first order in the perturbations, this condition applies to the equilibrium surface and writes:

$$[\hat{\mathbf{n}} \cdot \Delta\boldsymbol{\sigma}]_-^+ = \mathbf{0} \tag{2.23}$$

$$\text{with, } \Delta\boldsymbol{\sigma} = \delta\boldsymbol{\sigma} + \mathbf{x} \cdot \nabla\boldsymbol{\sigma}_0,$$

where $\Delta\boldsymbol{\sigma}$ and $\delta\boldsymbol{\sigma}$ are the Lagrangian and Eulerian increments of stress, respectively. The latter is equal to the sum of the isotropic increment of pressure and deviatoric tensor in the fluid, $\delta\boldsymbol{\sigma} = -\delta P\mathbb{1} + \boldsymbol{\tau}$, while the equilibrium stress tensor is purely isotropic: $\boldsymbol{\sigma}_0 = -P_0\mathbb{1}$. The two components of eq. (2.23) tangential to the boundary vanish identically in the absence of viscosity. At the interface with a solid boundary, eq. (2.23) is replaced by the simpler 'no-slip' condition:

$$[\mathbf{x}]_-^+ = \mathbf{0}. \tag{2.24}$$

In the absence of viscosity, this constraint is too strong and must be replaced by the weaker 'no-penetration' condition:

$$[\hat{\mathbf{n}} \cdot \mathbf{x}]_-^+ = 0. \tag{2.25}$$

2.7. *Tidal perturbations*

Tidal forcing enters the picture via the gravity potential:

$$\Phi = \Phi_0 + \delta\Phi(t), \tag{2.26}$$

where $\delta\Phi(t) = \delta\Phi(t)^{\text{ext}} + \delta\Phi(t)^{\text{int}}$ is the sum of the external time-varying part of the tidal potential plus the internal contribution from deformation and must satisfy eq. (2.14). The external forcing induced by a single object can be decomposed in terms of spherical harmonics and Fourier modes (e.g. Ogilvie 2014):

$$\delta\Phi(t)^{\text{ext}} = \sum_{n}\sum_{\ell=2}^{\infty}\sum_{m=-\ell}^{\ell} \delta\Phi^{\text{ext}}_{\ell,m,n} \mathrm{Y}_{\ell}^{m}(\theta,\varphi) e^{in\Omega_o t}, \tag{2.27}$$

where θ and φ are the colatitude and azimutal coordinates, and Ω_o is the mean orbital frequency. Tides with angular frequency $n\Omega_o$, and azimutal number m are perceived as oscillations at frequency $\omega = n\Omega_o - m\Omega$ in the planetary frame. ω corresponds to the response frequency of the fluid to the orbital forcing.

2.8. *Simplification: Spherical symmetry*

In principle the fluid configuration should satisfy the hydrostatic condition eqs. (2.3) and (2.4). If $\boldsymbol{\Omega} \neq \mathbf{0}$, the equilibrium shape is an ellipsoid of revolution around that axis (taken along the z-axis). If the equilibrium potential Φ_0 is not axisymmetric–as it is the case for objects tidally locked in low orbit around their parent body–the equilibrium ellipsoid is triaxial. Such ellipsoidal shapes do not readily lend themselves to spherical harmonics discretization. It may be tempting to resort to ellipsoidal harmonics. However, in addition to the practical difficulties associated with their use, they can only accommodate exact ellipsoids, i.e. those whose sections along their principal axes are true ellipses, and this precise type of figure can only exist for an object of homogeneous density (Chandrasekhar 1987). This situation does not lend itself well to the exploration of internal structures of differentiated objects, which typically starts with the adoption of a radial density profile. Theoretically, it is possible to approximate the equilibrium figure of such a body by means of a series expansion in flattening (Clairaut's method). Axisymmetric bodies can also be represented by a series of concentric Maclaurin ellipsoids (Hubbard 2013). These methods all rely on the use of complicated coordinate systems, making it all the more difficult to achieve the high numerical resolution required for viscous fluids. Moreoever, many partially fluid bodies such as terrestrial planets or icy moons have shapes far removed from hydrostatic equilibrium.

For all the above reasons, it is often valuable to stick to the simple spherical shape as a first order approximation. This assumption alone simplifies eq. (2.12) immensely by making $\mathbf{g}_0 = -g_0\hat{\mathbf{r}}$, where $\hat{\mathbf{r}}$ is the unit radial vector, thus making the buoyancy force radial and equal to $-N^2(\mathbf{x}\cdot\hat{\mathbf{r}})\hat{\mathbf{r}}$, while the Brunt-Viäsälä frequency reduces to its more common expression (De Boeck et al. 1992):

$$N^2 = -\frac{g_0^2}{c^2} - \frac{g_0}{\rho_0}\frac{d\rho_0}{dr}. \tag{2.28}$$

As we can see, two inputs are needed to calculate the fluid motion: a radial density profile, ρ_0, and another profile giving the evolution of the speed of sound, c with depth. As always, one recovers the incompressible limit by letting $c \to \infty$.

3. Free solutions

A detailed resolution of eq. (2.12) is beyond the scope of this short communication, we present only a general outline. For definiteness, we focus on the case where the fluid fills the whole sphere. Since we expect the tidal dissipation to be amplified by gravito-inertial modes, an important first step is to compute these modes. This is done by setting $\mathbf{x} = \tilde{\mathbf{x}}e^{\lambda t}$, and solving the resulting differential eigenvalue problem in $\hat{\mathbf{x}}$ and λ with the external forcing $\delta\Phi^{\text{ext}}$ set to zero.

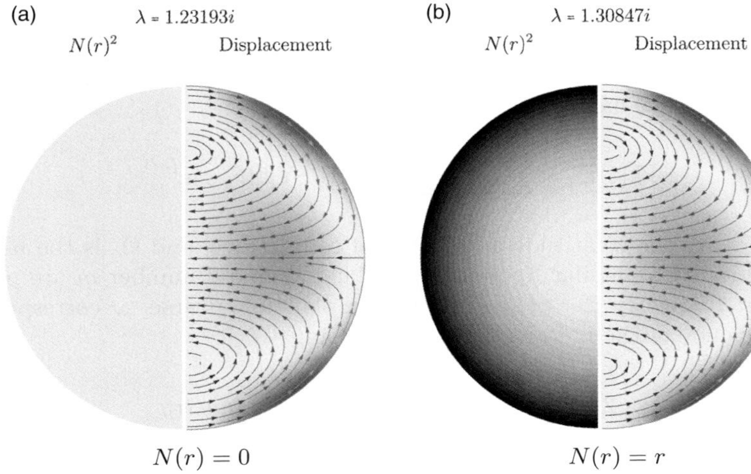

Figure 1. Example of free solutions to eq. (2.12) for two different buoyancy profiles ($m=2$). (a) Inertial mode. (b) gravito-inertial mode. The arrows represent the displacement in the meridional plane. They scale from blue to red based on the value of the displacement's norm. The background color represents the azimutal displacement with blue and red corresponding to a flow coming out and in the page, respectively. The left parts of each figure show the equipotentials of the function $N^2(r)$.

When buoyancy is set to zero, the spectrum is populated by inertial modes with frequencies $\omega = \mathrm{Im}(\lambda)$ satisfying $|\omega| < 2\Omega$. Solutions to the momentum equation can then be arrived at analytically (Greenspan 1968) based on the boundary condition eq. (2.25). Alternatively, one may discretise the domain and solve the resulting algebraic eigenvalue problem numerically (Rekier et al. 2019). The second option can be readily adapted to more complex boundary conditions and to the case where $N(r) \neq 0$. In such a case, inertial modes get mixed with gravity modes. These mixed modes can be classified depending on the relative values of ω, Ω, and the maximum value of $N(r)$ (Triana et al. 2021). Figure 1a shows an example of inertial mode with azimutal wave number $m=2$. Figure 1b shows its gravito-inertial mode counterpart when the Brunt-Väisälä frequency grows linearly from the centre.

4. Outlook

The shapes and frequencies of the solutions to eq. (2.12) strongly depend on the internal fluid structure. While on the one hand, this makes their computations more challenging, on the other hand it opens up the possibility of using these to probe planets and stars interior. Although viscous forces are expected to affect the frequencies of the modes very weakly due to their typical weakness compared to the dominant Coriolis force, taking viscosity into account is not only crucial in order to estimate tidal dissipation, it also contributes to regularize modes that are otherwise singular, by turning surfaces of singularities across the inviscid fluid into thin internal shear layers contributing to resonantly enhance tidal dissipation. The aim of the present work was to present the mathematical formalism governing the problem of tidal dissipation in some details. A more detailed account of its solution will be presented in a future work.

References

Chandrasekhar, S. (1987). *Ellipsoidal Figures of Equilibrium*. Dover Books on Mathematics. Dover.

Dahlen, F. A. and Tromp, J. (1998). *Theoretical Global Seismology.* Princeton University Press.

De Boeck, I., Van Hoolst, T., and Smeyers, P. (1992). The subseismic approximation for low-frequency modes of the Earth applied to low-degree, low-frequency g-modes of non-rotating stars. *Astronomy and Astrophysics*, 259:167–174.

Dintrans, B. and Rieutord, M. (2001). A comparison of the anelastic and subseismic approximations for low-frequency gravity modes in stars. *Monthly Notices of the Royal Astronomical Society*, 324(3):635–642.

Greenspan, H. P. (1968). *The Theory of Rotating Fluids.* Cambridge Monographs on Mechanics and Applied Mathematics. Cambridge U.P, London.

Hubbard, W. B. (2013). Concentric Maclaurin Spheroid Models Of Rotating Liquid Planets. *ApJ*, 768(1):43.

Ogilvie, G. I. (2014). Tidal dissipation in stars and giant planets. *Annual Review of Astronomy and Astrophysics*, 52(1):171–210.

Rekier, J., Trinh, A., Triana, S. A., and Dehant, V. (2019). Inertial modes in near-spherical geometries. *Geophysical Journal International*, 216(2):777–793.

Triana, S. A., Dumberry, M., Cébron, D., Vidal, J., Trinh, A., Gerick, F., and Rekier, J. (2021). Core Eigenmodes and their Impact on the Earth's Rotation. *Surv Geophys.*

Complex Planetary Systems II: Latest Methods for an Interdisciplinary Approach
Proceedings IAU Symposium No. 382, 2024
A. Lemaitre & A.-S. Libert, eds.
doi:10.1017/S1743921323004155

Secular dynamics of a coplanar, non-resonant planetary system, consisting of a star and two planets

Vladislav Sidorenko

Keldysh Institute of Applied Mathematics, Moscow, Russia
email: vvsidorenko@list.ru

Abstract. We consider an exoplanetary system consisting of a star and two planets. The masses of the planets are significantly less than the mass of the star. The evolution of the orbital motion of exoplanets is studied within the framework of a double averaged unrestricted three-body problem. The main attention is paid to coplanar configurations, when the star and planets move in a certain plane that preserves a constant position. The possibility of reversing the orbital motion of the inner planet is noted.

Keywords. Three-body problem, secular evolution, averaging technique

1. Introduction

In the general case, the motion of three gravitating points is rather complicated. The situation is simplified when we can apply the averaging technique to study their dynamics. Several cases are known when the averaged problem is integrable. The first such case was described by Harrington (1968): a point moves around a pair of points at a distance significantly greater than the distance between the points in the pair. The next case was found by Ziglin (1976): two points move in close orbits around the third, whose mass is much greater than their masses. From considerations of symmetry, we can conclude that if the motion of these points occurs in some constant plane, then the double averaged problem will be integrable for any ratio between the major semi-axes of their orbits. The study of this case was started by Pauwels (1983). Further investigations of the planar double averaged unrestricted problem were stimulated by the discovery of exoplanetary systems (e. g., Michtchenko and Malhotra (2004); Migaszewski and Gozdziewski (2008)). However, there is still no classification of the dynamical behavior of the system even by the number of stationary solutions.

2. Double Averaged Equations of Motion

The evolution of the coplanar motion of two planets can be described by the equations

$$\frac{d\varpi_\Delta}{dt} = \mu\left[\frac{\chi\sqrt{1-e_1^2}}{\bar{\mu}_1 e_1\sqrt{a}}\frac{\partial W}{\partial e_1} - \frac{\sqrt{1-e_2^2}}{\bar{\mu}_2 e_2}\frac{\partial W}{\partial e_2}\right] \tag{1}$$

$$\frac{de_1}{dt} = -\mu\chi\frac{\sqrt{1-e_1^2}}{\bar{\mu}_1 e_1\sqrt{a}}\frac{\partial W}{\partial \varpi_\Delta}, \quad \frac{de_2}{dt} = -\mu\frac{\sqrt{1-e_2^2}}{\bar{\mu}_2 e_2}\frac{\partial W}{\partial \varpi_\Delta}.$$

In system (1), indices 1 and 2 correspond to the inner planet and the outer planet, respectively. The variable $\varpi_\Delta = \varpi_1 - \varpi_2$ is the difference in the longitudes of the pericenters

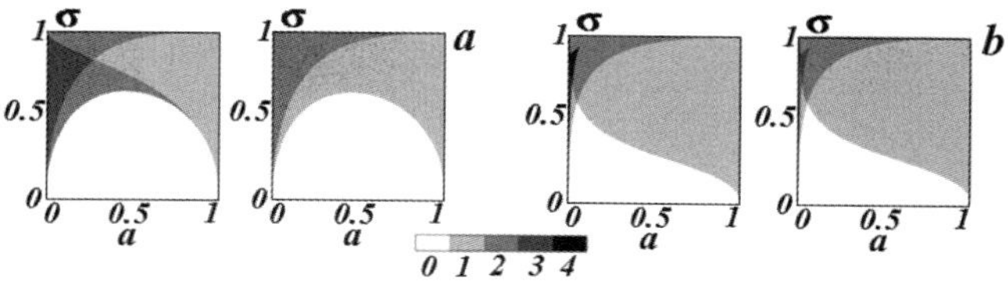

Figure 1. The total number of stationary solutions and the number of stable stationary solutions: a - $\bar{\mu}_1 = 0.25$, b - $\bar{\mu}_1 = 0.75$

of the planets, the variables e_1 and e_2 are the averaged values of the eccentricities of the osculating orbits of the planets. The units of measurement are chosen in such a way that the total mass of the star and planets is equal to 1, the average value of the major semi-axis of the outer planet is equal to 1, and, finally, the value of the gravitational constant is also equal to 1. The parameter μ characterizes the part of the mass of planets in the total mass of the system, parameters $\bar{\mu}_1$ and $\bar{\mu}_2$ set the ratio of the masses of the planets ($\bar{\mu}_1 + \bar{\mu}_2 = 1$). If the direction of rotation of the inner planet coincides with the direction of rotation of the outer planet, the factor χ in system (1) is equal to 1; in the case of retrograde motion of the inner planet $\chi = -1$.

The function W in (1) is a double averaged disturbing function. When constructing W numerically, it is convenient to use the fact that after averaging over the orbital motion of the first planet, the second averaging is equivalent to averaging over the motion in the field of the Gaussian ring specified by the force function W_{Gauss} (for details see Kondratyev (2012)):

$$W = \frac{\bar{\mu}_1\bar{\mu}_2}{4\pi^2}\int_0^{2\pi}\int_0^{2\pi}\frac{d\lambda_1 d\lambda_2}{|\mathbf{r}_1 - \mathbf{r}_2|} = \frac{\bar{\mu}_1\bar{\mu}_2}{2\pi}\int_0^{2\pi} W_{Gauss} d\lambda_2.$$

The parameter μ determines only the rate of secular evolution. The qualitative properties of motion depend on the ratio of the masses of the planets, on the value of the major semiaxis a of the inner planet, and on the value of the normalized angular momentum of the system

$$\sigma = \left[\chi\bar{\mu}_1\sqrt{a(1-e_1^2)} + (1-\bar{\mu}_1)\sqrt{1-e_2^2}\right] / \left[1 - \bar{\mu}_1(\sqrt{a} - 1)\right] .$$

3. Results of Investigation

We used equations (1) to study the secular evolution of the orbital motion of the planets. Similar to Michtchenko and Malhotra (2004), special attention was paid to apsidal resonances - stationary solutions of equations (1). The number of apsidal resonances for different values of the system parameters has been established, and their stability has been studied. Some examples are provided by Figure 1.

As it turned out, under certain conditions the direction of the orbital motion of the inner planet changes. Although this is a known effect (e.g., see Kaplan *et al.* (2008)), the scenarios of evolution with alternating forward and retrograde motion of the inner planet in the planar problem have not been considered before.

References

Harrington, R.S. 1968, *Astron. J.*, 73, 190
Ziglin, S.L. 1976, *Sov. Asron. Lett.*, 2, 69

Pauwels, T. 1983, *Cel. Mech.*, 30, 229
Michtchenko, T.A., Malhotra, R. 2004, *ICARUS*, 168, 237
Migaszewski, C., Gozdziewski, K. 2008, *MNRAS*, 388, 789
Kondratyev, B.P. 2012, *Solar System Research*, 46, 352
Kaplan, S.R., Levi, M., Montgomery, R. 2008, *DCDS, Ser. B*, 10, 569

Complex Planetary Systems II: Latest Methods for an Interdisciplinary Approach
Proceedings IAU Symposium No. 382, 2024
A. Lemaitre & A.-S. Libert, eds.
doi:10.1017/S1743921323003927

Chaotic diffusion in the GJ 876 exoplanetary system

Raphael Alves Silva[1], Cristian Beaugé[2] and Sylvio Ferraz-Mello[1]

[1]Instituto de Astronomia, Geofísica e Ciências Atmosféricas, Universidade de São Paulo (Brasil)
email: alvesraphael92@gmail.com

[2]Universidad Nacional de Córdoba, Observatorio Astronómico - IATE (Argentina)

Abstract. This is a study about the dynamical stability of the GJ 876 exoplanetary system. The phase space of initial conditions (ICs) is characterized using the MEGNO indicator of chaos and the Shannon entropy approach for estimations of diffusivity. The results are compared to analyze correlations between the chaotic layers and the instability timescales. The long-term dynamical behavior of the system is reminiscent of the *stable chaos*, at least within the system's lifetime.

Keywords. Shannon entropy, planetary instability, chaotic diffusion, Laplace resonance

1. Chaotic diffusion and instability times

Chaos and (macroscopic) *instability* constitute two different concepts that sometimes present intrinsic correlations among general dynamical problems. This may be the reason why there are some occasions in which both are wrongly used as synonyms. One can find systems where the existence of the first one does not implicate the occurrence of the second one, a scenario that has been reported in the case of the GJ 876 system (Batygin et al. 2015).

In our work, we applied the orbital parameters established in the work of Millholland et al. (2018) as the nominal condition of the real system (see Table 4, p. 7 therein), disregarding the dynamical contribution of the innermost planet due to both its reduced mass with respect to the others, and its proximity to the host star (Batygin et al. 2015; Martí et al. 2016). Hence, we modeled the system's architecture as a 4-body problem, with the Hamiltonian written using Jacobi angle-action variables (Alves Silva et al. 2021).

In Fig. 1, we plotted the results of our analysis. The first panel (left-hand frame) exhibits a MEGNO dynamical map, obtained from integrations of a grid of 300×300 ICs in the (a_3, e_3)-plane of the outermost mass. Each solution was integrated for a maximum timespan of $T_1 = 10^4$ yr (the yellow region indicates unstable orbits within times $t < T_1$). For each solution of such a map, the respective values of the MEGNO $\langle Y \rangle$ (Cincotta & Simó 2000) were computed numerically. We used them to obtain the respective Lyapunov times T_L of the trajectories, as it is indicated in the color bar.

The middle panel exhibits an instability dynamical map constructed from a grid of 200×200 ICs integrated within the same timespan T_1. At this opportunity, each solution was solved together with another five shadow trajectories to enhance statistically the outcomes. Then, for each one of this set of six trajectories, our numerical routine was able to compute the normalized Shannon entropy S associated with that orbital solution

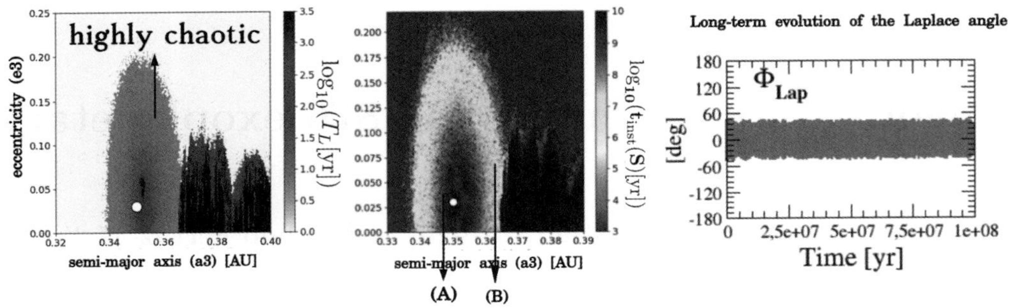

Figure 1. (*Left frame*): MEGNO map, with each solution associated to its respective Lyapunov time T_L. The nominal condition appears near the inner core of the resonance, where 10 yr $\lesssim T_L \lesssim$ 500 yr. (*Center frame*): instability time map, given in terms of the logarithm of $t_{\rm inst}(S)$: dark red regions are unstable in fast times. Dark blue region (A) is a domain of low diffusivity, with instability times of orders of tens of Gyr; (B) indicates domains where the instabilities manifest within tens of Myr. (*Right frame*): long-term integration of the nominal condition, displaying a stable evolution of the Laplace angle $\Phi_{\rm Lap}$ for 0.1 Gyr.

(see Cincotta et al. (2020a) and Cincotta et al. (2020b) for detailed explanations of the Shannon entropy approach). Using ergodic arguments, from the time derivative of S one can estimate a diffusion coefficient D_S, to which is possible to associate a macroscopic instability time† $t_{\rm inst}(S) = \frac{\Delta^2}{D_S}$, where Δ^2 denotes a given mean-square displacement, the squared distance between the initial and boundary values of actions-like variables (see Cincotta et al. (2020a,b); Alves Silva et al. (2021)).

In the right-hand panel, we show a long-term integration of the nominal condition (identified by the white dot in both previous frames) run for $T_2 = 0.1$ Gyr. The plot presents the evolution of the Laplace angle $\Phi_{\rm Lap} = \lambda_1 - 3\lambda_2 + 2\lambda_3$ (λ_i designates the mean longitude of the i-th body), since it is known that the GJ 876 system is currently inside a three-body resonant chain among its most massive planets (Batygin et al. 2015; Martí et al. 2016; Millholland et al. 2018). The system displayed stability within T_2, with $\Phi_{\rm Lap}$ oscillating around zero with an amplitude of ≈ 96 deg.

2. Discussion

Using numerical methods to solve the equations of motion, we obtained a chaotic profile of the phase space surrounding the nominal condition in terms of $\langle Y \rangle$, noticing that the system is immersed in a highly chaotic domain, for which T_L is no greater than hundreds of years. Despite the strong chaoticity, we verified that the system is expected to last over billions of years since it is located in a region of slow diffusion according to our Shannon entropy estimations. The most exciting characteristic of such measures is that they have run within the same timespan T_1 as the Lyapunov time simulations. The long-term behavior of the system was checked with integration of 0.1 Gyr, which is one-tenth of the instability times $t_{\rm inst}$ estimated for the small vicinity (A) of the nominal solution. In the current stage of our work, we are carefully analyzing the stable chaos scenario of the system.

3. Acknowledgements

This work was supported by grants from FAPESP (2016/13750-6) and CAPES.

† The instability time $t_{\rm inst}$ shall be interpreted as an expectation of the time within which dynamical instabilities would manifest on a large-scale, making the disruption of the original system likely.

Supplementary material

To view supplementary material for this article, please visit http://doi.org/10.1017/S1743921323003927

References

Alves Silva, R., Beaugé, C., Ferraz-Mello, S. et al. 2021, *A&A*, 652, A112.
Batygin, K., Deck, K.M., Holman, M.J. 2015, *ApJ*, 149, 167.
Cincotta, P.M., Simó, C. 2000, *A&AS*, 147, 205.
Cincotta, P. M., Giordano, C. M., Alves Silva, R., Beaugé, C., 2020. *CeMDA*, 133.
Cincotta, P. M., Giordano, C. M., Alves Silva, R., Beaugé, C., 2020. *Physica D.*
Martí, J.G., Cincotta, P.M., Beaugé, C. 2016, *MNRAS*, 460, 1094.
Millholland, S., Laughlin, G., Teske, J., Butler, R.P., et al. 2018, *ApJ*, 155, 106.

Complex Planetary Systems II: Latest Methods for an Interdisciplinary Approach
Proceedings IAU Symposium No. 382, 2024
A. Lemaitre & A.-S. Libert, eds.
doi:10.1017/S1743921323003605

Jovian encounter manifolds

Nataša Todorović and Nikola Knežević

Belgrade Astronomical Observatory, Volgina 7, 11000 Belgrade, Serbia
email: ntodorovic@aob.rs

Abstract. Recent numerical studies have shown that the entire solar system is permeated with arch-like structures originating from all planets. Particles placed on such arches experience planetary close encounters after only one or few orbital revolutions. In this work, we are interested how thece arches, which we associate to encounter manifolds of Jupiter, appear in three dimensions for higher inclinations.

Our results show that about 0.5% of the observed domain $[a, e, i] = [2\ AU, 11.5\ AU] \times [0, 0.7] \times [0°, 90°]$ is covered by the manifolds. For inclinations up to $\sim 5°$, the arch-like structures are almost unchanged compared to those initially observed in the orbital plane of Jupiter. At higher inclinations, the number of encounter orbits rapidly decreases to narrow domains where the manifolds stretch up to inclinations of 90° (and above) in a very steep manner.

Keywords. manifolds, Jupiter, encounters, chaos

1. Introduction

Chaos in the solar system has been investigated for more than a century from a variety of perspectives (Poincaré 1892; Sussman and Wisdom 1992; Laskar 1994; Batygin et al. 2015). It is generally accepted that chaos is present almost everywhere, but contrary to popular belief it is not completely disordered. Recent numerical studies have shown that the entire solar system is permeated with arch-like structures originating from all planets (Todorović et al. 2020). Particles placed on such arches experience planetary encounters after only one or few orbital revolutions, which is why they have been identified with the traces of stable invariant manifolds related to Lagrangian points L1 and L2 in the Sun-planet system (Koon *et al.* 2000; Gómez *et al.* 2004).

Invariant manifolds are crucial in our understanding of chaos, but also in the construction of natural space transfers (see e.g. Scantamburlo et al. (2022); Davis *et al.* (2010); Vaquero and Howell (2014); Conley (1968); Fitzgerald and Ross (2022); Daquin *et al.* (2016); Topputo et al. (2005)). However, their calculation is not trivial and is possible only in a narrow domain of three body energies in simplified dynamical models. The arches are numerically observed in the realm of high energies, they still do not have a proper mathematical interpretation and their general structure in phase space remains unknown.

The idea of this work is to investigate how the Jovian arc-like structures, initially obtained in the orbital plane of Jupiter (in the semi-major axis eccentricity (a, e) plane), appear in three dimensions at higher inclinations. In our discretized numerical treatment, we leave out the mathematical definitions and make a simple assumption that *encounter manifolds* (or just manifolds further in the text) are a collection of initial conditions leading to planetary close approaches in rapid times. Following this presumption we start our examination.

2. Methodology

The experiment was carried out following the analogy of dynamical mapping: we take a dense grid of equidistant initial conditions (ICs) in a given $a \times e \times i$ domain, track each particle for a fixed amount of time (for 100 years), and record their eventual entries into the close neighborhood of Jupiter. More precisely, ICs are selected inside a box defined with $a \times e \times i = [2AU, 11.5AU] \times [0, 0.7] \times [0°, 90°]$, with resolution of 0.001 AU in semi-major axis, 0.0014 in eccentricity and $1°$ in inclination, which makes a total number of 432 250 000 orbits.

The initial epoch was set to 7th January 2022 (JD 2459586.5) and the remaining orbital elements, the argument of pericenter ω, and the longitude of the node Ω were equal to the corresponding elements of Jupiter for the same epoch. The mean anomaly M was shifted for 60 degrees in front of the planet, because we construct the result around Jupiter's Lagrangian point L4, but also for the consistency with Todorović et al. (2020).

We have developed a Python program using the REBOUND N-body code (Rein and Liu 2012), where all parameter settings relevant to the computations are simplified and automatized (more details will be provided in future publications). The orbits were integrated using the WHfast scheme (Wisdom & Holman 1991; Rein and Tamayo 2015), with Jupiter as the only planet in the model. Besides observing close encounters, we also calculated FLIs Froeschlé et al. (1997); Froeschlé and Lega (2000); Froeschlé et al. (2000); Todorović (2017); Daquin *et al.* (2016); Guzzo and Lega (2015) and MEGNOs (Cincotta and Simó 2000; Hinse *et al.* 2010; Rein and Tamayo 2016; Kováčová et al. 2022) for each IC, but in this work, we limit our attention only to Jovian approaches smaller than 0.1 AU.

The computation lasted for about two months on the ARIS supercomputer† located in Athens (Greece).

3. Results and discussion

According to our results, about 0.5 per cent of the observed space is occupied by the encounter manifolds i.e. 2080251 (out of 432250000) particles approach Jupiter to a distance smaller than 0.1 AU in 100 years. Let us remark that this result refers only to the above-described selection of parameters since the number of encounter orbits constantly grows with the passage of time and changes for different approaching distances from the planet.

These 2 million particles are placed along certain surfaces (which we identified with encounter manifolds) presented in Figs. 1 and 2. For better visibility, we show the same data from two perspectives (on the two figures) and colour them based on inclination. ICs on lower inclinations are grey, on higher inclinations pink, and the ones in between are white. The given interval in $a \in [3.5, 11.5]$ AU is somewhat smaller than the observed domain, since the number of approaching orbits between 2 AU and 3.5 AU was negligible. Manifolds are not observed in the main belt.

Although we are in unknown territory, a certain geography is discernible. The arch-like structures initially observed in the orbital plane of Jupiter remain relatively unchanged up to inclinations of about $i \sim 5°$. For $i > 5°$ the number of ICs leading to Jovain approaches sharply decreases, but the manifolds survive up to $i = 90°$ (although in limited parts). At the top of the interval, the structures seem cut off, which suggests that they extend on even higher inclinations. The authors hypothesize that they could wrap around the Sun similar to how they wrap around the planet (see e.g. Lega and Guzzo (2016); Paez and Guzzo (2020)).

† https://doc.aris.grnet.gr/system/hardware/

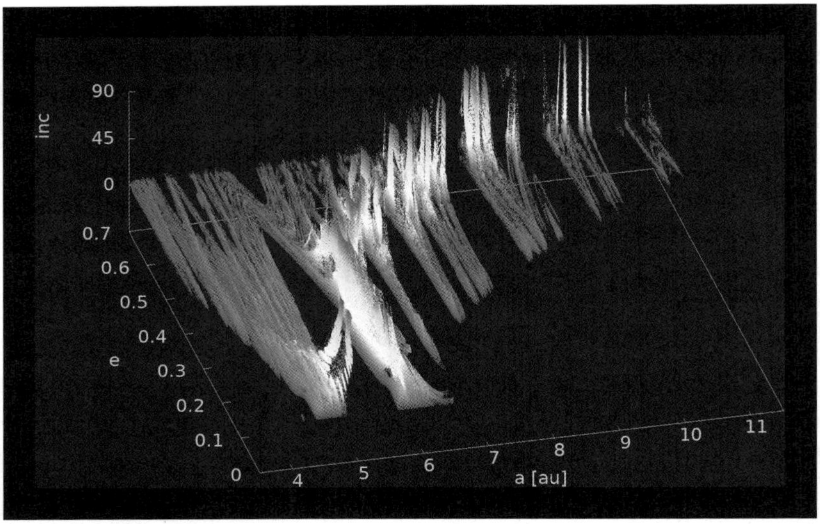

Figure 1. Encounter manifolds composed of initial conditions reaching Jupiter distance smaller than 0.1 AU in 100 years, in the region $a \times e \times i = [3.5, 11.5]$ $AU \times [0, 0.7] \times [0, 90]$ deg. The lowest inclination are coloured grey, the highest are pink, and those in between are white. The arch-like structures initially observed in the orbital plane of Jupiter, remain almost unchanged for inclination below 5 degrees. Manifolds reaching higher inclinations are limited to small areas around the Trojan island (the gap at $a \in 5.2$ AU) and along a narrow direction spreading diagonally through the whole domain. They extend up to 90 degrees (and above) in a steep manner almost vertically.

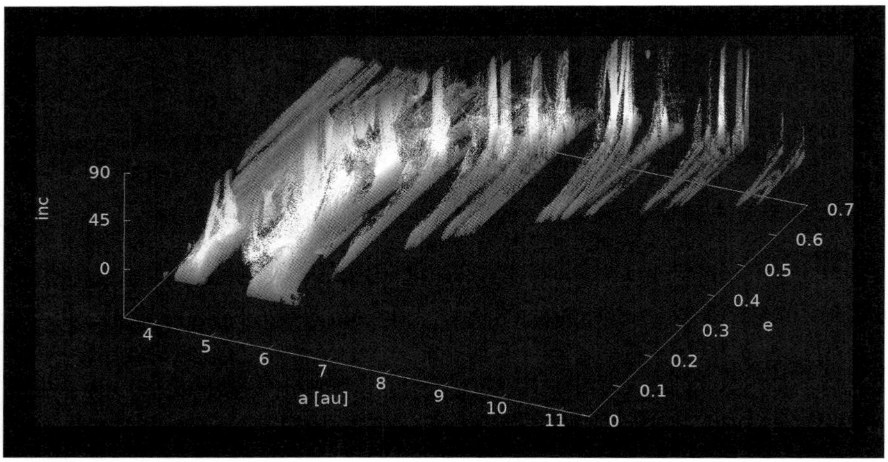

Figure 2. The same data set as in Fig.1 from another perspective. Manifolds occupy about 5 % of the observed space.

The limited parts from which manifolds stretch up to 90° inclinations are located around the stable Trojan island around ~ 5.2 AU (a closer look is given in Fig. 3), and along an unidentified diagonal form which extends in the entire (a, e) domain (more studies will be provided in future work).

In Fig.3 covering the interval between $a \in [4.5, 6.1]$, the layered complex manifold surfaces are better visible. As expected, encounter orbits are absent in the rounded gap of the Trojan island. But it is somewhat paradoxical that this region where

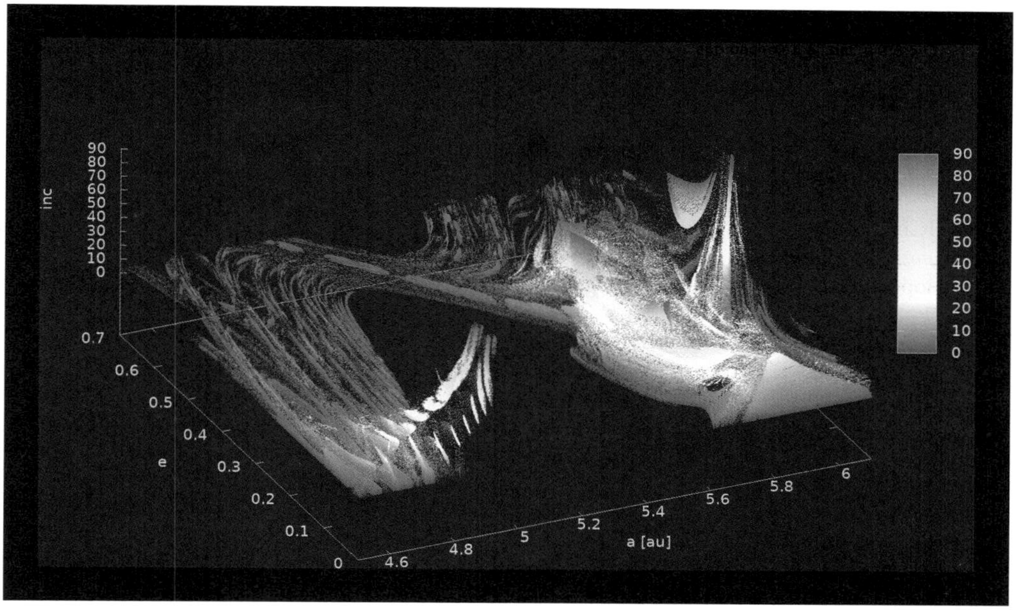

Figure 3. A closer look at the encounter manifolds for $a \in [4.5, 6.1]$ AU around the Jupiter Trojan island corresponding to the L4 point. ICs at low inclinations (coloured in grey) are vertically oriented and form the same structures as observed in Todorović et al. (2020). The white and pink particles lie on multi-layered complex surfaces. A somewhat larger representation of manifolds is notable on the right side of the island. Further research is needed to explain these results.

the stabilities are studied for times comparable to the lifetime of the solar system (Holt *et al.* 2020; Levison et al. 1997; Giorgilli and Skokos 1997; Di Sisto et al. 2014; Efthymiopoulos and Sándor 2005), is surrounded by highly unstable manifolds, whose orbits can hardly survive one century.

4. Conclusion

We conducted massive numerical investigations and detected Jovian encounter manifolds in three dimensions (a, e, i). Although the result has provided new insights into their global structure, such representation is still partial. Encounter manifolds are observed for only one planet, one set of remaining orbital elements, and one epoch. How the ensemble of all the manifolds would look and what is their theoretical explanation, is certainly a challenge for further investigations. It remains an open question whether we will ever see them in their entirety.

5. Acknowledgments

The authors are supported by the Ministry of Education, Science, and Technological Development of the Republic of Serbia (contract No. 451-03-47/2023-01/ 200002). The computational time was granted from the National Infrastructures for Research and Technology S.A. (GRNET S.A.) in the National HPC facility - ARIS - under hpce3021/HPC-EUROPA3 (INFRAIA-2016-1-730897) project, and hosted by Dr. K. Tsiganis from the Aristotle University of Thessaloniki. N.T. gratefully acknowledges the support of the IAU travel grant to attend the Meeting IAUS 382.

References

Batygin, K., Morbidelli, A., and Holman, M.J.: 2015, *The Astrophysical Journal* 799, 120

Cincotta, P.M., Giordano, C.M., and Simó, C.: 2003, *Physica D Nonlinear Phenomena* 182, 151

Cincotta, P.M. and Simó, C.: 2000, *Astronomy and Astrophysics Supplement Series* 147, 205

Conley, C.C.: 1968, *SIAM J. Appl. Math* **16**, 732

Daquin, J., Rosengren, A.J., Alessi, E.M., Deleflie, F., Valsecchi, G.B., and Rossi, A.: 2016, *Celestial Mechanics and Dynamical Astronomy* 124, 335

Davis, K.E., Anderson, R.L., Scheeres, D.J., and Born, G.H.: 2010, *Celestial Mechanics and Dynamical Astronomy* 107, 471

Efthymiopoulos, C. and Sándor, Z.: 2005, *Monthly Notices of the Royal Astronomical Society* 364

Fitzgerald, J. and Ross, S.D.: 2022, *Advances in Space Research* 70, 144

Froeschlé, C., Lega, E., and Gonczi, R.: 1997, *Celestial Mechanics and Dynamical Astronomy* 67, 41

Froeschlé, C. and Lega, E.: 2000, *Celestial Mechanics and Dynamical Astronomy* 78, 167

Froeschlé, C., Guzzo, M., and Lega, E.: 2000, *Science* 289, 2108

Giorgilli, A. and Skokos, C.: 1997, *Astronomy and Astrophysics* **317**, 254

Gómez, G., Koon, W.S., Lo, M.W., Marsden, J.E., Masdemont, J., and Ross, S.D.: 2004, *Nonlinearity* **17**, 1571

Guzzo, M. and Lega, E.: 2015, *Astronomy and Astrophysics* **579**, A79

Hinse, T.C., Christou, A.A., Alvarellos, J.L.A., and Goździewski, K.: 2010, *MNRAS* 404, 837

Holt, T. R., Nesvorný, D., Horner, J., King, R., Marschall, R., Kamrowski, M., and, ...: 2020, *Monthly Notices of the Royal Astronomical Society* 495, 408

Koon, W.S., Lo, M.W., Marsden, J.E., and Ross, S.D.: 2000, *Chaos* **10**, 427

Kováčová, M., Kornoš, L., and Matlovič, P.: 2022, *MNRAS* 509, 3842

Laskar, J.: 1994, *Astronomy and Astrophysics* 287, L9

Lega, E. and Guzzo, M.: 2016, *Physica D Nonlinear Phenomena* 325, 41

Levison, H.F., Shoemaker, E.M., and Shoemaker, C.S.: 1997, *Nature* 385, 42

Paez, R.I. and Guzzo, M.: 2020, *International Journal of Non Linear Mechanics* 120, 103417

Poincaré, H.: 1892, *Les méthodes nouvelles de la mécanique céleste, by Poincaré, Henri. Paris, Gauthier-Villars et fils, 1892.*

Rein, H. and Liu, S. F.: 2012, *Astronomy and Astrophysics* 537, A128

Rein, H. and Tamayo, D.: 2015, *MNRAS* 452, 376

Rein, H. and Tamayo, D.: 2016, *MNRAS* 459, 2275

Scantamburlo, E., Guzzo, M., and Paez, R.I.: 2022, *Acta Astronautica* **200**, 97

Di Sisto, R.P., Ramos, X.S., and Beaugé, C.: 2014, *Icarus* 243, 287

Sussman, G.J. and Wisdom, J.: 1992, *Science* 257

Todorović, N.: 2017, *MNRAS* 465, 4441

Todorović, N., Wu, D.,& Rosengren, A.J.: 2020, *Sci. Adv.*, 6, eabd1313

Topputo, F., Vasile, M., and Bernelli-Zazzera, F.: 2005, *Journal of the Astronautical Sciences* 53, 353

Vaquero, M. and Howell, K.C.: 2014, *Acta Astronautica* 94, 302

Wisdom, J. and Holman, M., 1991, *The Astronomical Journal*, 102, 1528

Complex Planetary Systems II: Latest Methods for an Interdisciplinary Approach
Proceedings IAU Symposium No. 382, 2024
A. Lemaitre & A.-S. Libert, eds.
doi:10.1017/S1743921323003617

On the predictability horizon in Impact Monitoring of Near Earth Objects

Giacomo Tommei[1] and Giovanni B. Valsecchi[2,3]

[1]University of Pisa
email: giacomo.tommei@unipi.it

[2]IAPS-INAF, Rome

[3]IFAC-CNR, Sesto Fiorentino

Abstract. The Impact Monitoring (IM) of Near Earth Objects (NEOs) is a fundamental part of the planetary defense strategy. Current NEO IM systems (Aegis, NEODyS and Sentry) scan the Confidence Region (CR) of each observed object looking for Virtual Impactors (VIs) with a time horizon of about 100 years. This procedure is performed regardless of the uncertainty with which the orbit of the object is known, and without considering whether a scattering encounter is present in the propagation time span. In view of the likely future increase of the IM workload due to higher future NEO discovery rates, it might be more reasonable to adapt the predictability horizon of the impacts to each object, taking into account the orbit uncertainty and the close encounters experienced. In this paper we discuss the problem of estimating a reasonable predictability horizon when multiple close encounters are present and start to address the problem proposing a formal mathematical definition of scattering encounter.

Keywords. Near Earth Objects, Impact Monitoring, dynamics, close encounters, scattering encounter

1. Introduction

The Impact Monitoring (IM) of Near Earth Objects (NEOs) is a relatively young field of research, considering that 25 years ago precise algorithms to compute the probability of an impact with the Earth did not exist; CLOMON, the first automatic IM system, was born in 1999 (Milani *et al.* 1999, 2000a; Chesley and Milani 2000; Tommei 2019, 2021).

Recent years have seen a strong interest in asteroid hazard:

- since 2015 three systems for the detection of imminent impactors (small asteroidal objects detected a few days before the possible impact with the Earth) have been developed: Scout at JPL/NASA (Farnocchia *et al.* 2015), NEORANGER at University of Helsinki (Solin and Granvik 2018) and NEOScan at University of Pisa/SpaceDyS (Spoto *et al.* 2018);
- the year 2020 saw the increase of IM operational systems: in addition to the two historical systems, CLOMON2 at University of Pisa/SpaceDyS and Sentry at JPL/NASA (Milani *et al.* 2005a; Roa *et al.* 2021), the European Space Agency (ESA) started its own system, Aegis;
- on 2022, September 26th, the DART spacecraft successfully impacted on Dimorphos, the moon of the Near Earth Asteroid (NEA) (65803) Didymos, showing the capability of deflecting a small body (Cheng *et al.* 2018, 2023; Rivkin and Cheng 2023);

• the European Space Agency (ESA) will launch in 2024 the Hera mission which will visit the Didymos system to investigate the effects of the DART impact and to characterize the properties of the system (Michel *et al.* 2022).

The IM science, in addition to being useful for the planetary defense, is a fascinating field of research involving astronomy, physics, mathematics and computer science. Current and future challenges in this activity include the continuous need to check and upgrade, when necessary, the dynamical model, including non-gravitational perturbations where appropriate, the extension of the time span covered by IM for objects with very well determined orbits and, last but not least, the likely huge increase of the workload that will be caused by future large-scale surveys. The related issue of the so called imminent impactors (Mochi and Tommei 2021) is out of the scope of the present paper.

Here we are interested in the dynamics of NEOs influenced by multiple close encounters with the Earth. This is the main reason why the long term computation of an orbit of a NEO is not a simple task.

In Section 2 we briefly summarize the problems arising in the determination of the orbits of NEOs and the solutions adopted to handle the orbital uncertainty. Section 3 is devoted to the problem of the predictability horizon after multiple close encounters; in particular, we give a formal mathematical definition of scattering encounter. In Section 4 we draw some conclusions.

2. NEO dynamics

As already said, the computation of the orbit of a NEO is not a simple task because, in many cases, the observations are few and the dynamics is strongly affected by close encounters with the Earth and other planets. Even if a preliminary orbit is available, the uncertainty can be large and the best way to proceed is to consider a set of orbits belonging to a Confidence Region (CR, a subset of the 6-dimensional space of the orbital elements) where the astrometric residuals are acceptable. Such orbits, obtained with an appropriate sampling of the CR, are called Virtual Asteroids (VAs, Milani *et al.* 2000b); they can have a very low Minimum Orbit Intersection Distance with our planet (MOID, Gronchi 2005; Gronchi and Tommei 2007) and thus be dangerous. The goal of IM is to understand whether the CR contains some Virtual Impactor (VI), that is a subset of initial conditions bringing to an impact with the Earth in the future.

Under appropriate conditions, it may happen that the CR contains multiple VIs at the same date in the future. What happens most often in this case is that each VI has a dynamical path to its collision different from those of the other VIs.

For example, when (99942) Apophis was discovered and the timing of its 13 April 2029 approach to the Earth was still rather badly constrained from the then available observations, there were at least three separate VIs for 13 April 2036, that is, after the Earth would have made 7 revolutions about the Sun following the 2029 encounter:

• via the 6/7 mean motion resonance, with Apophis making 6 revolutions about the Sun (Chesley 2006);

• via the 5/7 resonance, with Apophis making 5 revolutions;

• and via the 4/7 resonance, with Apophis making only 4 revolutions.

Clearly, between 2029 and 2036 the CR has wound up itself about the Sun multiple times, and the three VIs define disconnected collision subsets; in this case each VI is a connected component of the CR.

Once a *representative of a VI*, i.e. a particular initial condition that leads to a collision, has been identified, an Impact Probability (IP) needs to be computed: in general terms, the IP of a VI is proportional to the volume of the VI in the orbital elements space. If we are looking for small IPs, the sampling should be very dense; but the real problem is how to ensure completeness to the VI search, taking into account the computational costs

(Del Vigna *et al.* 2019a). One class of methods, including Monte Carlo (MC) and statistical ranging, uses random sampling of the CR to study the probabilistic distributions of the orbits through the swarm of VAs.

When we have to manage a large catalog of objects and small probabilities with a small number of VAs, it is more cost-effective to sample the CR with a geometrical object, such as a smooth manifold. At the end of 1990s in Pisa a class of 1-dimensional sampling methods based on the Line Of Variations (LOV) was developed (Milani *et al.* 2005b): the LOV is a differentiable curve, which can represent, in some cases, the spine of the CR; it is sampled generating a swarm of indexed VAs, allowing the interpolation between consecutive VAs. The sampling could be done uniformly using the curve parameter; recently, the possibility to collect points at fixed step in probability has been introduced (Del Vigna *et al.* 2020).

An important and expensive (in terms of computational costs) step in IM is the propagation of the VAs, a set of orbits of the order of thousands, for 100 years or more. Therefore, in the light of the expected increase in NEO discoveries (Jones *et al.* 2018; Cibin *et al.* 2016), we wonder whether it is possible to identify an algorithm that tailors the propagation horizon to each object. In the NEO population there are objects with very different types of motion, and for which the OD procedure results in very different orbital uncertainties. If the CR is either large due to poor observational record, or becomes large due to intervening encounters, there is the possibility for the VAs to follow different dynamical paths, going through different sequences of close encounters, so that the predictability horizon (after which the dynamics will be even more chaotic, Spoto and Milani 2016) becomes short.

An algorithm that would allow to identify the close encounter that causes the subsequent unpredictability of the orbit (the so called *scattering encounter*) would be useful both in the daily IM operations and also for the long term IM, hat involves the analysis of objects with a very well constrained orbits, including, if needed, also the determination of non-gravitational accelerations, like the Yarkovsky effect (Bedini and Tommei 2023). But before designing such an algorithm it is necessary to give a formal definition of scattering encounter, that will be presented in the next section.

3. The scattering encounter

In order to discuss the dynamics at close encounters, it is useful to introduce the analytical theory of close encounters and the *b*-plane (Öpik 1976; Greenberg *et al.* 1988; Valsecchi *et al.* 2003). The *b*-plane is the plane centred on the planet and perpendicular to the pre-encounter planetocentric velocity vector $\vec{U}$. In the theory, the encounter is modelled as an instantaneous rotation of $\vec{U}$, taking place when the small body crosses the *b*-plane; the crossing point has coordinates ξ, ζ. The magnitude of $\vec{U}$, together with the angles θ, ϕ defining its direction, are all given by simple functions of a, e, i, the pre-encounter semimajor axis, eccentricity and inclination of the orbit of the small body; the encounter rotates $\vec{U}$ into $\vec{U}'$, whose magnitude is the same, but its direction is given by the angles θ', ϕ' (θ', ϕ' are explicit functions of ξ, ζ), allowing the computation of the post-encounter values a', e', i' of the small body orbit. The two *b*-plane coordinates are chosen so that ξ is the local MOID, while ζ is related to the timing of the encounter, i.e. to whether the small body passes by the planet earlier or later with respect to the closest possible distance associated to the local MOID.

Using the so-called wire approximation (Valsecchi *et al.* 2003), we consider a continuous stream of particles, all with the same U, θ, ϕ, ξ (i.e. all with the same a, e, i and MOID), differing only for ζ; in this way, we can model the projection of the LOV onto the *b*-plane, and, using simple analytical formulae, we can identify the regions on the *b*-plane of a certain encounter that lead to post-encounter orbits characterized by a a given

Table 1. The encounter sequences of (410777) 2009 FD and (101955) Bennu; for details, see the text.

	(410777) 2009 FD				(101955) Bennu		
Date	$b(r_\oplus)$	$\sigma_\zeta(r_\oplus)$	$\dot\sigma_\zeta(r_\oplus/y)$	Date	$b(r_\oplus)$	$\sigma_\zeta(r_\oplus)$	$\dot\sigma_\zeta(r_\oplus/y)$
2015/10/29.4	983.7	0.0021		2005/09/20.4	777.9	0.0002	
2063/03/29.3	304.5	0.2158	$4.5{\cdot}10^{-3}$	2054/09/30.0	922.8	0.0017	$3.1{\cdot}10^{-5}$
2064/10/26.4	625.1	0.2428	$1.8{\cdot}10^{-2}$	2060/09/23.0	117.6	0.0024	$1.1{\cdot}10^{-4}$
2136/03/28.3	511.3	18.30	$2.5{\cdot}10^{-1}$	2080/09/22.0	365.4	1.15	$5.7{\cdot}10^{-2}$
2185/03/29.4	210.4	208.7	3.9	2135/09/25.4	47.2	12.5	0.21

semimajor axis and thus by a given period (Valsecchi *et al.* 2018). That is, we can find the regions of the pre-encounter b-plane leading to specific *resonant returns*.

In recent years there were, among others, a couple of successful attempts of long term IM for two remarkable objects, (410777) 2009 FD (Spoto *et al.* 2014) and (101955) Bennu (Chesley *et al.* 2014)†. In both cases the OD involved solving for seven parameters, the six orbital elements and the Yarkovsky acceleration, that is essential in the study of the long term dynamics of NEOs.

Table 1 summarizes the sequences of encounters with the Earth within 0.5 au for the two NEAs, that are presented side by side: on the left 2009 FD, on the right Bennu. In both cases there are 4 intervening encounters before the final, scattering one; the Table lists the date of closest approach for the nominal VA, the impact parameter $b = \sqrt{\xi^2 + \zeta^2}$, the 1-$\sigma$ uncertainty along ζ, in Earth radii, as well as its variation in time in the time span between the encounters, in Earth radii per year.

The encounters of 2009 FD take place alternatively at either node, and only the scattering encounter of 2185, in which also an Earth collision was possible at the time of Spoto *et al.* (2014), is a resonant return from 2136‡. In the case of Bennu, all the encounters take place at the same node, and the scattering encounter of 2135 does not involve an Earth collision; nevertheless, the scattering of nearby VAs in 2135 is so strong that the exploration of the subsequent impact hazard has to be done through statistical means (Chesley *et al.* 2014), a feature that has not changed with the new data acquired after 2014 (Farnocchia *et al.* 2021). For both asteroids, the sequence of encounters preceding the scattering one leads to a very large increase of σ_ζ, of about five orders of magnitude in less than two centuries.

What do these two examples tell us? Essentially that, if we want to perform a long term propagation, it is crucial to analyze the close encounters and, in particular, identify the scattering encounter; for very well determined orbits, like the ones just discussed, it is an encounter after which the impact hazard can only be computed statistically, since the dynamics of the object has become chaotic. Since for an encounter to be considered a scattering one depends on the subsequent uncertainty of the orbit, it can happen that for not well constrained orbits also a not so deep encounter could turn out to be a scattering one.

We now propose a topological definition of scattering encounter, which will help us to develop an operational definition.

Definition 1. A close encounter of a minor body with the Earth is said to be a **scattering encounter** if the post-encounter CR ceases to be simply connected.

† In both cases, subsequent observations have led to the exclusion of many VIs (Del Vigna *et al.* 2019b; Farnocchia *et al.* 2021).

‡ The value of σ_ζ in the Table refers to the nominal VA, and differs from the stretching given in Table 2 of Spoto *et al.* (2014) since the latter refers to the 2185 VI, at about 1 σ_ζ from the nominal; as Fig. 6 of Spoto *et al.* (2014) shows, on the b-plane of 2185 the stretching varies appreciably along the LOV.

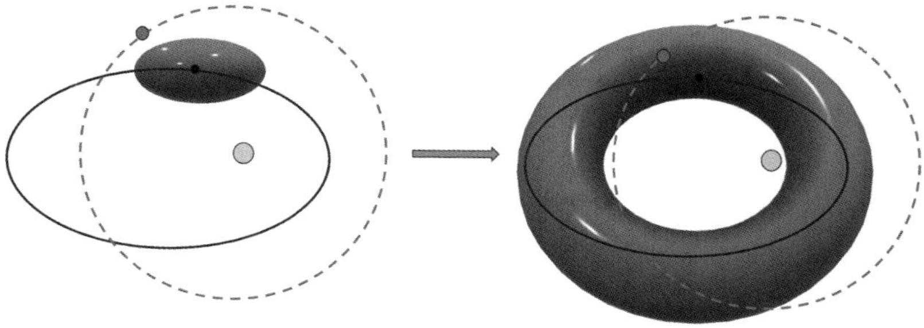

Figure 1. 3D representation of the effect of a scattering encounter.

Since it is not easy to visualize the situation in dimension 6, in Fig. 1 we have sketched the situation in dimension 3. The dashed red line is the Earth orbit, while the black continuous line is an hypothetical minor body orbit; on the left we have the nominal orbit of the asteroid and its CR, sketched with a blue ellipsoid before a close approach; we know that the geometrical shape could be more complex, like a bananoid for example (Vavilov 2020), but in any case a simply connected object. Due to encounters the CR grows, stretching along the orbit, until after the scattering encounter it changes completely the topology: from a simply connected confidence region, homotopic to a point, it becomes an object having a non-trivial fundamental group, like the torus in the figure. The underlying meaning is essentially that the uncertainty region, after a scattering encounter, contains orbits having periods in different resonances with the Earth, due to the large variation of the post-encounter semimajor axis throughout the CR.

4. Conclusions and future work

In this paper we introduced a topological formal definition for the scattering encounter of a minor body with the Earth (but the definition can be applied to close encounters with other planets) and we highlighted the need of finding an operational definition to be implemented in the IM automatic systems; this would be an important step to optimize the pipeline of such systems in view of the expected increase of their workload due to the forthcoming more capable NEA surveys.

Acknowledgments

Giacomo Tommei acknowledges the support from the Italian Space Agency through the grant *Monitoring Asteroids* (2022-33-HH.0).

References

Bedini L., Tommei G. 2023, submitted to *A&A*
Cheng A. F. *et al.* 2018, *Planet. Space Sci.* 157, 104
Cheng A. F. *et al.* 2023, *Nature* 616, 457
Chesley S. R. 2006, in : D. Lazzaro, S. Ferraz-Mello & J.A. Fernández (eds.), *Asteroids, Comets, Meteors 2005* Proc. IAU Symposium No. 229 (Cambridge: CUP), p. 215
Chesley S. R., Milani A. 2000, *Bull. Am. Astron. Soc.* 32, 862
Chesley S. R. *et al.* 2014, *Icarus* 235, 5
Cibin L., Chiarini M., Bernardi F., Ragazzoni R., Salinari P. 2016, *MemSAIt* 87, 197
Del Vigna A., Milani, A., Spoto F., Chessa A, Valsecchi G. B. 2019a, *Icarus* 321, 647
Del Vigna A., Roa J., Farnocchia D., Micheli M., Tholen D., Guerra F., Spoto F., Valsecchi G. B. 2019b, *A&A* 627, L11
Del Vigna A. 2020, *Celest. Mech. Dynam. Astron.* 132, 49
Del Vigna A., Guerra F, Valsecchi G. B. 2020, *Icarus* 351, 113966
Farnocchia D., Chesley S. R., Micheli M. 2015, *Icarus* 258, 18

Farnocchia D., Chesley S. R., *et al.* 2021, *Icarus* 369, 114594
Greenberg R., Carusi A., Valsecchi G. B. 1988, *Icarus* 75,1
Gronchi G. F. 2005, *Celest. Mech. Dynam. Astron.* 93, 295
Gronchi G. F., Tommei G. 2007, *Discret. Contin. Dynam. Syst. B* 7, 755
Jones R. L. *et al.* 2018, *Icarus* 303, 181
Michel P. *et al.* 2022, *PSJ* 3, 160
Milani A., Chesley S.R., Valsecchi G. B. 1999, *A&A* 346, L65
Milani A., Chesley S.R., Valsecchi G. B. 2000a, *Planet. Space Sci.* 48, 945
Milani A., Chesley S.R., Boattini A., Valsecchi G. B. 2000b, *Icarus* 145, 12
Milani A., Chesley S. R., Sansaturio M. E., Tommei G., Valsecchi G. B. 2005a, *Icarus* 173, 362
Milani A., Sansaturio M. E., Tommei G., Arratia O., Chesley S. R. 2005b, *A&A* 431, 729
Mochi M., Tommei G. 2021, *Universe* 7, 10
Öpik E. J. 1976, *Interplanetary Encounters* (Amsterdam: Elsevier)
Rivkin A., Cheng A. 2023, *Nature Communications* 14, 1003
Roa J., Farnocchia D,, Chesley S. R. 2021, *AJ* 162, 277
Solin O., Granvik M. 2018, *A&A* 616, A176
Spoto F., Milani A., Farnocchia D., Chesley S. R., Micheli M., Valsecchi G. B., Perna D., Hainaut O. 2014, *A&A* 572, A100
Spoto F., Milani A. 2016 *Celest. Mech. Dynam. Astron.* 124, 295
Spoto F., Del Vigna A., Milani A., Tommei G., Tanga P., Mignard F. , Carry B., Thuillot W., David P. 2018, *A&A* 614, A27
Tommei G. 2019, in *Proc. ESA NEO and DEBRIS Detection Conf., ESOC 22-24 January 2019*
Tommei G. 2021, *Universe* 7, 103
Valsecchi G.B., Milani A, Gronchi G. F., Chesley S. R. 2003, *A&A* 408, 1179
Valsecchi G.B., Alessi E. M., Rossi A. 2018. *Celest. Mech. Dynam. Astron.* 130, 8
Vavilov D. E. 2020, *MNRAS* 492, 4546

Complex Planetary Systems II: Latest Methods for an Interdisciplinary Approach
Proceedings IAU Symposium No. 382, 2024
A. Lemaitre & A.-S. Libert, eds.
doi:10.1017/S1743921323004192

Excitation of the obliquity of Earth-like planets via tidal forcing using the Andrade rheology

Ema F. S. Valente[1] **and Alexandre C. M. Correia**[1,2]

[1]CFisUC, Departamento de Física, Universidade de Coimbra, 3004-516 Coimbra, Portugal
email: emavalente@ua.pt

[2]IMCCE, Observatoire de Paris, PSL Université, 77 Av. Denfert-Rochereau, 75014 Paris, France

Abstract. Close-in planets undergo strong tidal effects with their host stars that modify their spins and orbits. Adopting a Maxwell rheology, it has been shown that for the 5/2 and 7/2 spin-orbit resonances, the obliquity of these planets can stabilise at a high value. Here, we show that these high obliquity metastable states can also be observed for the same spin-orbit resonances considering the Andrade rheology.

Keywords. planet-star interactions, dynamical evolution and stability, terrestrial planets

---

1. Introduction

Many exoplanets have been discovered in the habitable zone of low-mass stars, especially around M-dwarf stars. These stars comprise $\sim 70\%$ of all stars in the Milky Way (Bochanski et al. 2010), and the occurrence rate of Earth-sized planets is higher than around other main sequence stars (Tuomi et al. 2019). Furthermore, the detection of small planets is easier via radial velocity and transit techniques.

The search for habitable worlds around M-dwarf stars has some advantages. The habitable zone is closer to the star due to its low luminosity and these stars are the longest-living stars in the Universe, which could provide enough time for biological development. Yet, it has one major disadvantage. In the M-dwarfs planetary systems, the tidal interactions between the planet and the star are strong, which can lead the planet to develop synchronous rotation (the rotation period matches the orbital period) and zero obliquity (Hut 1980). This configuration is not ideal for the habitability.

Valente and Correia (2022) have studied the tidal evolution of these planetary systems and found that, in a two-body problem, the synchronous rotation can be prevented, and stabilisation on spin-orbit resonances with high obliquity states can occur. However, to handle the deformation of the planet, they only assume a Maxwell rheology. Here, we extend the work presented in Valente and Correia (2022) by adopting an Andrade rheological model (for a review of these and other main viscoelastic models, see Renaud and Henning 2018).

2. Model

We consider a system composed of a planet with mass m and a star with mass $M_\star$. The planet is an extended body with radius R and rotates with angular velocity $\boldsymbol{\omega} = \omega\, \mathbf{s}$.

It can be deformed under the action of tides, which gives rise to the tidal potential (e.g. Lambeck 1980),

$$V(\mathbf{r}) = -k_2 \frac{\mathcal{G} M_\star}{R} \left(\frac{R}{r}\right)^3 \left(\frac{R}{r_\star}\right)^3 P_2\left(\frac{\mathbf{r}\cdot\mathbf{r}_\star}{r\, r_\star}\right) , \tag{1}$$

where $\mathcal{G}$ is the gravitational constant, $\mathbf{r}$ is the distance measured from the planet's centre of mass, $\mathbf{r}_\star$ is the position of the star, $P_2(x) = (3x^2 - 1)/2$ is a Legendre polynomial, and k_2 is the second Love number for potential, that depends on the internal structure of the planet and is a function on the frequency of the perturbation, σ.

We obtain the spin and orbital equations of motion from the torque, $\mathbf{T} = -M_\star \mathbf{r}_\star \times \nabla V(\mathbf{r}_\star)$, and from the power, $\dot{E} = -M_\star \dot{\mathbf{r}}_\star \cdot \nabla V(\mathbf{r}_\star)$. Tidal effects slowly modify the spin and the orbit of the planet in a timescale much longer than the orbital and precession periods of the system. Thus, we can average the torque and the power over the mean anomaly and the argument of pericentre as

$$\langle \mathbf{T} \rangle = T_p\, \mathbf{p} + T_q\, \mathbf{q} + T_s\, \mathbf{s} \quad \text{and} \quad \langle \dot{E} \rangle = n\, T_E , \tag{2}$$

where n is the mean motion, and T_p, T_q, T_s, T_E are expressed in Hansen coefficients, which depend solely on the eccentricity (their full expressions are given in Valente and Correia 2022). $(\mathbf{p}, \mathbf{q}, \mathbf{s})$ is a set of orthogonal vectors, where $\mathbf{p} = \mathbf{k} \times \mathbf{s}$, $\mathbf{q} = \mathbf{s} \times \mathbf{p}$, and $\mathbf{k}$ is the unit vector normal to the orbital plane (see Correia and Valente 2022).

To handle the deformations of the planet, we need to adopt some rheological model for the second Love number k_2. Unlike Valente and Correia (2022), we choose the Andrade model. An Andrade solid is represented by three mechanical elements combined in series: a spring, a dashpot, and a spring-pot (e.g. Andrade 1910). The second Love number for the Andrade model is given by (e.g. Efroimsky 2012):

$$k_2(\sigma) = \frac{k_f}{1 + \tilde{\mu}(\sigma)} , \tag{3}$$

where

$$\tilde{\mu}(\sigma) = \left(\frac{\tau}{\tau_e} - 1\right) \left[1 - i(\sigma\tau_e)^{-1} + (i\sigma\tau_A)^{-\alpha}\, \Gamma(1+\alpha)\right]^{-1} \tag{4}$$

is the effective rigidity. Here τ_A is the anelastic relaxation time, τ_e is the Maxwell relaxation time, and $\tau = k_f \tau_e / k_e$ is a global relaxation time, where k_f and k_e are the fluid and the elastic Love numbers, respectively. The anelastic relaxation time can also be related to the Maxwell relaxation time, for simplicity we assume that $\tau_A = \tau$ (Efroimsky 2012). The relaxation times are not well determined for the Earth due to uncertainties in the effective viscosity, and so we treat τ as a free parameter. The α parameter is the time exponent of the Andrade transient creep strain law and is an empirical adjustable parameter whose value depends on the material. We usually have $0.2 \leq \alpha \leq 0.4$ for rocky planets (Castillo-Rogez et al. 2011), but $\alpha = 1$ for the Maxwell model. The dissipation of the system is only given by the imaginary part of Eq. (3).

In the following, we explore the secular tidal evolution of the planet (Eq. (2)) for different values of α, eccentricities, and τ values, in order to observe possible configurations where the Earth-like planets can have non-synchronous rotation rates and high obliquity states, as was observed with the Maxwell rheology (see Valente and Correia 2022).

3. Spin dynamics

Independently of the tidal model that we adopt, the final state of tidal evolution corresponds to circular orbits, zero obliquity, and synchronous rotation (Hut 1980). However, the spin evolves much faster than the orbit (Valente and Correia 2022), and so in this

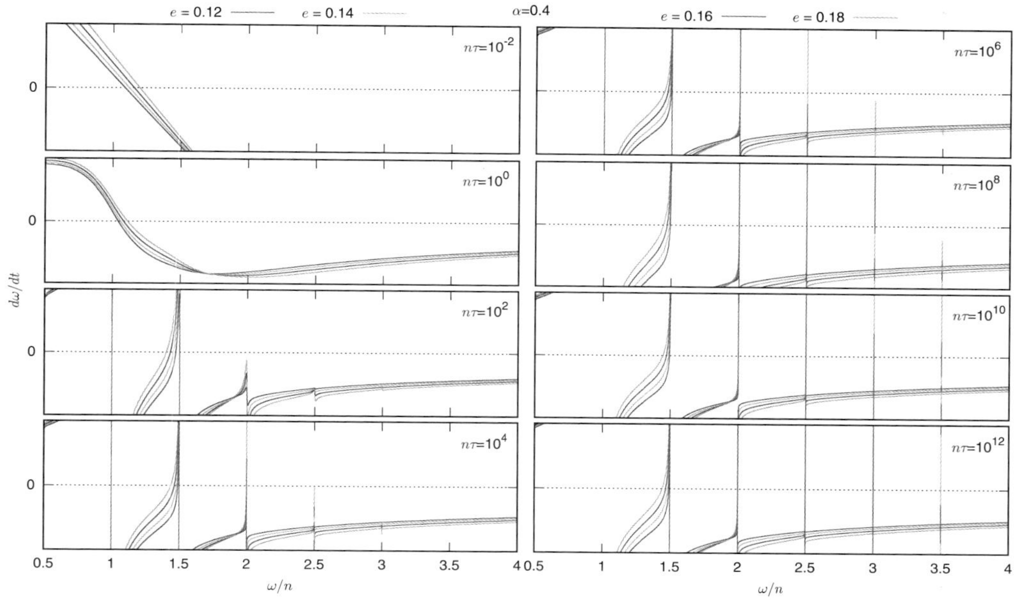

Figure 1. Variation of $d\omega/dt$ with ω/n, assuming the Andrade rheology with $\alpha = 0.4$, for different values of $n\tau$ and four different eccentricities: $e = 0.12$ (red), $e = 0.14$ (green), $e = 0.16$ (blue), and $e = 0.18$ (orange). We adopt here $\theta = 0$.

section, we assume that the eccentricity and semi-major axis are constant. We thus only consider the spin evolution given by the rotation rate, ω, and the obliquity, θ,

$$\frac{\dot{\omega}}{\omega} = -\frac{T_s}{C\omega} \quad \text{and} \quad \dot{\theta} \approx \frac{T_q}{C\omega} \sin\theta \,, \tag{5}$$

respectively, where $C = \xi m R$ is the principal moment of inertia of the planet, and ξ is an internal structure constant. Also, we limit our analysis to positive rotation rates ($\omega > 0$). Our main goal is to describe the possible equilibria and evolutionary paths for different tidal regimes (different τ values).

3.1. *Equilibrium rotation*

The final evolution of the spin has zero obliquity, which is an equilibrium point (Eq. (5)). Therefore, we initially assume $\theta = 0$ (planar case). In Fig. 1, we plot, as an example, the evolution of the rotation rate for $\alpha = 0.4$ as a function of ω/n for different values of eccentricities and $n\tau$. The interception with the horizontal dotted line $\dot{\omega} = 0$ gives all the equilibria for the rotation rate. Negative slopes correspond to stable equilibria, where the rotation can remain locked for a given value of the eccentricity. We adopt eccentricity values between $e = 0.12$ and $e = 0.18$, and vary $n\tau$ with the power 10^N up to $N = 12$. These values are not exactly the same as in Valente and Correia (2022) for the Maxwell model, because in the nonlinear regime, for the same $n\tau$ value, the Andrade model dissipates more energy.

For $n\tau \lesssim 1$, we observe that a single stable equilibrium point exists, close to synchronous rotation, which is known as the pseudo-equilibrium rotation. For $n\tau \gg 1$, additional stable equilibria appear. They cluster around semi-integer values, known as spin-orbit resonances, because $\dot{\omega}$ has extrema at these values. When the amplitude of $\dot{\omega}$ surpasses the line $\dot{\omega} = 0$, the rotation can be locked in that spin-orbit resonance. Here, we also observe that by increasing $n\tau$ and/or the eccentricity, the number of spin-orbit equilibria increase.

Although we only show here the example for $\alpha = 0.4$, by changing this parameter, we note a difference in the first capture in a given spin-orbit resonance for the same $n\tau$ value. By comparing our Fig. 1 with Fig.1 of Valente and Correia (2022), for instance, for $n\tau = 10^4$ and $e = 0.16$, in our Fig. 1 ($\alpha = 0.4$), the first capture occurs in the 2/1 resonance, while in Fig.1 of Valente and Correia (2022) (which is equivalent to $\alpha = 1$), the first capture occurs in the 3/1 resonance. Therefore, we conclude that the value of α modifies the captures in spin-orbit resonances. This is why we need to show higher values of $n\tau$, because by decreasing the value of α, the capture in a given spin-orbit resonance can only occur for higher values of $n\tau$.

In Fig. 1, we also observe that for the same $n\tau$ value, stabilisation can occur at different spin-orbit resonances, depending on the eccentricity. Here, we assume a constant eccentricity, yet it also evolves slowly with time (Valente and Correia 2022). By decreasing the eccentricity value, the higher-order spin-orbit equilibria become unstable, and the rotation is sequentially captured in lower-order resonances.

3.2. *Equilibrium obliquity*

Despite $\theta = 0$ being an equilibrium point (Eq. (5)), it is not necessarily stable. For $T_q > 0$, the equilibrium point can be unstable, and the obliquity is allowed to grow. To find these additional stable equilibrium states, we compute numerically the rotation and obliquity rates (Eq. (5)), determining the pair $(\omega/n, \theta)$ that simultaneously satisfy $T_q = T_s = 0$.

In Fig. 2, we compute the high obliquity stable equilibria for $\alpha = 0.2$, 0.3, 0.4, 0.5 and 0.6, and also the high obliquity state equilibria for the Maxwell rheology ($\alpha = 1$). We observe that for $\omega/n \approx 5/2$ and $\omega/n \approx 7/2$ also exist high obliquity states as in the case with the Maxwell model (Valente and Correia 2022). However, for $\alpha = 0.2$, the high obliquity state only exists for $\omega/n \approx 5/2$. These states do not directly depend on the value of $n\tau$, but they become unstable below some critical eccentricity, which depends on $n\tau$. We observe that as we increase the $n\tau$ value, the critical eccentricity becomes smaller. We mark with a dot the critical eccentricity for $n\tau = 10^N$, where the value of N is given next to the dot. By increasing the value of α, the high obliquity state curves of the 5/2 and 7/2 resonances widen, englobing more eccentricity values and $n\tau$ values. Also, by increasing α, high obliquities states exist for lower values of $n\tau$. Indeed, by increasing α, we approach the regime of the Maxwell rheology, and the curve tends to be equal to Fig. 2 f. An important difference is that for lower values of α, the maximal obliquity that can be attained is smaller. For instance, for $\alpha = 0.3$, the obliquity can reach nearly 15° for the 5/2 resonance and 11° for the 7/2 resonance, while for $\alpha = 0.6$, the obliquity can reach nearly 62° for the 5/2 resonance and 49° for the 7/2 resonance.

4. Aplication to Proxima b

We now apply our model to the long-term tidal evolution of Proxima Centauri b, a planet with mass $m = 1.17\ M_\oplus$, radius $R = 1.08\ R_\oplus$, and semi-major axis $a = 0.049$ au, which is in the habitable zone of a M-dwarf star with mass $M_\star = 0.12\ M_\odot$ (Anglada-Escudé et al. 2016). For the structure and Love numbers, we adopt those of the Earth, $\xi = 0.330$, $k_f = 0.933$ and $k_e = 0.295$ (Yoder 1995). The initial conditions of the spin are a rotation period of 24 h and the obliquity starts at $\theta = 10°$. We only show configurations where the high obliquity equilibria occur when α is in the range of $0.2 \le \alpha \le 0.4$ (the usually accepted range for rocky planets).

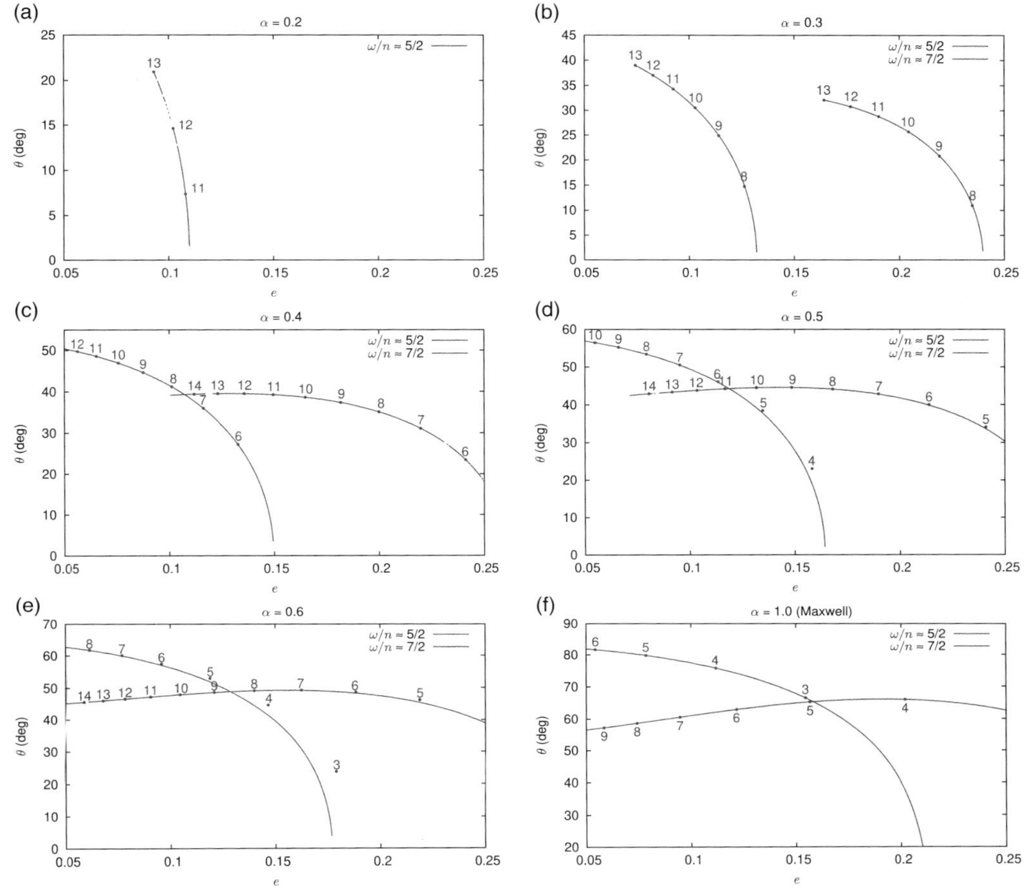

Figure 2. High obliquity stable equilibria as a function of the eccentricity for values of α between 0.2 and 0.6, and for the Maxwell model ($\alpha = 1$). The dots mark the critical eccentricity for $n\tau = 10^N$, where the N-value is given next to the dot. These states are found numerically by determining the pair $(\omega/n, \theta)$ that simultaneously satisfy $T_q = T_s = 0$.

The results of the simulations are shown in Fig. 3. For each panel, we show the eccentricity (top), the ω/n ratio (middle), and the obliquity (bottom) as a function of time. We do not present the full tidal evolution of the planet because we are only interested in showing that high obliquity states occur even for the Andrade rheology.

In Fig. 3 a, we show the evolution with $n\tau = 10^6$ for $\alpha = 0.4$ and the initial eccentricity is $e = 0.14$. At the beginning of the evolution, the rotation evolves rapidly towards the 3/1 resonance, where it stays briefly until destabilisation occurs. Then, it is captured in the 5/2 spin-orbit resonance. Here, a high obliquity state is attained for a very short time (~ 0.3 Gyr) with $\theta \approx 25°$. This is only possible for $n\tau > 10^5$ as we have seen in Fig. 2 c. The rotation subsequently evolves into the 2/1 spin-orbit resonance, where the obliquity drops to zero.

In Fig. 3 b we adopt $n\tau = 10^9$ for $\alpha = 0.3$ and the initial eccentricity is $e = 0.12$. Here, the rotation evolves directly into the 5/2 spin-orbit resonance. We observe the slow growth of the obliquity when the planet is captured in this resonance. This high obliquity equilibrium remains stable for ~ 3 Gyr until the destabilisation of the 5/2 spin-orbit

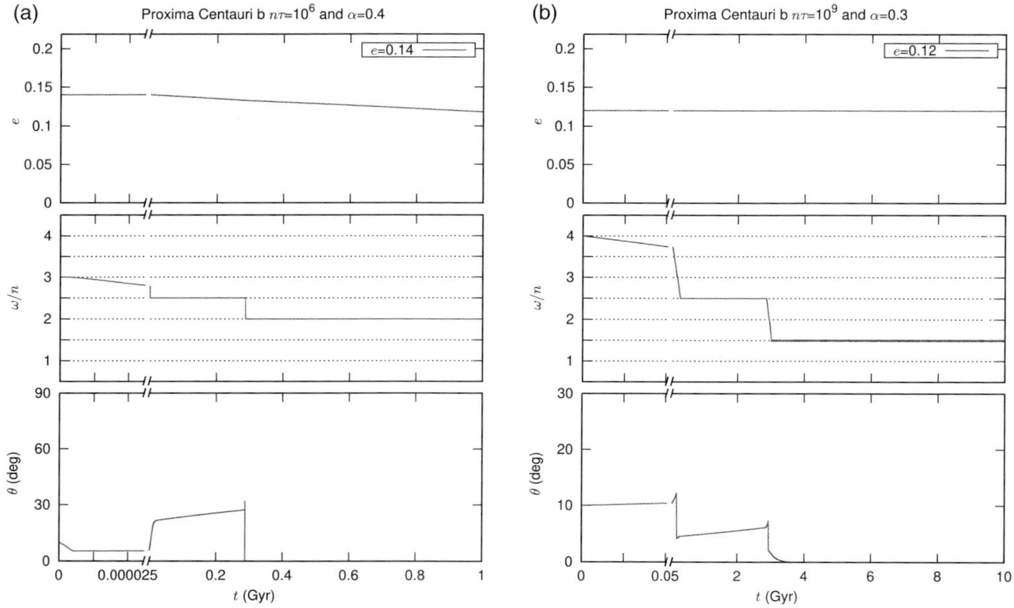

Figure 3. Secular evolution over time of Proxima b for a) $\alpha = 0.4$ with $n\tau = 10^6$ and b) $\alpha = 0.3$ with $n\tau = 10^9$; with initial eccentricities a) $e = 0.14$ and b) $e = 0.12$. We show the eccentricity (top), the ω/n ratio (middle) and the obliquity (bottom). Note the break in the time scale.

resonance occurs. Then, the rotation is captured in the 3/2 resonance and the obliquity descends again to zero. This behavior is also in agreement with Fig. 2 b.

5. Conclusion

In this work, we show that when we consider the Andrade rheology instead of the Maxwell one, rocky planets in the habitable zone of low-mass stars can still develop high obliquity states for the specific 5/2 and 7/2 spin-orbit resonances, as described in Valente and Correia (2022). However, these high obliquity states are less likely, because they are only possible for some particular values of the eccentricity and unrealistic large values of $n\tau$. The obliquity also attains lower values and lasts for shorter periods of time, since for the Andrade model the dissipation is higher than in the Maxwell model for the same $n\tau$ value.

Acknowledgements. This work was supported by grant SFRH/BD/137958/2018, and by projects CFisUC (UIDB/04564/2020 and UIDP/04564/2020), GRAVITY (PTDC/FIS-AST/7002/2020), and ENGAGE SKA (POCI-01-0145-FEDER-022217), funded by COMPETE 2020 and FCT, Portugal. The numerical simulations were performed at the Laboratory for Advanced Computing at University of Coimbra (https://www.uc.pt/lca).

References

Andrade, E. N. d. C. 1910, On the Rigidity of the Earth. *Proc. R. Soc. Lond. A*, 84, 1–12.

Anglada-Escudé, G., Amado, P. J., Barnes, J., Berdiñas, Z. M., Butler, R. P., Coleman, G. A. L., de La Cueva, I., Dreizler, S., Endl, M., Giesers, B., Jeffers, S. V., Jenkins, J. S., Jones, H. R. A., Kiraga, M., Kürster, M., López-González, M. J., Marvin, C. J., Morales, N., Morin, J., Nelson, R. P., Ortiz, J. L., Ofir, A., Paardekooper, S.-J., Reiners, A., Rodríguez, E., Rodríguez-López, C., Sarmiento, L. F., Strachan, J. P., Tsapras, Y., Tuomi, M., & Zechmeister, M. 2016, A terrestrial planet candidate in a temperate orbit around Proxima Centauri. *Nature*, 536(7617), 437–440.

Bochanski, J. J., Hawley, S. L., Covey, K. R., West, A. A., Reid, I. N., Golimowski, D. A., & Ivezić, Z. 2010, The luminosity and mass functions of low-mass stars in the galactic disk. ii. the field. *The Astronomical Journal*, 139(6), 2679–2699.

Castillo-Rogez, J. C., Efroimsky, M., & Lainey, V. 2011, The tidal history of Iapetus: Spin dynamics in the light of a refined dissipation model. *Journal of Geophysical Research (Planets)*, 116(E9), E09008.

Correia, A. C. M. & Valente, E. F. S. 2022, Tidal evolution for any rheological model using a vectorial approach expressed in Hansen coefficients. *Celestial Mechanics and Dynamical Astronomy*, 134(3), 24.

Efroimsky, M. 2012, Bodily tides near spin-orbit resonances. *Celestial Mechanics and Dynamical Astronomy*, 112, 283–330.

Hut, P. 1980, Stability of tidal equilibrium. *Astronomy and Astrophysics*, 92, 167–170.

Lambeck, K. 1980,. *The Earth's Variable Rotation: Geophysical Causes and Consequences.* Cambridge University Press.

Renaud, J. P. & Henning, W. G. 2018, Increased Tidal Dissipation Using Advanced Rheological Models: Implications for Io and Tidally Active Exoplanets. *Astrophysical Journal*, 857(2), 98.

Tuomi, M., Jones, H. R. A., Butler, R. P., Arriagada, P., Vogt, S. S., Burt, J., Laughlin, G., Holden, B., Shectman, S. A., Crane, J. D., Thompson, I., Keiser, S., Jenkins, J. S., Berdiñas, Z., Diaz, M., Kiraga, M., & Barnes, J. R. 2019,. Frequency of planets orbiting m dwarfs in the solar neighbourhood.

Valente, E. F. S. & Correia, A. C. M. 2022, Tidal excitation of the obliquity of Earth-like planets in the habitable zone of M-dwarf stars. *Astronomy and Astrophysics*, 665, A130.

Yoder, C. F. Astrometric and geodetic properties of Earth and the Solar System. In *Global Earth Physics: A Handbook of Physical Constants* 1995,, pp. 1–31. American Geophysical Union, Washington D.C.

Complex Planetary Systems II: Latest Methods for an Interdisciplinary Approach
Proceedings IAU Symposium No. 382, 2024
A. Lemaitre & A.-S. Libert, eds.
doi:10.1017/S174392132300409X

Comparison of Mars rotation angle models

Marie Yseboodt[1], **Rose-Marie Baland**[1] **and Sébastien Le Maistre**[1,2]

[1]Royal Observatory of Belgium, Avenue circulaire 3, Brussels, Belgium
email: m.yseboodt@oma.be,

[2]UCLouvain, Louvain-la-Neuve, Belgium

Abstract. We compare published solutions for the rotation angle $W(t)$ which describes the location of the prime meridian of Mars with respect to the ICRF equator in IAU recommendations. If the model for W includes a very long period term, we transform it into a quadratic polynomial with updated epoch value and rate, resulting in a difference in the mean epoch value up to 200 km. The mean and true epoch rotation angles are about 800 mas (13 m) apart in J2000 and should not be confused with each other in order to accurately locate the prime meridian. We identify two groups of radio-science solutions for W, which can be distinguished by the prime meridian location they used as a priori that differ from each other by about 100 m at J2000.

Keywords. Mars rotation, geodesy, dynamics, Viking, longitude, radio science, prime meridian

1. Introduction

In order to describe the orientation of the spin axis of a planet, the IAU Working Group on Cartographic Coordinates and Rotational Elements (WGCCRE) uses the equatorial coordinates (right ascension α and declination δ) orienting the planet with respect to the equator of the International Celestial Reference Frame (ICRF), see for example Archinal *et al.* (2018). A third angle, denoted W and which has a diurnal periodicity, is used to position the prime meridian with respect to the node of the planet equator over the ICRF equator. For Mars, Yseboodt *et al.* (2023) showed that if a precision of about 5 mas is targeted in the 1970-2030 time interval, α, δ and W should be modeled by the sum of a quadratic polynomial and of a periodic series. The main periodicities correspond to harmonics of the orbital (i.e. annual) period.

To describe the orientation and rotation of Mars, the radio-science community more commonly uses Euler angles, including the obliquity ε and node longitude ψ defined with respect to the planet mean orbit at a chosen epoch. The rotation angle positioning the prime meridian with respect to the node of Mars equator over Mars mean orbit is denoted ϕ. This set of angles $(\varepsilon, \psi, \phi)$ can be precisely converted into equatorial coordinates (α, δ, W), and vice versa, using the analytical method proposed by Yseboodt *et al.* (2023). This method allows us to compare different rotation models whatever the set of angles used.

Thanks to the large number of spacecraft missions sent to the planet Mars in the past 50 years, numerous rotation solutions have been published. The latest IAU-like Mars solutions (in Earth equatorial coordinates and denoted here "IAU angles") were obtained by following a method which first converts the periodic series of Euler angles estimate in the time domain, then numerically transforms them into IAU angles time series, and finally fits the IAU angles periodic series using frequency analysis techniques (Jacobson 2010; Kuchynka *et al.* 2014 and Jacobson *et al.* 2018). In these solutions, the expression for each IAU angle includes a very long period ($\gtrsim 70,000$ years) term that

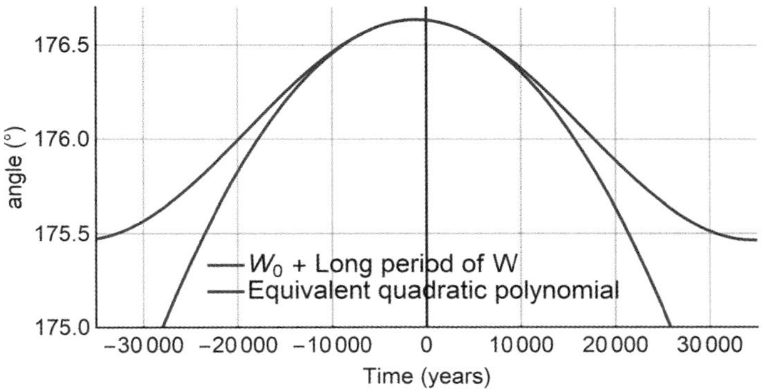

Figure 1. Locally (close to $t = 0$ here), a very long period term (in purple) can mimic the behavior of a quadratic function (in blue).

results from the lack of the quadratic term in the model used for the fit. Indeed, by choosing adequate amplitude, phase and frequency, long period signals can well mimic quadratic signals over an interval of a few decades around J2000. This is not anecdotal for the definition of a rotation model since adding a long period modulation instead of a quadratic term largely and artificially alters the epoch value and rate of each angle.

In Section 2, we transform the very long period term in W of the published IAU-like models into a quadratic function. We discuss the difference between the true and mean epoch value of W in Section 3. Then, we compare different published solutions for the rotation angle in Section 4. Finally we discuss some implications regarding the definition of the prime meridian of Mars.

2. Quadratic polynomial versus long period term

As explained in Yseboodt *et al.* (2023) and mentioned above, when no quadratic term is present in a IAU-like model obtained from a numerical fit, a very long period signal is needed to mimic the expected behavior on an interval of a few tens of years around the reference epoch, as shown in Fig. 1. This is demonstrated below where a long period modulation in the angle W plus a degree 1 polynomial is showed to be equivalent, to the second order in frequency f, to a degree 2 polynomial with no periodic terms:

$$\begin{aligned} W_0 + \dot{W}_0\, t + A \sin(f\, t + \varphi_0) &\approx (W_0 + A \sin\varphi_0) + (\dot{W}_0 + A\, f \cos\varphi_0)\, t - \frac{A\, f^2 \sin\varphi_0}{2}\, t^2 \\ &\approx W'_0 + \dot{W}'_0\, t + W_Q\, t^2 \qquad (1) \end{aligned}$$

Adding a long period modulation instead of a quadratic term significantly changes the epoch value and the rate at the initial epoch. As shown in Eq. (1), the "real" epoch value (W'_0) and rate ($\dot{W}'_0$) are then both split into a sum of a term coming from the degree 1 polynomial (W_0 and $\dot{W}_0$, respectively) and a term coming from the long period variations ($A \sin\varphi_0$ and $A\, f \cos\varphi_0$, respectively). The same reasoning applies to the α and δ angles.

Published solutions for the spin angle of Mars at t_0 are reported in the third column of Table 1. Corrected values (W'_0) are provided for the three solutions that include very long period terms (column 4). Since Jacobson (2010) is the transformation of Konopliv *et al.* (2006), by construction $W'_{0\,(Ja10)} = W_{0\,(Ko06)}$. The same applies for $W'_{0\,(IAU15)} = W_{0\,(Ku14)}$ and for $W'_{0\,(Ja18)} = W_{0\,(Ko16)}$. For Jacobson *et al.* (2018), the correction is as high as 3.3° (or 200 km at the Martian surface) for W_0 and 100 mas/year for the rate of W. For Kuchynka *et al.* (2014), the shift is about 34 km, see first panel of Figure 2.

Table 1. Comparison of the rotation angle W values at $t_0 = J2000$ for a selection of published rotation models (column 2). We transform the published solutions of Konopliv *et al.* (2006, 2016); Kuchynka *et al.* (2014) expressed in Euler angles into Earth equatorial coordinates by using the method described in Yseboodt *et al.* (2023). The fourth column provides the corrected epoch value (W_0') for these three solutions. The fifth and sixth columns present the values of the periodic variations at epoch ($\Delta W(t_0)$) and the true epoch values ($W(t_0)$). The last columns present the difference between $W(t_0)$ and the true epoch value of the IAU15 solution, in degree and meter.

	Reference	W_0 Mean value (deg)	W_0' (deg)	$\Delta W(t_0)$ Periodic (deg)	$W(t_0)$ True value (deg)	$W - W_{IAU15}(t_0)$ (deg)	(m)
IAU82	Davies *et al.* (1983)	176.655	-	0	176.655	0.022940	1357.1
IAU00	Seidelman *et al.* (2002)	176.630	-	0	**176.630**	-0.002060	**-121.9**
Ko06	Konopliv *et al.* (2006)	176.630239	-	0.000217	176.630456	-0.001604	-94.9
Ja10	Jacobson (2010)	176.076538	176.630239	0.553918	176.630456	-0.001604	-94.9
Ku14	Kuchynka *et al.* (2014)	176.631818	-	0.000241	176.632059	0	-0.1
IAU15	Archinal *et al.* (2018)	176.049863	176.631819	0.582197	**176.632060**	0	0
Ko16	Konopliv *et al.* (2016)	176.631896	-	0.000248	176.632145	0.000085	5.0
Ja18	Jacobson *et al.* (2018)	173.308792	176.631896	3.323352	176.632145	0.000085	5.0
LM23	Le Maistre *et al.* (2023)	176.631969	-	0.000216	176.632186	0.000126	7.5

3. Mean and true epoch rotation angles

When periodic terms are included in the definition of W, a distinction between the **mean** (W_0) and **true** ($W(t_0)$) epoch values can be made. Before IAU15 (Archinal *et al.* 2018), the successive IAU reference solutions for the orientation and rotation of Mars did not include periodic terms. This means that for the IAU00 solution for instance, the degree zero term in the time polynomial for the spin angle, W_0, is both the mean and true epoch value of W at the same time. The link between these two quantities is simply written here as:

$$W(t_0) = W_0 + \Delta W(t_0) \tag{2}$$

where the periodic series $\Delta W(t_0)$ can reach 0.00021-0.00025° (i.e. 780-890 mas or 12.4-14.3 m at the surface) at the J2000 epoch (only considering short period terms). A selection of published numerical values of these periodic variations is given in the fifth column of Table 1, while the corresponding true epoch values ($W(t_0)$) are given in the sixth column.

The radio-science community has long been ahead of the IAU WGCCRE regarding the addition of periodic terms in the model of orientation and rotation of Mars. The "Pathfinder model" of Folkner *et al.* (1997) already included nutation and seasonal spin variations (see also Konopliv *et al.* 2006). For the sake of consistency with the IAU solution, Konopliv *et al.* (2006) fixed the value of ϕ_0 so that the longitude of the Body Frame X-axis almost coincides (only 20 cm difference) with the one of the IAU00 solution at the J2000 epoch. This requires coinciding mean epoch values on one side with true epoch values on the other.

4. Comparison of published prime meridian positions

The prime meridian of Mars is defined as that meridian which passes through the center of the 500 m diameter Airy-0 crater (de Vaucouleurs *et al.* 1973). The last two estimations of its location (and therefore of W_0) are those of Duxbury *et al.* (2001) and Kuchynka *et al.* (2014). The first one (Duxbury *et al.* 2001), used for the IAU00 solution, is the average of various estimations based on different methods and mainly based on MOLA observations. Because the range of values covers about 500 m on the Martian surface, the uncertainty on the adopted $W_0 = 176.630°$ is about 250 m ($\sim 0.004°$). The

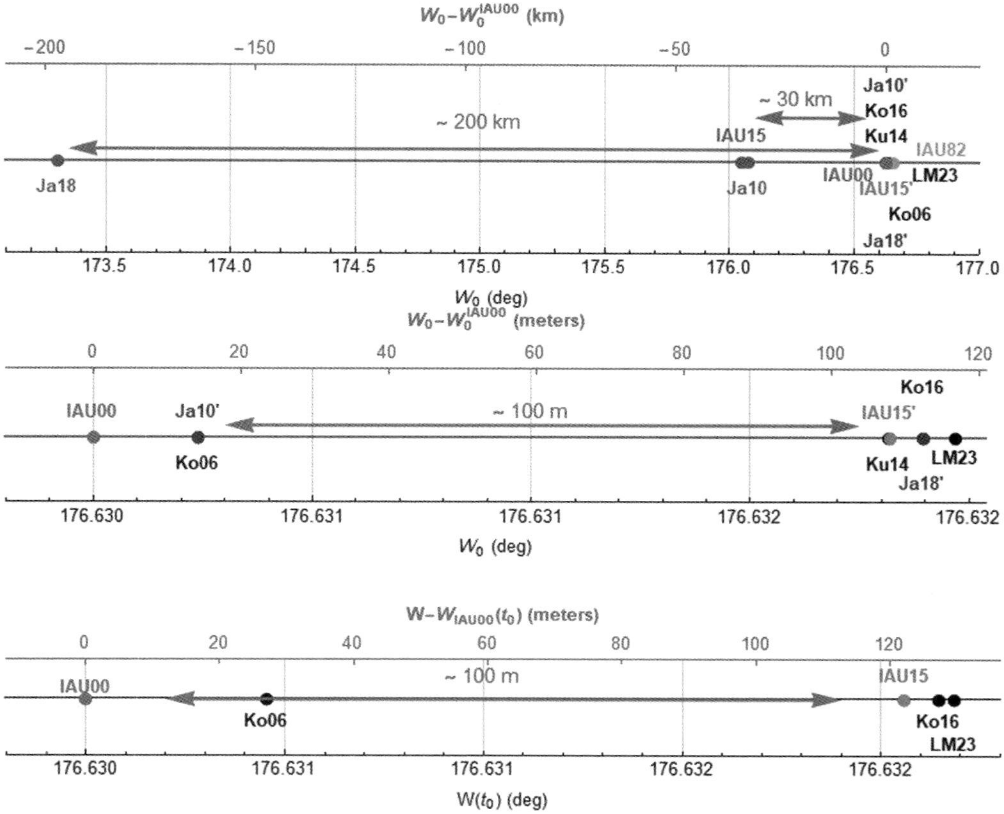

Figure 2. Comparison of the rotation angle W at the J2000 epoch for different published solutions. The names with $'$ are models with the very long period integrated into the epoch value. The first two panels show the mean epoch value W_0 while the last panel shows the true epoch value $W(t_0)$, taking into account the periodic variations at t_0. The red axis and values show the comparison to W_{IAU00} in meters. Since there are large and small differences between all the numerical values, 2 different intervals are shown: about 200 km for the first panel and about 120 meters for the other panels.

second one, estimated by Kuchynka *et al.* (2014) and used as the IAU15 solution, equals $W_0' = 176.631819°$ (after correction from long-period bias as explained in Sec. 2 and shown in Table 1) with an alleged precision of the order of 10 meters. This estimation of W_0 is based on Mars Global Surveyor's Mars Orbiter Camera (MOC) data, used to determine the cartographic longitude of Viking 1 lander relative to Airy-0, using pyramidal georegistration of lower to higher resolution imagery.

Note that the difference in X-axes longitudes between the IAU00 and IAU15 models is of about 100 m at J2000, within the error bar on the epoch value of W of the IAU00 model. The difference between the IAU82 and IAU00 solutions for W_0 was about 1.3 km. The IAU82 solution, issued with the second report of the WGCCRE, was the first providing the epoch value W_0 at the J2000 epoch and with respect to the J2000 ICRF.

The difference between the location of the prime meridian of the IAU00 and IAU15 solutions remains in the pure radio-science estimates of W_0, because radiometric data have no sensitivity to Airy-0 crater location. Thus, two groups of solutions for W epoch value can be identified. In the first group, Konopliv *et al.* (2006, 2011) choose their ϕ_0 value so that their prime meridian coincides with that of the IAU00 solution at the J2000

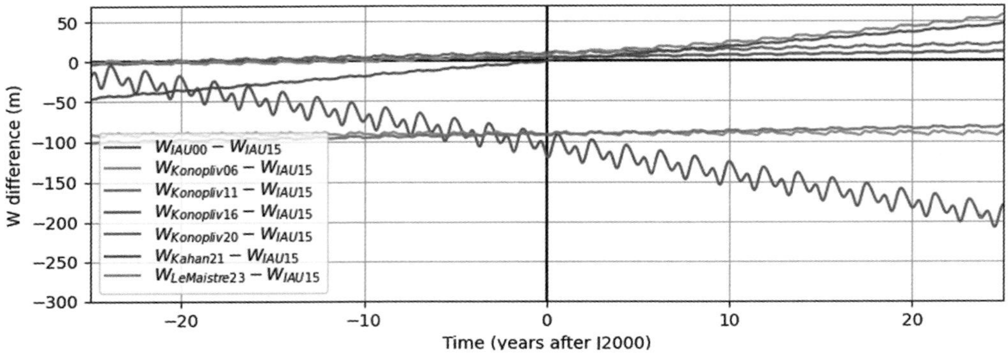

Figure 3. Difference between rotation angle $W(t)$ solutions and $W_{IAU15}(t)$, as a function of time, in meters.

epoch. In the second group, the prime meridian location is close to that of IAU15, either by fixing ϕ_0 value (Le Maistre *et al.* 2023) or by fixing the Viking 1 longitude to the Kuchynka *et al.* (2014) cartographic estimation, as done for example by Konopliv *et al.* (2016, 2020); Kahan *et al.* (2021). These two groups of solutions can be clearly identified on the left and right sides of the second panel of Figure 2 and in the last two columns of Table 1.

We also compare some published solutions at different times, see Figure 3. The difference between the IAU00 and IAU15 solutions clearly shows periodic variations, because the IAU00 solution did not have any periodic variations, unlike IAU15. Close to J2000, we see the two groups of solutions mentioned above: the first one close to IAU00, and the second one close to IAU15. However, the distinction between the groups is less clear away from J2000, mainly due to differences in the rotation rates.

5. Summary and discussion

Future IAU solutions shall replace the artificial terms at very long period by degree 2 polynomials in order to avoid large bias in the definition of the mean epoch values and rates and, consequently in the prime meridian location. Such a bias in the rotation angle $W(t)$ reaches ~ 200 km in Jacobson *et al.* (2018) solution and 30 km in Jacobson (2010) solution and in the current IAU solution (i.e. IAU15) adopted from Kuchynka *et al.* (2014) (see Figure 2). Even though they accurately represent the temporal evolution of the angles within a few decades around J2000, models with long period terms lead to confusion as to the location of the prime meridian. For example, if one decided to ignore all the periodic terms (short and long) of these models, the prime meridian would be very far from reality.

The exact location of the Airy-0 crater at the J2000 epoch is not defined by the mean epoch rotation angle W_0, but by the true epoch rotation angle $W(t_0)$, the difference being the rotation variations evaluated at J2000. As the periodic variations in W were not taken into account by Duxbury *et al.* (2001), their estimation of W_0 is possibly biased by up to a few tens of meters, which is negligible compared to the 250 m error coming from MOLA observations.

Following IAU recommendations, the radio-science teams generally adjust their solutions so that their prime meridian is close to that of the IAU model at the J2000 epoch. This is why the recent radio-science solutions are divided into two groups, one following IAU00 and one following IAU15, which are based on two different estimates for the mean epoch value W_0. Only a work based on cartographic product and images more precise

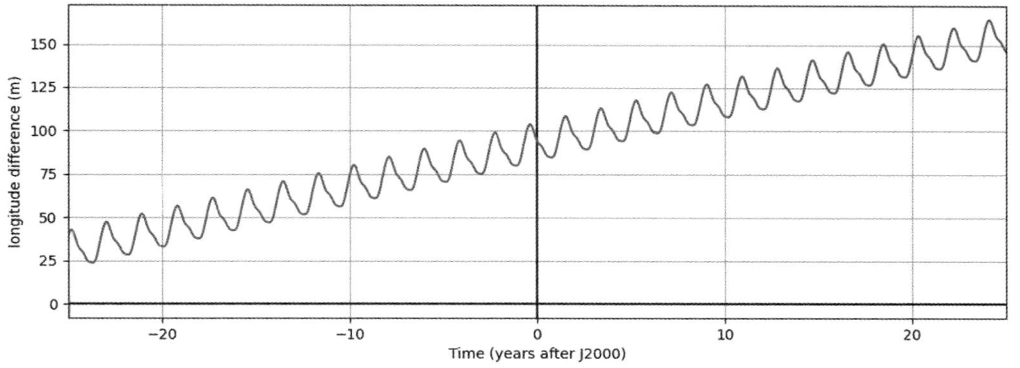

Figure 4. Comparison of the prime meridian position (see equation 3) between IAU00 and IAU15, as a function of time, in meters.

than used before can give a better estimation of W_0, which would then be used to give birth to a third group of Mars rotation models, hopefully close to the IAU15 group.

The intersection between the prime meridian and Mars true equator defines the X-axis of Mars body frame, reference for the cartographic longitude. Since any change in the IAU orientation angles displaces the true equator, and therefore the node from which W is measured, the location of the prime meridian is also affected by any difference in α and δ, besides the difference in W. The difference in longitude between the X-axes of two different orientation and rotation models expressed in IAU angles, $(\alpha_1, \delta_1, W_1)$ and $(\alpha_2, \delta_2, W_2)$ can be expressed, correct up to the first order in angles' differences, by

$$\Delta\lambda_{PrimeMer} = (W_1 - W_2) + (\alpha_1 - \alpha_2)\sin\delta \quad (3)$$

$$= (\phi_1 - \phi_2) + (\psi_1 - \psi_2)\cos\epsilon \quad (4)$$

where δ can be replaced either by the true epoch value of δ_1 or δ_2, or even by one of the mean epoch values, provided that the two models are not expressed with long period terms. From Eq. (3), we see that the prime meridian location changes when W changes, but also when α changes. Close to J2000, this $(\alpha_1 - \alpha_2)\sin\delta$ term equals -27 m for the difference IAU00-IAU15, which add to the 122 m difference in their W, leading to a difference in the prime meridian location of 95 meters at J2000 (see Figure 4).

Acknowledgments

This work was financially supported by the Belgian PRODEX program managed by the European Space Agency in collaboration with the Belgian Federal Science Policy Office.

References

B.A. Archinal, C.H. Acton, *et al.*, 2018, *Cel. Mec. Dyn. Astr.*, 130.

M.E. Davies, V.K. Abalakin, *et al.*, 1983, *Cel. Mec. Dyn. Astr.*, 29.

G. de Vaucouleurs, M.E. Davies, F.M. Sturms Jr., 1973, *JGR*, 78.

T. Duxbury, R.L. Kirk, B.A. Archinal, G.A. Neumann, 2001, `https://astrogeology.usgs.gov/groups/ISPRS`.

W. M. Folkner, C. F. Yoder, *et al.*, 1997, *Science*, 278.

R.A. Jacobson, 2010, *A. J.*, 139.

R.A. Jacobson, A.S. Konopliv, R.S. Park, W.M. Folkner, 2018, *Planetary and Space Science*, 152.

D.S. Kahan, W.M. Folkner, *et al.*, 2021, *Planetary and Space Science*, 199.

A.S. Konopliv, C.F. Yoder, E.M. Standish, D.-N. Yuan, and W.L. Sjogren, 2006, *Icarus*, 182.

A.S. Konopliv, S.W. Asmar, *et al.*, 2011, *Icarus*, 211.
A.S. Konopliv, R.S. Park, and W.M. Folkner, 2016, *Icarus*, 274.
A.S. Konopliv, R.S. Park, *et al.*, 2020, *GRL*, 47.
P. Kuchynka, W.M. Folkner, *et al.*, 2014, *Icarus*, 229.
S. Le Maistre, A. Rivoldini *et al.*, 2023, *Nature* 619.
P.K. Seidelmann, V.K. Abalakin, M. Bursa *et al.*, 2002, *Cel. Mec. Dyn. Astr.* 82.
M. Yseboodt, R.M. Baland, and S. Le Maistre, 2023, *Cel. Mec. Dyn. Astr.*, 135.

Complex Planetary Systems II: Latest Methods for an Interdisciplinary Approach
Proceedings IAU Symposium No. 382, 2024
A. Lemaitre & A.-S. Libert, eds.
doi:10.1017/S1743921323004751

Collision outcomes due to planetesimal and planetary embryo interactions in inclined binary star systems

Maximilian Zimmermann and Elke Pilat-Lohinger

University of Vienna, Departement of Astrophysics
email: maximilian.zimmermann@univie.ac.at

Abstract. In the final phase of terrestrial planet formation, planetary embryos and planetesimals are the building blocks for the growth of rocky planets. In this investigation, we study the dynamical behaviour of a circumstellar disk in an inclined binary star system. The disk consists of 2000 planetesimals and 25 embryos and is distributed between 1 and 4 au around the primary star. To compute the gravitational interaction of the whole system, we use our recently developed GPU N-body code GANBISS. GANBISS treats all collision as perfect merging and delivers the impact parameters that will be used to distinguish between different collision outcomes.

Keywords. Methods: numerical - Planets and satellites: formation - Binaries: close, inclined - Celestial Mechanics

1. Numerical methods and setup

GANBISS is written in CUDA C and runs on most modern NVIDIA GPUs. It uses the Bulirsch-Stoer method to solve the equations of motion. It is designed to simulate the dynamical evolution of a planetesimal-embryo disk in a binary star system. The disk can handle up to 10000 interacting objects in a simulation. For details see Zimmermann & Pilat-Lohinger (2023). In this study, all simulated binary configurations consist of two equal mass stars ($M_1 = M_2 = 1M_{\odot}$). The separation of the two stars is fixed to 30 au and the eccentricity and inclination are varied as shown in table 1. For comparison a planar configuration has been studied in addition. The dynamical interaction of all bodies was integrated over 1 Myr.

2. Results

From the perfect merging events of the N-body simulations, we obtained a wide range of collision parameters (impact velocity and angle). In further treatment, these parameters were used to classify collisions according to the Leinhardt & Stewart (2012) model (see figure 1 and table 1). Collisions were categorized as merging events ("pm", "gm", "pa" in table 1), hit-and-run collisions ("hr" in table 1), or destructive collisions ("pe", "cat" in table 1). When over 90 % of the initial mass is lost, it's termed super-catastrophic. A comparison of inclined systems to the planar case revealed that significantly more destructive collisions occur in the inclined cases: approximately 15 % for $i_b = 20°$ and 32.5 % for $i_b = 45°$, compared to just one in the planar case. Therefore, mass-growing collisions are less common in inclined binary star configurations, where about 18.5 % for $i_b = 20°$ and about 10 % for $i_b = 45°$ are mass-growing collisions, in contrast to the planar case where this fraction is around 64.5 % of all collisions. The largest fraction of

Table 1. Calculated collision outcomes for the different binary star configurations according to Leinhardt & Stewart (2012). Abbreviations are the following ones: pm = perfect merging; gm = graze-and-merge; hr = hit-and-run; pa = partial accretion; pe = partial erosion; cat = super-catastrophic.

a_b [au]	e_b	i_b [°]	pm	gm	hr	pa	pe	cat	# collisions
30	0.0	0	1018	104	686	126	1	0	1935
30	0.0	20	82	10	1028	193	107	135	1555
30	0.2	20	109	9	1050	191	99	137	1595
30	0.0	45	14	1	876	154	146	366	1557
30	0.2	45	13	2	897	141	135	367	1555

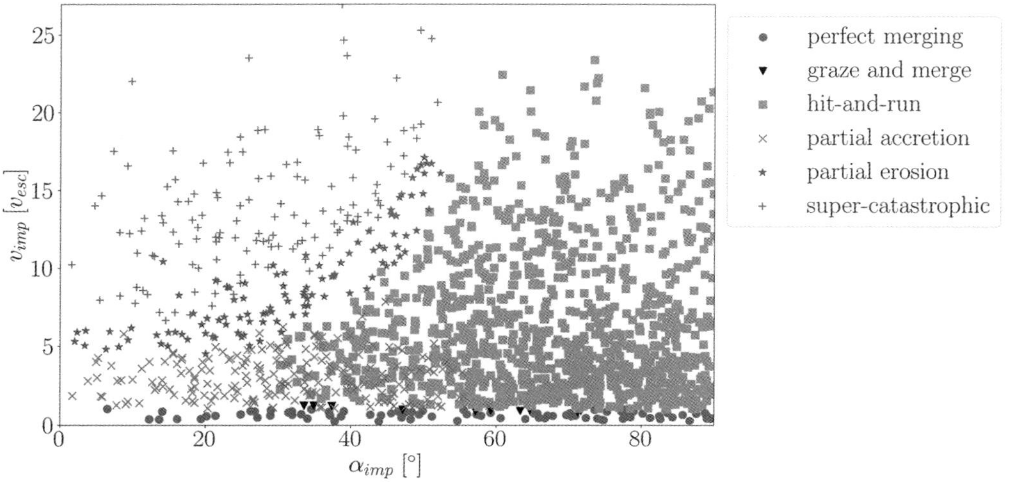

Figure 1. Distribution of the impact parameters (impact angle α_{imp} and impact velocity v_{imp}) for the binary configuration: $a_b = 30$ au; $e_b = 0.0$ and $i_b = 20°$. Note that $\alpha_{imp} = 0°$ corresponds to head-on collisions while 90° matches grazing collisions, and v_{imp} is in terms of the mutual escape velocity v_{esc}. Each dot represents one collision of the simulation. The color-coding refers to the calculated collision outcome when applying the analytic model from Leinhardt & Stewart (2012) to the resulting N-body simulation collisions.

collisions in the inclined cases are the hit-and-run events (~ 60 %). For the planar case the hit-and-run events are ~ 35 % of all collisions.

Note that the overlapping of different collision outcomes in figure 1 results from the fact that the collision outcomes have been determined for various mass-ratios of the colliding objects.

3. Conclusion

We simulated a disk of planetesimals and planetary embryos in S-type motion in different configurations of binary star systems (variation of i_b and e_b) and studied the growth of the disk objects via perfect merging. We tracked the impact parameters of all collisions to acquire better results for the growth of rocky planets following Leinhardt & Stewart (2012). Our results showed an increase in destructive collision events for inclined binary systems. Overall, the largest fraction of collision outcomes are hit-and-run collisions in the inclined configurations and the fraction of mass growing collisions decreases for increasing i_b.

A more realistic collision handling – e.g. using smooth particle hydrodynamics simulations (SPH) Maindl *et al.* (2013) – is needed for N-body simulations in order to investigate the late stage of terrestrial planet formation in stellar systems.

4. Acknowledgements

The authors want to acknowledge the support by the Austrian FWF - project P33351-N and S11608-N16.

The computational results presented have been achieved using the Vienna Scientific Cluster (Projects 71637, 71686, 70320).

References

Zimmermann, M. and Pilat-Lohinger, E. 2023, CeMDA 135

Maindl, T. I., Schäfer, C., Speith, R., Süli, A., Forgács-Dajka, E. and Dvorak, R. 2013 Astronomische Nachrichten, 334

Leinhardt, Z. and Stewart, S. T., 2012, ApJ, 745

Author index